THE ADAM FAMILY OF PROTEASES

## PROTEASES IN BIOLOGY AND DISEASE

SERIES EDITORS:

NIGEL M. HOOPER, *University of Leeds, Leeds, United Kingdom*

UWE LENDECKEL, *Otto-von-Guericke University, Magdeburg, Germany*

---

Volume 1
PROTEASES IN TISSUE REMODELLING OF LUNG AND HEART
Edited by Uwe Lendeckel and Nigel M. Hooper

Volume 2
AMINOPEPTIDASES IN BIOLOGY AND DISEASE
Edited by Nigel M. Hooper and Uwe Lendeckel

Volume 3
PROTEASES IN THE BRAIN
Edited by Uwe Lendeckel and Nigel M. Hooper

Volume 4
THE ADAM FAMILY OF PROTEASES
Edited by Nigel M. Hooper and Uwe Lendeckel

# THE ADAM FAMILY OF PROTEASES

Edited by

Nigel M. Hooper
*University of Leeds, U.K.*

and

Uwe Lendeckel
*University of Magdeburg, Germany*

Springer

A C.I.P. Catalogue record for this book is available from the Library of Congress.

ISBN-10 0-387-25149-9 (HB)
ISBN-13 978-0-387-25149-3 (HB)
ISBN-10 0-387-25151-0 (e-book)
ISBN-13 978-0-387-25151-6 (e-book)

Published by Springer,
P.O. Box 17, 3300 AA Dordrecht, The Netherlands.

*www.springeronline.com*

*Printed on acid-free paper*



Printed in the Netherlands.

# Contents

# Contributing Authors

**Reidar Albrechtsen**
Institute of Molecular Pathology, University of Copenhagen, 2100 Copenhagen, Denmark

**Dominique Alfandari**
Department of Veterinary and Animal Sciences, University of Massachusetts, Amherst, Massachusetts, USA.

**Suneel S. Apte**
Department of Biomedical Engineering and Orthopaedic Research Center, Cleveland Clinic Foundation, 9500 Euclid Avenue, Cleveland, OH 44195, USA

**Elizabeth Arner**
Joint Biology, Pfizer Global Research and Development, St Louis Research Laboratories, 700 Chesterfield Parkway West, St Louis, MO, USA

**Joaquín Arribas**
Medical Oncology Research Program. Vall d'Hebron University Hospital Research Institute. Psg. Vall d'Hebron 119-129 Barcelona 08035, Spain

**Jörg W. Bartsch**
Developmental Biology and Molecular Pathology, W7, Bielefeld University, 33615 Bielefeld, Germany

**Carl P. Blobel**
Arthritis and Tissue Degeneration Program and Cell Biology Program, Hospital for Special Surgery at Weil Medical College of Cornell University, New York, NY, USA

**Lance Bridges**
Department of Cell Biology, University of Virginia, USA

**Chunghee Cho**
Department of Life Science, Gwangju Institute of Science and Technology, Gwangju 500-712, Republic of Korea

**Douglas DeSimone**
Department of Cell Biology, University of Virginia, USA

**Eva Engvall**
The Burnham Institute, La Jolla, California 92037, USA

**Anne M. Fourie**
Johnson & Johnson Pharmaceutical Research and Development, San Diego, USA

**Daniel S. Greenspan**
Department of Pathology and Laboratory Medicine, University of Wisconsin, Madison, USA

**Dieter Hartmann**
Vlaams Interuniversitair Instituut voor Biotechnologie, VIB 4; and Department  for Human Genetics, Katholieke Universiteit Leuven, Campus Gasthuisberg, Herestraat 49, B-3000 Leuven, Belgium

**Keisuke Hoiruchi**
Arthritis and Tissue Degeneration Program and Cell Biology Program, Hospital for Special Surgery at Weil Medical College of Cornell University, New York, NY, USA

**Shoichi Ishiura**
Department of Life Sciences, Graduate School of Arts and Sciences, The University of Tokyo, Japan

**Tiebang Kang**
Department of Chemisty and Biochemistry and Institute of Molecular Biophysics, Florida State University, Tallahassee, Florida, USA

**Carine Le Goff**
Department of Biomedical Engineering and Orthopaedic Research Center, Cleveland Clinic Foundation, 9500 Euclid Avenue, Cleveland, OH 44195, USA

**Anne-Marie Malfait**
Joint Biology, Pfizer Global Research and Development, St Louis Research Laboratories, 700 Chesterfield Parkway West, St Louis, MO, USA

**Silvia Naus**
Developmental Biology and Molecular Pathology, W7, Bielefeld University, 33615 Bielefeld, Germany

**Robert G. Newcomer**
Department of Chemisty and Biochemistry and Institute of Molecular Biophysics, Florida State University, Tallahassee, Florida, USA

**Andrea Rittger**
Developmental Biology and Molecular Pathology, W7, Bielefeld University, 33615 Bielefeld, Germany

**Soraya Ruiz-Paz**
Medical Oncology Research Program. Vall d'Hebron University Hospital Research Institute. Psg. Vall d'Hebron 119-129 Barcelona 08035, Spain

**Paul Saftig**
Biochemical Institute, Christians-Albrecht-Universität Kiel, Olshausenstr. 40, 24018 Kiel, Germany

**Qing-Xiang Amy Sang**
Department of Chemisty and Biochemistry and Institute of Molecular Biophysics, Florida State University, Tallahassee, Florida, USA

**Uwe Schlomann**
Developmental Biology and Molecular Pathology, W7, Bielefeld University, 33615 Bielefeld, Germany

**Monika Tomczuk**
Department of Cell Biology, University of Virginia, USA

**Micky Tortorella**
Joint Biology, Pfizer Global Research and Development, St Louis Research Laboratories, 700 Chesterfield Parkway West, St Louis, MO, USA

**Han-Mou Tsai**
Hematology Division, Montefiore Medical Center, Albert Einstein College of Medicine, New York, USA

**Wei-Man Wang**
Department of Pathology and Laboratory Medicine, University of Wisconsin, Madison, USA

**Ulla M. Wewer**
Institute of Molecular Pathology, University of Copenhagen, 2100 Copenhagen, Denmark

**Judith White**
Department of Cell Biology, University of Virginia, USA

**Dirk Wildeboer**
Developmental Biology and Molecular Pathology, W7, Bielefeld University, 33615 Bielefeld, Germany

**Tyra Wolfsberg**
National Human Genome Research Institute, NIH, USA

**Yun-Ge Zhao**
Department of Chemisty and Biochemistry and Institute of Molecular Biophysics, Florida State University, Tallahassee, Florida, USA

# Preface

Over the last few years a new family of zinc metalloproteases have been discovered, mainly through advances in genome sequencing. This family has been given the name ADAM for 'a disintegrin and metalloprotease', reflecting the fact that these proteins have multiple domains including a protease domain and a disintegrin domain. A branch of the family, called ADAMTS, also have thrombospondin-like motifs. The role of these proteases in a diversity of biological processes is gradually coming to light and this volume, the fourth in the *Proteases in Biology and Disease* series, aims to provide the first comprehensive review of the roles of ADAM and the related ADAMTS proteases in biology and disease.

The book begins with an introduction to the ADAM family by Judith White and colleagues who review the history and phylogeny of the ADAMs as well as structural and functional aspects of their major domains. This chapter hopefully provides the reader with an appreciation of the widespread, varied, and fascinating means by which ADAMs affect key cell surface events, including cell signalling, cell adhesion, and cell migration. In Chapter 2, Keisuke Hoiruchi and Carl Blobel focus on recent insights that have emerged from studies of knockout mice for ADAM proteases that are widely expressed, namely ADAMs 8, 9, 10, 12, 15, 17 and 19. This includes studies from multiple, as well as single, knockouts. The availability of viable knockout mice for some of the ADAMs sets the stage for a more comprehensive analysis of potential functions of these proteins in physiological and pathological processes. Jörg Bartsch and colleagues in Chapter 3 describe work on ADAM8 which is involved in immune responses. Whereas embryonal development in ADAM8 deficient mice appears normal, its upregulation under inflammatory conditions like that seen in chronic neurodegeneration, after administration of lipopolysaccharides and in allergic asthma, seems to reflect a specific function of ADAM8 in cytokine response.

In Chapter 4, Shoichi Ishiura describes the molecular aspects of ADAM9 and the role of this protease in ectodomain shedding, matrix degradation and cell-to-cell contact. ADAM10 is the subject of Chapter 5 by Paul Saftig and Dieter Hartmann. ADAM10 has emerged as a key player in the Notch pathway in both flies and nematodes and as a candidate Amyloid Precursor Protein α-secretase in mammals. However, the list of ADAM10 substrates has expanded rapidly, and at present includes adhesion molecules, guideposts of cell migration and axon navigation and key signaling factors of the immune system. An especially fascinating aspect of ADAM10 function is the high proportion of substrates that are further processed by I-Clips, like γ-secretase, in regulated intramembrane proteolysis. The next chapter by Ulla Wewer and colleagues describes the two alternatively spliced forms of ADAM12. This ADAM interacts with integrin and syndecan adhesion receptors via the disintegrin and cysteine-rich domains, and influences cell shape, cytoskeleton, and the organization of the extracellular matrix. ADAM12 is expressed mainly during development and differentiation, in remodelling tissues, and in fast growing tissues such as placenta and malignant tumors, suggesting that the protease may have direct implications for clinical medicine.

In Chapter 7, Dominique Alfandari describes the role of ADAM13 in development. ADAM13 is expressed in Cranial Neural Crest cells and somites during neurulation and subsequent tailbud formation in *Xenopus laevis*. ADAM17, possibly the most intensively studied ADAM, is the subject of Chapter 8 by Joaquín Arribas and Soraya Ruiz-Paz. Although the participation of ADAM 17 in the proteolytic release of the ectodomains of different cell surface proteins has been well established, critical questions such as how its metalloprotease activity is regulated or how its substrates are recognized remain to be answered. In Chapter 9, Tiebang Kang and colleagues describe the domain structure, regulation, processing and functions of ADAM19. The disintegrin and cysteine-rich domains serve to regulate the proteolytic activity of ADAM19, while complex signalling pathways under the control of molecules such as protein kinase C, calcium, and calmodulin regulate ADAM19 expression and activity.

ADAM28 is the subject of Chapter 10 by Anne Fourie. Both membrane-bound and secreted isoforms of ADAM28 have been identified and the protease is activated by autocatalytic removal of the pro-domain. The physiological functions of ADAM28 are not known, but its expression pattern, together with its substrate and integrin binding selectivity, suggest potential roles in spermatogenesis, lymphocyte maturation and function, inflammation and cancer. In Chapter 11, Chunghee Cho provides an account of those mammalian ADAMs with testis-specific or -predominant expression. Uncovering the *in vivo* functions of the testicular ADAM proteases present in both mice and humans should provide insights into the mammalian reproductive system involving protease-mediated events. Daniel Greenspan and Wei-Man Wang provide an overview of the ADAMTS proteases in Chapter 12 and also specifically describe the properties of ADAMTS2, the procollagen III N-protease. There are 19 known ADAMTS proteases in vertebrates, and defects in a number of these are implicated as causal in diseases that include dermatosparaxis, osteoarthritis, inflammatory joint disease and thrombotic thrombocytopenic purpura. ADAMTS proteinases are also involved in growth, organogenesis and fertility in a broad spectrum of species that range from humans to worms.

ADAMTS3 and ADAMTS14 are the subject of Chapter 13 by Carine Le Goff and Suneel Apte. ADAMTS3 and ADAMTS14 belong to the procollagen aminopropeptidase subfamily of ADAMTS proteases that also includes ADAMTS2. These enzymes appear to have co-evolved with their substrates, the major fibrillar collagen types I, II and III by gene duplication from a primitive precursor. In Chapter 14, Anne-Marie Malfait and colleagues describe ADAMTS4 and ADAMTS5, otherwise known as aggrecanases. Work from a number of groups has begun to provide insight into the molecular basis for the role of these proteases in aggrecan catabolism and knowledge continues to accumulate on the expression pattern of these proteases in different tissues and their potential role in normal physiological mechanisms and in disease. In the final chapter, Han-Mou Tsai describes ADAMTS13. Studies on the homeostasis of von Willebrand factor, a plasma glycoprotein that mediates platelet adhesion and aggregation at the site of vessel injury, led to the discovery that it is cleaved in the circulation by a zinc metalloprotease, ADAMTS-13, in a shear stress dependent manner. In thrombotic thrombocytopenic purpura, autoimmune inhibitors of the protease or genetic mutations of the *ADAMTS-13* gene cause a severe deficiency of ADAMTS-13 in plasma.

We trust that this fourth volume in the *Proteases in Biology and Disease* series will prove to be a timely and useful source of information on ADAM family proteases for all those researchers involved in studying any aspect of the biology of this emerging family of zinc metalloproteases. Finally, we would like to thank all the authors for their scholarly contributions and apologize to them for editorial changes in the interests of consistency.

*Nigel M. Hooper and Uwe Lendeckel*

# Chapter 1

# INTRODUCTION TO THE ADAM FAMILY

Judith White[1], Lance Bridges[1], Douglas DeSimone[1], Monika Tomczuk[1] and Tyra Wolfsberg[2]
[1]*Department of Cell Biology, University of Virginia,* [2]*National Human Genome Research Institute, NIH, USA*

**Abstract**: ADAMs (proteins containing A Disintegrin and A Metalloprotease domain) are multidomain and multifunctional proteins that are emerging as key regulators of critical events that occur at the cell surface. Many ADAMs (roughly half) are active metalloproteases, and several of these (e.g. ADAMs 10, 17, and 19) exert important functions *in vivo*, for example in development of the heart and brain. The best-characterized *in vivo* activity of ADAM proteases is as ectodomain sheddases. By shedding cell surface proteins (e.g. cytokines and growth factors), ADAMs initiate extracellular signaling events (e.g. signaling through epidermal growth factor receptors). ADAM-mediated ectodomain shedding (e.g. of Notch) can also set the stage for important intracellular signaling events. ADAMs have also been reported to shed surface proteins involved in both cell-cell and cell-matrix adhesion. The disintegrin and cysteine-rich domains of ADAMs exhibit adhesive activities in tissue culture-based studies. The important roles that several proteolytically inactive ADAMs play in development (ADAMs 2, 3, 14, and 23) suggest that ADAM adhesive activities may be relevant to their function. In this chapter, we first review the history and phylogeny of the ADAMs as well as structural and functional aspects of their major domains. We next review how ADAMs function as ectodomain sheddases, how their protease activities may be regulated, and how ADAMs may function in modulating cell adhesion and cell migration. We end with a very brief discussion of the role of ADAMs in development and disease and conclude by posing some questions for future research. Our goal is to give an appreciation for the widespread, varied, and fascinating means by which ADAMs affect, or may affect, key cell surface events: cell signaling, cell adhesion, and cell migration.

**Key words**: Disintegrin, metalloprotease, sheddase, signaling, cell adhesion.



*N. M. Hooper and U. Lendeckel (eds.), The ADAM Family of Proteases,* 1-28.

# 1. HISTORY, GENERAL FEATURES, AND PHYLOGENY

The ADAM family was unearthed during an analysis of a critical cell-cell interaction, the union between sperm and egg that initiates zygotic development (Blobel *et al.* 1992; Wolfsberg *et al.* 1993; Wolfsberg *et al.* 1995). The name ADAM denotes two of the key ADAM domains, A Disintegrin and A Metalloprotease domain, as well as their historic roots. ADAMs are present in all multicellular organisms, and ADAM-like sequences have been noted in filamentous fungi as well as in the unicellular eukaryote, *Schizosaccharomyces pombe* (but not in *Saccharomyces cerevisiae*) (Nakamura *et al.* 2004b). All ADAMs contain (from N- to C-terminus): pro, metalloprotease, disintegrin, cysteine-rich, spacer, transmembrane, and cytoplasmic tail domains (Fig. 1).

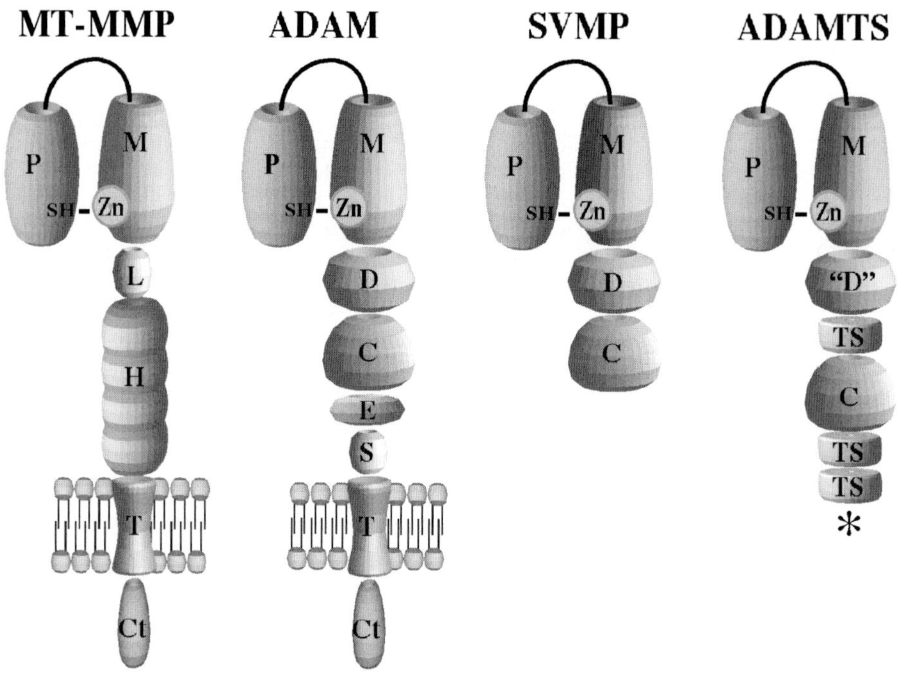


*Figure 1.* Domains of ADAMs and Relatives. ADAMs possess a Pro, P; Metalloprotease, M; Disintegrin, D; Cysteine-Rich, C; EGF-like module, E; Spacer, S; Transmembrane, T; and Cytoplasmic tail, Ct. The P and M domains share features with their relatives (e.g. a Zn binding catalytic site), but downstream domains differ. Relatives depicted are a membrane type (MT) MMP and two other adamalysins, a P-III SVMP and an ADAMTS protein. L, linker; H, hemopexin domain; TS, thrombospondin repeat. * denotes that ADAMTS proteins vary in length in their C-terminal regions, which contain variable numbers of TS repeats.

For the majority, domains downstream of the pro domain are of similar length and share cysteine alignment. ADAMs 10 and 17 differ from the others in several respects (e.g. size of metalloprotease domain and features of their cysteine-rich and spacer regions). One version of each ADAM is transmembrane-anchored, but alternative splicing has been shown to generate secreted forms of several ADAMs.

*Table 1.* ADAM protease substrates

| **ADAM** | **[a]Substrate** |
|---|---|
| ADAM 8 | [b]CD23, [c]CHL1, [f]ADAM8 (pro domain) |
| ADAM 9 | [b]proHB-EGF, kit ligand-1, p75 neurotrophin receptor (NTR), [d]APP, oxytocinase, [e]collagen XVII, fibronectin |
| ADAM 10 | [b]proTNFα, proEGF, proHB-EGF, probetacellulin, proTGFα, IL-6R, Notch, Delta 1, fractalkine (CX3CL1), CXCL16, ephrin-A2, [c]CD44, L1 adhesion molecule, [d]APP, Prion Protein Cellular (PrPc) |
| ADAM 12 | [b]proHB-EGF, [d]IGFBP-3, IGFBP-5, oxytocinase |
| ADAM 13 | [e]fibronectin, [f]ADAM 13 (cysteine-rich domain) |
| ADAM 15 | [b]proamphiregulin, proTGFα, CD23, [e]collagen IV |
| ADAM 17 | [b]proTNFα, proHB-EGF, proTGFα, CD30, CD40, p55 TNFR, p75NTR, TrkA, TRANCE/RANKL, proamphiregulin, proepiregulin, proneuregulinα2c, c-Kit, erbB4/HER4, GHR, IL-IR, IL-6R, IL-15Rα, M-CSFR, Notch1, fractalkine (CX3CL1), [c]CD44, L-selectin, MUC1, VCAM-1, GP1b-α, [d]APP, Ebola Glycoprotein, PrPc, Vibrio cholera cytolysin, [e]collagen XVII |
| ADAM 19 | [b]proTNFα (in TACE deficient cells), neuregulin, TRANCE/RANKL, [f]ADAM 19 (cysteine-rich domain) |
| ADAM 28 | [b]CD23, [d]IGFBP-3, [f]ADAM 28 (pro domain) |
| ADAM 33 | [a]kit-ligand 1, [d]APP |

[a]Peptide and model substrates (eg α2-macroglobulin) are not included; substrate listing as follows: [b]signalling ligands and receptors, [c]adhesion molecules, [d]other, [e]extracellular matrix components, [f]auto-catalytic. Primary references can be found in Seals and Courtneidge 2003; Becherer and Blobel 2003; White 2003; or Tomczuk 2004.

ADAMs are members of the adamylsin family of the metzincin subclan of Zn-dependent metalloproteases, which also includes astacins and matrix metalloproteases (MMPs). The adamalysin family also includes snake venom metalloproteases (SVMPs) and ADAMTS proteins (ADAMs containing thrombospondin repeats) (Fig. 1). In contrast to SVMPs and ADAMTS proteins, only ~60% of ADAMs contain a catalytic site signature motif (HExGHxxGxxHD); all of the latter that have been tested are proteolytically active (Table 1). Excluding pseudogenes, public databases list 20 genes encoding human ADAMs and 37 genes encoding mouse ADAMs; 8 of the latter have only been found to date as predictions. As seen in Fig. 2, there appear to be several ADAM subfamilies, and the ability to be or not to be proteolytically active appears to be conserved among subfamily members. Of the mouse ADAM genes, 22 are reported as testis specific (or testis predominant, e.g. testases such as ADAM24). Several of these are intronless multi-copy genes, that are not present (or only present as pseudogenes) in the human genome (Fig. 2 and Chapter 11).

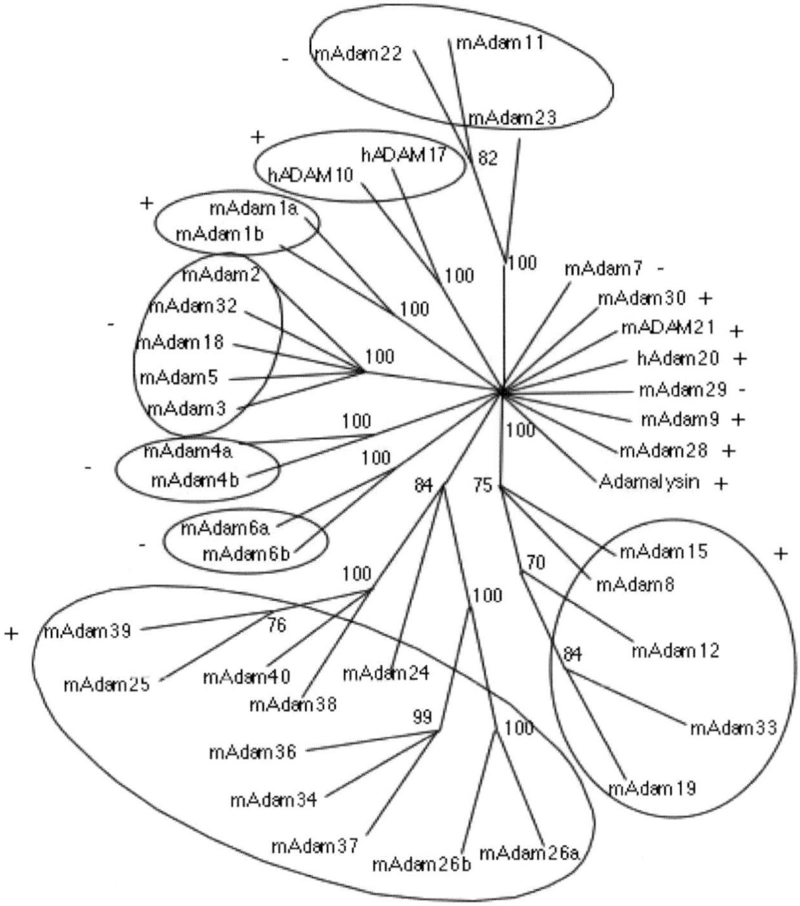


*Figure 2.* Phylogenetic Analysis of ADAM Metalloprotease Domains. Numbers on the nodes indicate the bootstrap value, the number of times that fork was generated in 100 separate analyses. Branch lengths are not significant. Related groups of ADAMs are circled, and `+` and `-` indicate the presence or absence of the catalytic active site motif. Note that there are no human testases (bottom, left group), and no mouse Adam20. To generate the tree, sequences of mouse Adams (mAdams) were aligned to human ADAM20 (hADAM20) and the SVMP Adamalysin with MUSCLE (http://www.drive5.com/muscle/). Since ADAMs 10 and 17 cannot be aligned to other ADAMs based on sequence alone, they were incorporated into the alignment via a structure-based alignment of human ADAMs 10 and 17 with Adamalysin (Maskos *et al.* 1998). Trees were constructed using algorithms in PHYLIP

(http://evolution.genetics.washington.edu/phylip.html). The sequences were subjected to 100 bootstrap replicates, and the parsimony method PROTPARS was used to calculate phylogenies. Nodes with bootstrap values of 69 or less were discarded. The tree diagram was generated with TREEVIEW (http://taxonomy.zoology.gla.ac.uk/rod/treeview.html). Analysis of additional sequences groups the ADAMTSs with ADAMs 10 and 17. If nodes with lower bootstrap values are retained, additional groupings are evident. For example, ADAMs 4a, 4b, 6a, 6b 20, 21, 29, and 30 join the existing testase group. Trees constructed using a distance method show a grouping of ADAMS 4a/4b with 6a/6b with a bootstrap cutoff of 70. The reported testis specific/predominant ADAMs are all of those in the branches including ADAMs 2, 4, 6, and 24, as well as ADAMs 20 (in humans), 21, 29, and 30.

The ADAM family was inaugurated with the characterization of ADAMs 1 and 2 (Blobel *et al.* 1992; Wolfsberg *et al.* 1993), which had been implicated in fertilization (Primakoff *et al.* 1987). Other early milestones on the family timeline were studics implicating ADAMs in myoblast fusion (Yagami-Hiromasa *et al.* 1995), in cell differentiation during neurogenesis (Pan and Rubin, 1997; Rooke *et al.* 1996), in axon extension (Fambrough *et al.* 1996), and in the shedding of important signaling ligands (Black *et al.* 1997; Moss *et al.* 1997). Soon thereafter it was demonstrated, using knockout mice, that both proteolytically active (Peschon *et al.* 1998) and proteolytically inactive (Cho *et al.* 1998) ADAMs are involved in (predicted) developmental processes. Another early milestone was the presentation of the first high resolution structure of an ADAM catalytic domain (Maskos *et al.* 1998). ADAMs have clearly grown beyond their roots, and they have emerged as key regulators of many important events that occur at the cell surface. For recent reviews, see (Becherer and Blobel, 2003; Seals and Courtneidge, 2003; White, 2003).

# 2. DOMAINS

## 2.1 Metalloprotease Domain

ADAM proteases can shed or cleave a large array of substrates (Table 1), and ADAM-mediated shedding is clearly very important *in vivo* (Peschon *et al.* 1998). ADAM metalloprotease domains are preceded by a pro domain, which is thought to work as an intramolecular chaperone controlling folding of the catalytic domain. The pro domain maintains the protease domain in a latent state, and it may influence intracellular trafficking (Anders *et al.* 2001; Gonzales *et al.* 2004). Most ADAM pro domains are removed by a proprotein convertases (e.g., furin) in the secretory pathway. Two (ADAMs 8 and 28) have been shown to undergo autocatalytic pro domain removal (Howard *et al.* 2000; Schlomann *et al.* 2002). Like those of most metzincins,

ADAM pro domains contain a 'cysteine switch box' and are thought to be activated by a cysteine-switch mechanism. Recent work has indicated, however, that additional features of the pro domain assist in maintaining the catalytic domain of ADAM17 (TACE) in its latent state (Gonzales *et al.* 2004).

The structures of the metalloprotease domains of ADAM17 (TACE; PDB# 1BKC) and ADAM33 (PDB# 1R54) are known at high resolution (Maskos *et al.* 1998; Orth *et al.* 2004). Both contain a central five-stranded β-sheet surrounded by five α-helices and a conserved methionine below the active site. As anticipated by its larger size (~50 residues), the ADAM17 catalytic domain has several projecting loops not seen in ADAM 33.

ADAMs are major, albeit not exclusive, proteases that execute the critical function of ectodomain shedding (Becherer and Blobel, 2003). In this manner ADAMs provide extracellular signaling ligands (section 3.1) and they help provide intracellular signaling molecules (section 3.2). Substrates that can be shed by ADAMs (Table 1) include signaling ligands (e.g. the potent cytokine TNF-α and EGF family members), receptors for signaling ligands (e.g. for EGF family members, TNF-α, nerve growth factors, and growth hormone), and cell adhesion molecules (e.g. CD44, L1, and L-selectin). Shedding can be constitutive, but can generally be enhanced by natural stimuli such as ligands for G protein coupled receptors (GPCRs). The critical questions of which ADAMs shed which biologically active proteins in which specific cells at which stages of development or under which physiological conditions are beginning to be addressed. For example, a recent study using fibroblasts from mice null for specific ADAMs showed that ADAM10 preferentially sheds two EGF family members, whereas ADAM 17 preferentially sheds four others (Sahin *et al.* 2004). Other recent studies have shown that ADAMs 10, 15, and 17 shed different EGF ligands in different cell types (Schafer *et al.* 2004a; Schafer *et al.* 2004b).

The natural substrates shed by ADAMs encompass type I, type II and GPI-anchored proteins including ones with monomeric and oligomeric ectodomains. Although ADAMs display cleavage site specificity towards peptide substrates, they cleave broader ensembles of sites in cell membrane-anchored substrates. Point mutations at the cleavage sites of membrane-anchored substrates generally have minimal effect. Hence the emerging picture is that cleavage sites in membrane-anchored substrates are selected based on their distance from the membrane (~10-20 amino acids), secondary structure, and accessibility (Black *et al.* 2003; Franzke *et al.* 2004). Sequences distal to the cleavage site may also be important (Hattori *et al.* 2000; Zhao *et al.* 2001). Given the steric constraints imposed on a membrane-anchored protease to cleave a membrane-anchored substrate, it is reasonable that secreted ADAMs preferentially cleave different substrates than their membrane-anchored counterparts (Reddy *et al.* 2000; Schlomann *et al.* 2002; Wakatsuki *et al.* 2004).

In addition to signaling ligands, cell adhesion molecules, and receptors, other proteins located in extracellular spaces can be cleaved by ADAMs. For example, ADAM12S (secreted ADAM12; Chapter 6) cleaves insulin-like growth factor binding proteins 3 and 5. In addition there is a suggestion that an ADAM cleaves TGF-α at a membrane distal site (in addition to its juxtamembrane site) during processing to its bioactive form (Hinkle *et al.* 2003) And, based on the actions of metalloprotease inhibitors, or by analogy with SVMPs, ADAMs may be involved in shedding or cleaving other important proteins (see legend to Table 2 in White (2003)). An interesting group of (potential) substrates encompasses molecules involved in cell-matrix interactions such as syndecans (Holen *et al.* 2001), integrins (Matsui *et al.* 2000), CD44 (Nagano *et al.* 2004; Nakamura *et al.* 2004a), and ECM components such as fibronectin (FN), laminin, and collagen (Matsui *et al.* 2000). Although cleavage of ECM proteins is generally considered to be the purview of MMPs, ADAMTS proteins cleave several important ECM components (aggrecan, versican, and von Willibrand factor) (Apte, 2004; also see Chapters 12-15), and four ADAMs have been shown, *in vitro*, to cleave ECM substrates (Table 1). For example, ADAM13 generates an ~90 kD product of FN and can 'remodel' a FN substrate (Alfandari *et al.* 2001).

Many challenging questions remain about ADAM metalloproteases. As mentioned above, one of the most pressing questions is deciphering which ADAMs precisely cleave which biologically relevant substrates in which cells at which stages of development or under which set of conditions. This question encompasses possible overlapping functions with ADAM-TS proteins and MMPs (e.g. MT-MMPs). Despite differences in their active site structures (Solomon *et al.* 2004), as reflected by differing inhibitor profiles (Baker *et al.* 2002), ADAMs and MMPs can act upon overlapping sets of substrates. Other pressing questions include: Where does shedding occur (in the secretory pathway, at the cell surface, in lipid rafts)? How do ancillary domains (disintegrin, cysteine-rich, cytoplasmic tail) contribute to proteolytic specificity or activity? Do non-proteolytic ADAMs act as endogenous inhibitors (e.g., by competing for substrates) or as facilitators (e.g., by interacting with endogenous inhibitors)? How do cells balance ADAM-mediated shedding of signaling ligands (e.g. TNF-α) with shedding of their cognate receptors (e.g. TNF-α receptor) to titrate a signaling response? How are various ADAM domains organized in three dimensions such that the protease domain can access varied cleavage sites (e.g. ones close to the membrane and in the extracellular space)? Does the ability of some ADAMs to cleave ECM components relate to their *in vivo* functions? Several of these questions will be addressed in section 3.

## 2.2 Disintegrin Domain

ADAMs are unique among membrane-anchored proteins (and non-venomous proteins) in having a disintegrin domain. While other proteins such as *Dictyostelium* AmpA (Varney *et al.* 2002) and ADAMTS proteins (Apte, 2004), have disintegrin-like domains, domains with a characteristic cysteine alignment and characteristic disintegrin loops are only seen in ADAMs and SVMPs (Fig. 1). PII SVMP disintegrins associate with RGD-interacting (αIIBβ3, αvβ3, α5β1) or 'leukocyte' (α4β1, α4β7, α9β1) or α1β1 integrins (Marcinkiewicz, 2004), with sequences near the centers of their disintegrin loops (e.g. RGD, MLD, KTS) being important for function. Other venomous proteins have been identified that interact with α2β1 (Marcinkiewicz, 2004) or α3β1, α6β1, and α7β1 integrins (Eble *et al.* 2003). Like those of PIII SVMPs, most ADAM disintegrin loops contain 14 amino acids with the 'canonical' disintegrin loop motif shown at the top of Table 2. The disintegrin loops of ADAMs 10 and 17 vary from this motif. Interestingly, the disintegrin domain is the most highly conserved domain between the *S. pombe* ADAM-like protein and its relatives in higher eukaryotes (Nakamura *et al.* 2004b), and it displays the canonical disintegrin loop motif.

Early hints that ADAM disintegrin domains can interact with integrins came from studies in the fertilization system (for review, see Kaji and Kudo (2004)). Since that time, recombinant disintegrin domains from fourteen ADAMs have been shown to support adhesion mediated by a variety of integrins: α4β1, α5β1, α6β1, α7β1, α9β1, αvβ3, and α4β7 (Table 2). No ADAM disintegrin domain has yet been shown to interact with an I domain-containing integrin. Mutations of residues within the disintegrin loop have been shown to impair adhesive activity. The sequence $RX_6DLPEF$ functions as an α9β1-interacting motif (Eto *et al.* 2002), but ADAMs with variant loop sequences can also interact with α9β1 (Table 2). A given ADAM disintegrin domain can interact with multiple integrins (Tomczuk *et al.* 2003), and different residues within the loop appear to predominate in interactions with different integrins (Bridges *et al.* 2004). Sequences outside of the disintegrin loop also appear to contribute to adhesive activity (Bridges *et al.* 2003). Several reports indicate that full length membrane-anchored ADAMs can interact with membrane-anchored integrins (Bax *et al.* 2004; Eto *et al.* 2000; Kawaguchi *et al.* 2003).

Major questions regarding ADAM disintegrin domains are whether and how the interactions with integrins observed in model systems relate to *in vivo* functions (section 3.4). If such connections can be made, then it will be important to further decipher the sequence requirements and integrin selectivity of ADAM disintegrin domains, to gain information on (cell type specific) regulation of ADAM-integrin interactions, and to determine the major consequences of ADAM-integrin pairing.

**Table 2. ADAM Disintegrin-Integrin Interactions**

| ADAM | Disintegrin Loop Sequence | Integrins |
|---|---|---|
| Canonical | CRXXXXXCDXXEXC | |
| mADAM 1 | CRPAEDVCDLPEYC | α9β1 |
| mADAM 2 | CRLAQDECDVTEYC | α4β1, α6β1, α9β1 |
| mADAM 3 | CRKSKDQCDFPEFC | α4β1, α6β1, α7β1, α9β1 |
| hADAM 7 | CRPAKDECDFPEMC | α4β1, α9β1, α4β7 |
| mADAM 9<br>hADAM 9 | CRGKTSECDVPEYC<br>CRGKTSECDVPEYC | α6β1<br>α9β1, αvβ5 |
| mADAM 12<br>hADAM 12 | CRGSSNSCDLPEFC<br>CRDSSNSCDLPEFC | α7β1, α9β1<br>α9β1 |
| mADAM 15<br>hADAM 15 | CRPPTDDCDLPEFC<br>CRPTRGDCDLPEFC | α9β1<br>α5β1, α9β1, αvβ3 |
| hADAM 17 | CQEAINATCKGVSYC | α5β1 |
| hADAM 23 | CRDAVNECDITEYC | αvβ3 |
| hADAM 28 | CRPAKDECDLPEMC | α4β1, α9β1, α4β7 |
| hADAM 33 | CRQAMGDCDLPEFC | α9β1 |

Residues conserved among all known mammalian orthologues are shaded; arrowheads denote residues that have been implicated in integrin recognition. Primary references can be found in White 2003, Bridges and Bowditch 2004 or Tomczuk 2004.

## 2.3 Cysteine-Rich Domain

All ADAMs contain a cysteine-rich domain before their transmembrane domains. Several observations suggest that the cysteine-rich domain is important for the biological activities of some ADAMs (see also sections 3.3 and 3.4). Recombinant constructs containing the ADAM12 cysteine-rich domain can interact with syndecans and promote β1 integrin-mediated cell spreading (Thodeti *et al.* 2003). A recombinant construct containing the disintegrin and cysteine-rich domains of ADAM13 binds to fibronectin using its cysteine-rich domain (Gaultier *et al.* 2002). Elements in the disintegrin and/or cysteine-rich domain of ADAM17 influence protease activity (Lee *et al.* 2003; Reddy *et al.* 2000). The cysteine-rich domain of *Xenopus* ADAM13 is critical for specifying a protease-directed activity *in vivo* (Smith *et al.* 2002; Chapter 7). ADAM14 (unc-71) is a non-proteolytic ADAM that is involved in cell migration in *C. elegans*. Some of the loss of function alleles of ADAM14 map to its cysteine-rich domain (Huang *et al.* 2003).

## 2.4 Cytoplasmic Tail

ADAM cytoplasmic tails vary in length and composition. Several are subject to alternative splicing and many contain one or more SH3 binding sites or sites for serine, threonine or tyrosine phosphorylation (Seals and Courtneidge, 2003). Numerous adapter proteins have emerged as ADAM cytoplasmic tail binding partners (Table 3).

*Table 3.* ADAM cytoplasmic tail interactions

| ADAM | Cytoplasmic Binding Partners |
|---|---|
| ADAM 9 | Endophilin I, MAD2β, PACSIN3, PKCδ, SH3PXI, SRC |
| ADAM 10 | Lck, MAD2, PACSIN3 |
| ADAM 12 | α-actinin-1, α-actinin-2, FISH, Grb2, PACSIN3, PI3 kinase, PKCδ, Src, YES, Eve-1 |
| ADAM 13 | Ab1, PACSIN2, Src |
| ADAM 15 | Ab1, endophilin I, FISH, Fyn, Grb2, Hck, Lck, MAD2, MAD2β, PACSIN3, SH3PXI, Src |
| ADAM 17 | Erk, MAD2, PTPH, SAP97 |
| ADAM 19 | ArgBP1, Beta cop, FISH, MAD2β, PACSIN3, Ubiquitin |
| ADAM 22 | 14-3-3ζ |

Primary references can be found in Seals and Courtneidge 2003 or Tomczuk 2004.

Phosphorylation of, or binding of adapter proteins to, the cytoplasmic tail could influence ADAM maturation (e.g. Pro domain removal), trafficking (e.g. to the cell surface), membrane localization (e.g. to rafts), association with cytoskeletal elements, or proteolytic activity. In section 3.3 we discuss evidence that these types of cytoplasmic tail modifications influence protease activity. In addition, there are suggestions that interactions between the cytoplasmic tail of ADAM12 and α-actinins-1 and -2 influence myoblast fusion (Cao *et al.* 2001; Galliano *et al.* 2000). And, binding of 14-3-3ζ to the cytoplasmic tail of ADAM22 appears to influence cell adhesion and spreading on FN, an effect that may be mediated by β1 integrins (Zhu *et al.* 2003).

## 3. MODELS FOR BIOLOGICAL ACTIVITIES

ADAMs shed the ectodomains of a wide variety of substrates (Table 1). ADAM-mediated shedding can evoke extracellular signaling (section 3.1) or it can precede cleavage events that evoke intracellular signaling (section 3.2). ADAMs can also influence cell-cell and cell-matrix interactions (section 3.4).

### 3.1 ADAM Shedding in Extracellular Signaling: The EGF Ligand Paradigm

In the simplest mode an ADAM constitutively sheds a membrane-anchored substrate by cleaving a site in its juxtamembrane region. The released ligand (e.g., a growth factor or cytokine) can then initiate an autocrine or paracrine extracellular signaling pathway by binding to cognate cell surface receptors (e.g., growth factor and cytokine receptors). Shedding of cognate receptors (e.g. TNF receptors) could inhibit or titrate the extracellular signaling response. ADAM-mediated shedding can be activated by experimental (e.g. phorbol esters) and natural (e.g. ligands for GPCRs) stimulants. A very important mechanism underlying mitogenic responses is GPCR transactivation of EGF receptors (Prenzel *et al.* 1999). ADAMs 10, 12, 15, and 17 have been implicated in various GPCR/EGFR transactivation events (Asakura *et al.* 2002; Lemjabbar and Basbaum, 2002; Sahin *et al.* 2004; Schafer *et al.* 2004a; Schafer *et al.* 2004b; Yan *et al.* 2002). For GPCR/EGFR transactivation to occur, GPCR ligands activate ADAMs to shed EGF family members from their membrane-bound pro forms. The released EGF family member (e.g. HB-EGF) then activates its cognate EGFR (Fig. 3A). Different ADAMs may participate in constitutive vs. stimulated shedding, and in a given cell type different ADAMs may preferentially shed different EGF family members (Sahin *et al.* 2004; Schafer *et al.* 2004a; Schafer *et al.* 2004b). Other substances (than GPCR ligands) can activate ADAMs: TNF was shown to induce ADAM17 to produce TGF-α, which activated an EGFR response linked with hepatocyte replication (Argast *et al.* 2004). Similarly, tobacco smoke has been shown to activate ADAM17 to produce amphiregulin, leading (via EGFR) to proliferation of lung epithelial cells (Lemjabbar *et al.* 2003) or to produce TGF-α leading (via EGFR) to increased mucin production (Shao *et al.* 2004). Protein kinases have been implicated in the activation of ADAM sheddases: PKC and MAP kinases in response to phorbol esters, and MAP kinases (Erk and p38 MAP kinase) in response to different natural stimuli (Black, 2003). Another way in which the effect of ADAM shedding might be augmented is by reducing the volume of intercellular spaces, for example as a result of compressive stress (Tschumperlin *et al.* 2004). In this manner the net concentration of an EGF ligand (shed constitutively) could be enhanced, thereby upregulating EGFR signaling (Tschumperlin *et al.* 2004).

## 3.2 ADAM Shedding in Intracellular Signaling: The Notch Paradigm

ADAMs also participate in intracellular signaling. This was first shown by documenting the role of kuzbanian (fly ADAM10) in the lateral inhibition that leads to selective differentiation of neuronal cells through the Notch signaling pathway in *Drosophila* (Pan and Rubin, 1997; Rooke *et al.* 1996). And, ADAM10 has proven to be critical for Notch signaling in mice (Hartmann *et al.* 2002). Notch is a cell surface receptor. Its large ectodomain can be shed by ADAM10. Following ADAM10-mediated shedding, presenilin (Wolfe and Kopan, 2004) cleaves a site in the transmembrane domain of Notch. The resultant fragment leaves the membrane and travels to the nucleus to influence gene transcription (Fig. 3B). Recent work suggests that other ADAM-mediated shedding events precede similar intramembrane cleavages that result in intracellular signaling. New work documents ADAM-mediated Notch-like processing for two Notch ligands, Delta and Jagged (Bland *et al.* 2003; LaVoie and Selkoe, 2003; Six *et al.* 2003), for the cell-matrix adhesion molecule, CD44 (Murakami *et al.* 2003) (see section 3.4), and for HB-EGF (Nanba *et al.* 2003). Other important molecules that may be subject to ADAM-initiated Notch-like proteolytic cascades are Erb-B4 (Ni *et al.* 2001), neuregulin (Bao *et al.* 2003), and N-cadherin (Marambaud *et al.* 2003).

## 3.3 Regulation of Protease Activity

As discussed above ADAM sheddases can be activated by natural (e.g. GPCR ligands, TNF, tobacco smoke) or experimental (phorbol esters,

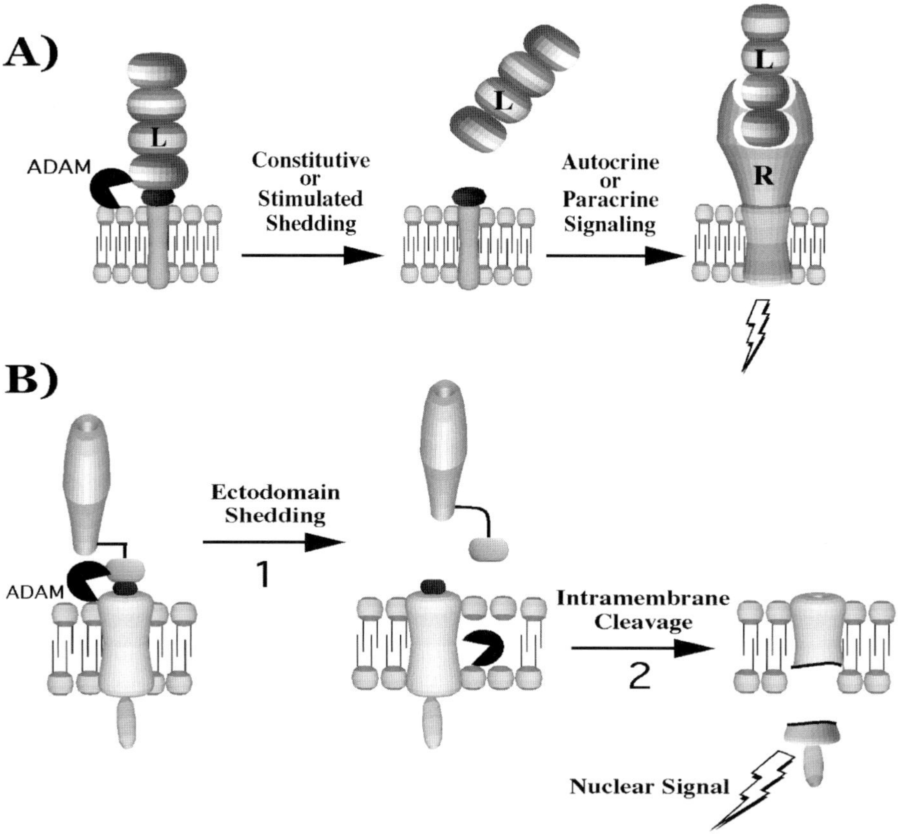


*Figure 3.* ADAM Proteases in Extracellular (A) and Intracellular (B) Signaling. (A) ADAM cleavage of a membrane-anchored ligand (L) such as Pro-HB-EGF at a juxtamembrane site produces a soluble ligand that can subsequently bind, in an autocrine or paracrine manner, to a cognate receptor (R) propagating a cellular signal. (B) A cell surface receptor (e.g. Notch) or ligand (e.g. Delta) is initially processed at a juxtamembrane site in its ectodomain by an ADAM protease (1) leading to a subsequent cleavage at an intramembrane site (2). Following intramembrane cleavage, a C-terminal fragment is translocated to the nucleus activating and recruiting various transcription factors.

pervanadate) stimuli. However, the mechanisms that regulate activation and other aspects of ADAM proteolytic function are ill defined. Several observations suggest that ADAM proteolytic function can be influenced by domains downstream of their metalloprotease domains (Fig. 1). First, isolated ADAM metalloprotease domains or ectodomains cleave different substrates than their full length (membrane-anchored) counterparts (Reddy *et al.* 2000; Schlomann *et al.* 2002; Wakatsuki *et al.* 2004). Second, ADAMs are subject to alternative splicing to generate different cysteine-rich sequences, different (or no) cytoplasmic tails, or secreted forms. Third, certain ADAMs can 'self-shed' to generate secreted forms (Table 1). Hence downstream ADAM domains are likely to be very important for overall proteolytic activity.

### 3.3.1 Roles of disintegrin and cysteine-rich domains

In addition to their potential roles in cell adhesion and cell migration (section 3.4), several studies indicate that ADAM adhesive domains (disintegrin and cysteine-rich) can influence ADAM proteolytic function (reviewed in Bridges and Bowditch, 2004; White, 2003). In one such study shedding of the ectodomain of IL-1R-II by ADAM17 was found to require its cysteine-rich region (Reddy *et al.* 2000). And, the cysteine-rich domain of ADAM13 was found to be the major determinant specifying an ADAM13 protease-mediated developmental event in *Xenopus* embryos; the disintegrin domain was found to enhance the effect (Smith *et al.* 2002). The disintegrin domain of ADAM10 is not, however, needed for its activity as an amyloid precursor protein α-secretase (Fahrenholz *et al.* 2000). The role of downstream ADAM cysteine-rich (and disintegrin) domains may be analogous to the role of domains downstream of the metalloprotease domains of some MMPs and ADAMTS proteins (Overall, 2002; Soejima *et al.* 2003; Somerville *et al.* 2003). As proposed for the former, downstream ADAM domains might help select substrates (Hattori *et al.* 2000), they might help the ADAM get to its site of action, or they might influence its interaction with inhibitors, such as TIMPs (Lee *et al.* 2003). A recent study has shown that TIMP3 functions as an endogenous regulator of ADAM17 (Mohammed *et al.* 2004).

### 3.3.2 Roles of the cytoplasmic tail

ADAM cytoplasmic tails have been proposed to modulate protease activity through interactions with adapter proteins or by becoming phosphorylated. ADAM tails cannot, however, be obligatorily required since (stimulated) shedding of certain ADAM substrates appears to occur equally well with ADAMs that lack their cytoplasmic tails (Reddy *et al.* 2000; Wakatsuki *et al.* 2004).

Although the mechanisms are not yet clear, several studies have indicated that interactions between ADAM cytoplasmic tails and adapter proteins are relevant to overall ADAM protease activity. An interaction between PACSIN2 and the cytoplasmic tail of ADAM13 was shown to interfere with two ADAM protease-mediated developmental events (Cousin *et al.* 2000). Conversely, by binding to the cytoplasmic tail of ADAM12, PACSIN3 was reported to positively regulate HB-EGF shedding in response to phorbol ester and angiotensin II (Mori *et al.* 2003). Another ADAM12 interacting protein, Eve-1, was also shown to positively regulate stimulated shedding of several EGF ligands, including HB-EGF (Tanaka *et al.* 2004). Two proteins that interact with the ADAM17 cytoplasmic tail, protein-tyrosine phosphatase-1 (Zheng *et al.* 2002) and SAP97 (Peiretti *et al.* 2003), have been shown to influence the ability of ADAM17 to shed several ligands.

The role of phosphorylation of ADAM cytoplasmic tails for protease function is under active investigation. It was suggested that Erk-mediated phosphorylation of a threonine in the cytoplasmic tail of ADAM17 is needed for efficient shedding of TrkA, a nerve growth factor receptor (Diaz-Rodriguez *et al.* 2002). Another study showed that although the cytoplasmic tail of ADAM17 is phosphorylated (and dephosphorylated) on serines in response to growth factor stimulation, neither modification was needed for induced shedding of TGF-$\alpha$ (Fan *et al.* 2003). Hence phosphorylation of the cytoplasmic tail cannot be a general mechanism for activating ADAM17. Moreover, as noted above, the cytoplasmic tail of ADAM17 is not needed for phorbol ester-stimulated shedding of several physiological ligands (Black *et al.* 2003; Reddy *et al.* 2000). Remarkably, phorbol esters can activate the intrinsic activity of tail-less (membrane-anchored) ADAM17, and this effect occurs without changes in either the cell surface expression levels of ADAM17 or processing of ADAM17 to its mature form. The authors suggested that phorbol ester-activation of tail-less ADAM17 occurs by engaging an activator or releasing an inhibitor (Doedens *et al.* 2003). The cytoplasmic tail of ADAM15 has also been considered as a functional target for phophorylation. The ADAM15 cytoplasmic tail binds more of some, but less of other, adapter proteins when it is phosphorylated (Poghosyan *et al.* 2002). Whether, and if so how, such phosphorylation events regulate the protease activity of ADAM15 remains to be determined.

### 3.3.3 Other Possible Regulatory Mechanisms

Other post-translational mechanisms have been invoked to explain regulation of ADAM sheddase activities. These include regulation of Pro domain removal (Nagano *et al.* 2004), cell surface expression (Hougaard *et al.* 2000), or localization to lipid rafts (Wakatsuki *et al.* 2004). With regards to the latter possibility, two studies have suggested that low cholesterol, which could influence membrane properties, promotes some ADAM-mediated shedding events (Kojro *et al.* 2001; Matthews *et al.* 2003). And the tetraspanin integrin-associated protein CD9, which may help foster protein localization to special membrane domains, has been found in a complex with ADAM10 and a substrate, HB-EGF (Yan *et al.* 2002). With regards to the possibility of regulated pro domain removal, a study on the mechanism of CD44 shedding (see section 3.4) suggested that calcium influx leads to association of calmodulin with the pro form of ADAM10 and conversion of more ADAM10 to its mature form. In the same study, phorbol ester was shown to elicit greater conversion of ADAM17 to its mature form and greater shedding of CD44 (Nagano *et al.* 2004).

## 3.4 ADAMs in Cell Adhesion and Migration

ADAMs may influence cell adhesion and cell migration by virtue of their proteolytic or adhesive activities. Below we consider some of the ways in which ADAMs may modulate (either positively or negatively) such events. Both proteolytically active ADAMs (Alfandari *et al.* 2001; Martin *et al.* 2002) as well as a proteolytically inactive ADAM (Huang *et al.* 2003) have been implicated in cell migration events.

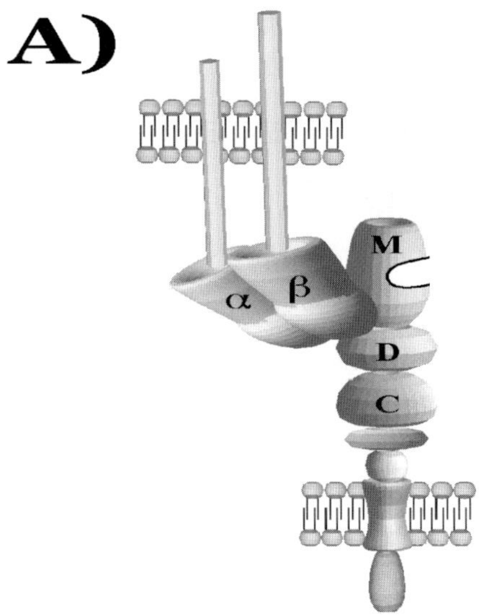

**B)**

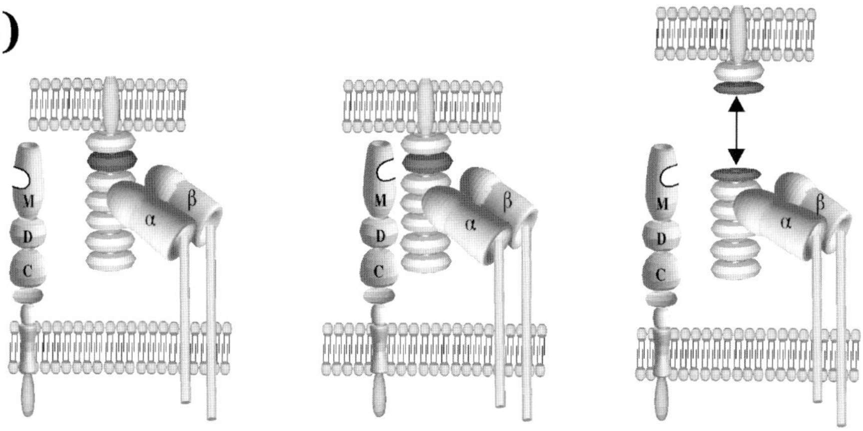


**C)**

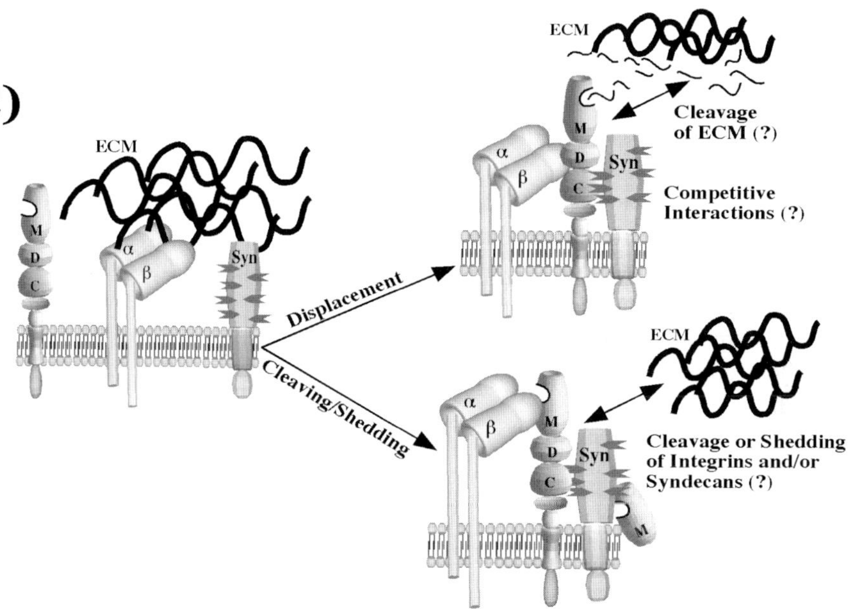


*Figure 4.* Models for ADAMs as Modulators of Cell-Cell and Cell-Matrix Interactions. ADAMs are depicted as in Fig. 1. Integrins are shown as α and β subunits, extracellular matrix component, ECM; syndecan, Syn. (A) Interaction of an ADAM with an integrin facilitates a positive cell-cell interaction. (B) ADAM shedding of an adhesion counter receptor (e.g VCAM) terminates (i.e. negatively regulates) integrin mediated cell-cell contacts. (C, top) ADAM association with integrins and/or syndecans could competitively displace (double-headed arrow) integrin and/or syndecan mediated cell-ECM interactions. Binding of the ADAM to, or possibly cleavage of, ECM components may accompany this process. (C, bottom) Alternatively, the ADAM could attenuate cell-ECM contacts by cleaving or shedding the integrin or syndecan.

### 3.4.1 ADAMs in Cell-Cell and Cell-Matrix Adhesion

ADAMs may foster cell-cell interactions (Fig. 4A) by binding to integrins on neighboring cells (Eto *et al.* 2000; Herren *et al.* 2001). Conversely, they may inhibit cell-cell adhesion by shedding cell-cell adhesion molecules (Fig. 4B) such as L1, CHL1, VCAM, and selectins (Table 1). An interesting scenario is the case of two prochemokine ligands, fractalkine (CX3CL1) and CXCL16. In membrane-anchored form these proteins promote cell-cell adhesion. Shedding of their ectodomains breaks cell-cell contacts, and the shed ectodomains now act as a chemoattractants (Abel *et al.* 2004; Hundhausen *et al.* 2003). Another interesting case is shedding of cell-anchored eprin-A2, which breaks its association with its cell-anchored receptor, thereby leading to axon repulsion (Hattori *et al.* 2000).

ADAMs may foster cell-matrix adhesion by binding to ECM components (Gaultier *et al.* 2002) or by activating β1 integrins (Thodeti *et al.* 2003). They may inhibit cell-matrix adhesion by interacting with integrins (through their disintegrin domains) and/or with syndecans (through their cysteine-rich domains), thereby weakening cell-matrix interactions (Fig. 4C). Future work is needed to assess whether cell-anchored ADAMs, integrins and syndecans interact functionally "in cis" through the implicated ADAM domains and whether such interactions affect cell-matrix interactions. ADAMs may also modify cell-matrix adhesion by shedding cell-matrix adhesion molecules (Fig. 4C) such as syndecans (Holen *et al.* 2001) and CD44 (Nagano *et al.* 2004; Nakamura *et al.* 2004a). And as suggested by the ability of an SVMP, ADAMs could negatively impact cell-matrix adhesion by cleaving an integrin.

### 3.4.2 ADAMs in Cell Migration

ADAMs could influence cell migration by modulating cell-cell or cell-matrix adhesion by any of the means discussed in the preceding section, and they could do so by acting in either the migrating cell or in cells in the migratory pathway. To begin a migratory event, cells must often break cell-cell contacts (e.g. as in epithelial to mesenchymal transitions) and favor cell-matrix contacts. And, there must be an optimum level of cell-matrix adhesion for cell migration to proceed (Palecek *et al.* 1997). In this context it is intriguing that a non-proteolytically-active ADAM in *C. elegans* (unc71; ADAM14) functions (cell non-autonomously) in cell migration, and that it works in concert with integrins (and a netrin) to do so (Huang *et al.* 2003).

ADAMs could also promote cell migration by shedding ligands that stimulate or direct cell migration or, as described above, that influence cell-cell and cell-matrix interactions. Numerous substances known or suggested to be shed by ADAMs could participate in such events. These include EGF family members, chemokines, ephrins, CD44, syndecans, Kit, or receptors for EGF family members (e.g. Erb B4), Kit, ephrins, or chemoattractants. For example, shedding of syndecans (which may be mediated by both MT-MMPs and ADAMs) is well known to promote cell migration (Holen *et al.* 2001; Kajita *et al.* 2001). Shed L1 also stimulates cell migration, and it has been proposed to do so by interacting with integrins (Mechtersheimer *et al.* 2001).

ADAMs could also promote cell migration by cleaving ECM components. This could be by cleavages that release promigratory growth factors bound to ECM, by directly shedding promigratory substances from the cell surface, or by cleaving ECM components (e.g. laminin and FN) so as to reveal cryptic sites. ADAMs could also locally degrade or remodel ECM components so as to help clear a pathway for cell migration. An ability to cleave an ECM molecule may be facilitated by an ability of the ADAM to interact with integrins and syndecans (Fig. 4C). Two proteolytically active ADAMs that can cleave matrix molecules *in vitro* have been implicated in cell migration events (Alfandari *et al.* 2001; Martin *et al.* 2002). It will be interesting to determine if the action of these ADAMs on ECM molecules contributes to their function in cell migration.

## 4. ROLE IN DEVELOPMENT AND DISEASE

ADAMs have now been implicated in a multitude of developmental events including fertilization, neurogenesis, myogenesis, adipogenesis, bone development, neural crest cell migration, branching morphogenesis in the lung, and development of the heart and other organ systems (Becherer and Blobel, 2003; Seals and Courtneidge, 2003). Mice lacking (or encoding defective) genes for twelve individual ADAMs have been characterized (reviewed in Becherer and Blobel, 2003; Kaji and Kudo, 2004; Novak, 2004; Seals and Courtneidge, 2003) (also see Chapter 2). Some of these mice display strong defects such as infertility (ADAMs 1-3), early embryonic lethality (ADAM10), perinatal lethality (ADAMs 17 and 19), or ataxia (ADAMs 22 and 23). Two show milder defects, problems with adipogenesis and myogenesis for ADAM12, (Chapters 2 and 6) and defects in neovascularization (upon challenge) for ADAM 15 (Chapter 2). Two (ADAM9$^{-/-}$ and ADAM11$^{-/-}$) have no reported defects (Chapter 2). Due to their possible overlapping activities, it will be interesting to unravel the phenotypes of mice lacking multiple ADAMs.

The only ADAM that has been genetically linked with a human disease is ADAM33, which was linked to asthma and bronchial hyper-responsiveness in Caucasians (Lind *et al.* 2003; Van Eerdewegh *et al.* 2002). ADAMs have, however, been implicated in numerous disease states. Studies have associated ADAMs with tumor cell proliferation (Lemjabbar *et al.* 2003; Schafer *et al.* 2004a; Schafer *et al.* 2004b) and with tumor cell migration (Murai *et al.* 2004; Schafer *et al.* 2004a). Moreover, altered expression levels have implicated ADAMs in cancers of multiple tissues and organs including prostate, breast, kidney, liver, neurons, and glia (for example see Costa *et al.* 2004). In addition to cancer, ADAMs (notably ADAM17) have been implicated in numerous inflammatory diseases including rheumatoid arthritis and inflammations associated with CNS diseases (Black *et al.* 2003). ADAMs have been ascribed a protective role in Alzheimer's disease by virtue of their α-secretase activity (Allinson *et al.* 2003; Postina *et al.* 2004). ADAM12 may participate in the biology of muscular dystrophy and cardiac diseases (see Chapter 6), and ADAM 8 may participate in neurodegenerative diseases (Schlomann *et al.* 2000) (also see Chapter 3). The former is only a sampling of the diseases that ADAMs have been implicated in and the list is sure to grow. One of the challenges, of course, is to ascertain whether it will be possible to design inhibitors of specific ADAMs that could aid in combating some of these diseases without damaging normal physiological processes (Moss and Bartsch, 2004).

## 5. CONCLUDING REMARKS

ADAMs are important in several developmental and disease processes, and some clearly function in these events as ectodomain sheddases. As discussed above, many important questions remain regarding ADAM metalloproteases in biology and pathology. Of paramount importance are continuing efforts to elucidate which ADAMs (and other metalloproteases) shed which key substrates in specific biological contexts and deciphering mechanisms that regulate ADAM proteases. Moreover, it will be very important to elucidate how ADAMs that lack a protease active site sequence function. Several ADAMs in this group play important roles in development: ADAMs 2 and 3 in fertilization, *C. elegans* ADAM14 in cell migration, and ADAM23 in the brain. A key question about non-proteolytically-active ADAMs is whether any of their *in vivo* functions depend on adhesive activities ascribed to their disintegrin or cysteine-rich domains in tissue culture studies. Recall that ADAMs are unique among cell surface proteins in possessing a disintegrin domain and unusual in their capacity to display both proteolytic and adhesive activities. It will therefore be rewarding to see if there is meaningful interplay between ADAM proteolytic and adhesive domains to enact important biological functions. A large related area of research will be to assess the role of ADAMs in cell adhesion and cell migration. ADAMs could function in such events through proteolytic activity (e.g. shedding of adhesion or pro-migratory molecules), adhesive activity (e.g. ADAM-integrin or ADAM-syndecan associations), or through their ability to associate with other cellular proteins (e.g., cytoplasmic adaptor proteins). The future will surely continue to dazzle us with how these elegant and multifaceted proteins regulate important events that occur at the cell surface.

## ACKNOWLEGMENTS

Work in the authors' laboratories was supported by grants from the NIH (GM48739 to JMW and DE14365 to DWD). We thank Jay Fox, Cezary Marcinkiewicz, and Marcos Milla for helpful discussions. Due to space limitations we were not able to cite many important references.

## REFERENCES

Abel, S., Hundhausen, C., Mentlein, R., Schulte, A., Berkhout, T. A., Broadway, N., Hartmann, D., Sedlacek, R., Dietrich, S., Muetze, B., Schuster, B., Kallen, K. J., Saftig, P., Rose-John, S., and Ludwig, A. (2004). The transmembrane CXC-chemokine ligand 16 is induced by IFN-gamma and TNF-alpha and shed by the activity of the disintegrin-like metalloproteinase ADAM10. *J Immunol* **172,** 6362-72.

Alfandari, D., Cousin, H., Gaultier, A., Smith, K., White, J. M., Darribere, T., and DeSimone, D. W. (2001). Xenopus ADAM 13 is a metalloprotease required for cranial neural crest-cell migration. *Curr Biol* **11,** 918-30.

Allinson, T. M., Parkin, E. T., Turner, A. J., and Hooper, N. M. (2003). ADAMs family members as amyloid precursor protein alpha-secretases. *J Neurosci Res* **74,** 342-52.

Anders, A., Gilbert, S., Garten, W., Postina, R., and Fahrenholz, F. (2001). Regulation of the alpha-secretase ADAM10 by its prodomain and proprotein convertases. *Faseb J* **15,** 1837-9.

Apte, S. S. (2004). A disintegrin-like and metalloprotease (reprolysin type) with thrombospondin type 1 motifs: the ADAMTS family. *Int J Biochem Cell Biol* **36,** 981-5.

Argast, G. M., Campbell, J. S., Brooling, J. T., and Fausto, N. (2004). Epidermal growth factor receptor transactivation mediates tumor necrosis factor-induced hepatocyte replication. *J Biol Chem* **279,** 34530-6.

Asakura, M., Kitakaze, M., Takashima, S., Liao, Y., Ishikura, F., Yoshinaka, T., Ohmoto, H., Node, K., Yoshino, K., Ishiguro, H., Asanuma, H., Sanada, S., Matsumura, Y., Takeda, H., Beppu, S., Tada, M., Hori, M., and Higashiyama, S. (2002). Cardiac hypertrophy is inhibited by antagonism of ADAM12 processing of HB-EGF: metalloproteinase inhibitors as a new therapy. *Nat. Med.* **8,** 35-40.

Baker, A. H., Edwards, D. R., and Murphy, G. (2002). Metalloproteinase inhibitors: biological actions and therapeutic opportunities. *J Cell Sci* **115,** 3719-27.

Bao, J., Wolpowitz, D., Role, L. W., and Talmage, D. A. (2003). Back signaling by the Nrg-1 intracellular domain. *J Cell Biol* **161,** 1133-41.

Bax, D. V., Messent, A. J., Tart, J., Van Hoang, M., Kott, J., Maciewicz, R. A., and Humphries, M. J. (2004). Integrin {alpha}5{beta}1 and ADAM-17 Interact in Vitro and Co-localize in Migrating HeLa Cells. *J Biol Chem* **279,** 22377-22386.

Becherer, J. D., and Blobel, C. P. (2003). Biochemical properties and functions of membrane-anchored metalloprotease-disintegrin proteins (ADAMs). *Curr Top Dev Biol* **54,** 101-23.

Black, R. A., Doedens, J. R., Mahimkar, R., Johnson, R., Guo, L., Wallace, A., Virca, D., Eisenman, J., Slack, J., Castner, B., Sunnarborg, S. W., Lee, D. C., Cowling, R., Jin, G., Charrier, K., Peschon, J. J., and Paxton, R. (2003). Substrate specificity and inducibility of TACE (tumour necrosis factor alpha-converting enzyme) revisited: the Ala-Val preference, and induced intrinsic activity. *Biochem Soc Symp* **70,** 39-52.

Black, R. A., Rauch, C. T., Kozlosky, C. J., Peschon, J. J., Slack, J. L., Wolfson, M. F., Castner, B. J., Stocking, K. L., Reddy, P., Srinivasan, S., Nelson, N., Boiani, N., Schooley, K. A., Gerhart, M., Davis, R., Fitzner, J. N., Johnson, R. S., Paxton, R. J., March, C. J., and Cerretti, D. P. (1997). A metalloproteinase disintegrin that releases tumour-necrosis factor- alpha from cells. *Nature* **385,** 729-733.

Bland, C. E., Kimberly, P., and Rand, M. D. (2003). Notch-induced proteolysis and nuclear localization of the Delta ligand. *J Biol Chem* **278,** 13607-10.

Blobel, C. P., Wolfsberg, T. G., Turck, C. W., Myles, D. G., Primakoff, P., and White, J. M. (1992). A potential fusion peptide and an integrin ligand domain in a protein active in sperm-egg fusion. *Nature* **356,** 248-252.

Bridges, L. C., and Bowditch, R. D. (2004). ADAM-Integrin interactions; Potential integrin regulated ectodomain shedding activity. *Current Pharmaceutical Design,* in press.

Bridges, L. C., Sheppard, D. and Bowditch, R. D. (2004). ADAM disintegrin-like domain recognition by the lymphocyte integrins alpha4beta1 and alpha4beta7. *Biochem J,* in press.

Bridges, L. C., Hanson, K. R., Tani, P. H., Mather, T., and Bowditch, R. D. (2003). Integrin alpha4beta1-dependent adhesion to ADAM 28 (MDC-L) requires an extended surface of the disintegrin domain. *Biochemistry* **42,** 3734-41.

Cao, Y., Kang, Q., and Zolkiewska, A. (2001). Metalloprotease-disintegrin ADAM 12 interacts with alpha-actinin-1. *Biochem. J.* **357,** 353-361.

Cho, C., Bunch, D. O., Faure, J. E., Goulding, E. H., Eddy, E. M., Primakoff, P., and Myles, D. G. (1998). Fertilization defects in sperm from mice lacking fertilin beta. *Science* **281,** 1857-9.

Costa, F. F., Verbisck, N. V., Salim, A. C., Ierardi, D. F., Pires, L. C., Sasahara, R. M., Sogayar, M. C., Zanata, S. M., Mackay, A., O'Hare, M., Soares, F., Simpson, A. J., and Camargo, A. A. (2004). Epigenetic silencing of the adhesion molecule ADAM23 is highly frequent in breast tumors. *Oncogene* **23,** 1481-8.

Cousin, H., Gaultier, A., Bleux, C., Darribere, T., and Alfandari, D. (2000). PACSIN2 is a regulator of the metalloprotease/disintegrin ADAM13. *Dev Biol* **227,** 197-210.

Diaz-Rodriguez, E., Montero, J. C., Esparis-Ogando, A., Yuste, L., and Pandiella, A. (2002). Extracellular signal-regulated Kinase phosphorylates tumor necrosis factor alpha-converting enzyme at threonine 735: a potential role in regulated shedding. *Mol Biol Cell* **13,** 2031-44.

Doedens, J. R., Mahimkar, R. M., and Black, R. A. (2003). TACE/ADAM-17 enzymatic activity is increased in response to cellular stimulation. *Biochem Biophys Res Commun* **308,** 331-8.

Eble, J. A., Bruckner, P., and Mayer, U. (2003). Vipera lebetina venom contains two disintegrins inhibiting laminin-binding beta1 integrins. *J Biol Chem* **278,** 26488-96.

Eto, K., Huet, C., Tarui, T., Kupriyanov, S., Liu, H. Z., Puzon-McLaughlin, W., Zhang, X. P., Sheppard, D., Engvall, E., and Takada, Y. (2002). Functional classification of ADAMs based on a conserved motif for binding to integrin alpha 9beta 1: implications for sperm-egg binding and other cell interactions. *J Biol Chem* **277,** 17804-17810.

Eto, K., Puzon-McLaughlin, W., Sheppard, D., Sehara-Fujisawa, A., Zhang, X.-P., and Takada, Y. (2000). RGD-independent binding of integrin a9b1 to the ADAM 12 and -15 disintegrin domains mediates cell-cell interaction. *J Biol Chem* **275,** 34922-34930.

Fahrenholz, F., Gilbert, S., Kojro, E., Lammich, S., and Postina, R. (2000). Alpha-secretase activity of the disintegrin metalloprotease ADAM 10. Influences of domain structure. *Ann N Y Acad Sci* **920,** 215-22.

Fambrough, D., Pan, D., Rubin, G. M., and Goodman, C. S. (1996). The cell surface metalloprotease/disintegrin Kuzbanian is required for axonal extension in *Drosophila Proc Nat'l Acad Sci USA* **93,** 13233-13238.

Fan, H., Turck, C. W., and Derynck, R. (2003). Characterization of growth factor-induced serine phosphorylation of tumor necrosis factor-alpha converting enzyme and of an alternatively translated polypeptide. *J Biol Chem* **278,** 18617-18627.

Franzke, C. W., Tasanen, K., Borradori, L., Huotari, V., and Bruckner-Tuderman, L. (2004). Shedding of collagen XVII/BP180: structural motifs influence cleavage from cell surface. *J Biol Chem* **279,** 24521-9.

Galliano, M. F., Huet, C., Frygelius, J., Polgren, A., Wewer, U. M., and Engvall, E. (2000). Binding of ADAM12, a marker of skeletal muscle regeneration, to the muscle-specific actin-binding protein, alpha -actinin-2, is required for myoblast fusion. *J. Biol. Chem.* **275,** 13933-13939.

Gaultier, A., Cousin, H., Darribere, T., and Alfandari, D. (2002). ADAM13 disintegrin and cysteine-rich domains bind to the second heparin- binding domain of fibronectin. *J Biol Chem* **277,** 23336-44.

Gonzales, P. E., Solomon, A., Miller, A. B., Leesnitzer, M. A., Sagi, I., and Milla, M. E. (2004). Inhibition of the tumor necrosis factor-alpha-converting enzyme by its pro domain. *J Biol Chem* **279,** 31638-45.

Hartmann, D., de Strooper, B., Serneels, L., Craessaerts, K., Herreman, A., Annaert, W., Umans, L., Lubke, T., Lena Illert, A., von Figura, K., and Saftig, P. (2002). The disintegrin/metalloprotease ADAM 10 is essential for Notch signalling but not for alpha-secretase activity in fibroblasts. *Hum Mol Genet* **11,** 2615-24.

Hattori, M., Osterfield, M., and Flanagan, J. G. (2000). Regulated cleavage of a contact-mediated axon repellent. *Science* **289,** 1360-1365.

Herren, B., Garton, K. J., Coats, S., Bowen-Pope, D. F., Ross, R., and Raines, E. W. (2001). ADAM15 overexpression in NIH3T3 cells enhances cell-cell interactions. *Exp Cell Res* **271,** 152-60.

Hinkle, C. L., Mohan, M. J., Lin, P., Yeung, N., Rasmussen, F., Milla, M. E., and Moss, M. L. (2003). Multiple metalloproteinases process protransforming growth factor-alpha (proTGF-alpha). *Biochemistry* **42,** 2127-36.

Holen, I., Drury, N. L., Hargreaves, P. G., and Croucher, P. I. (2001). Evidence of a role for a non-matrix-type metalloproteinase activity in the shedding of syndecan-1 from human myeloma cells. *Br J Haematol* **114,** 414-21.

Hougaard, S., Loechel, F., Xu, X., Tajima, R., Albrechtsen, R., and Wewer, U. M. (2000). Trafficking of human ADAM 12-L: retention in the trans-Golgi network. *Biochem Biophys Res Commun* **275,** 261-267.

Howard, L., Maciewicz, R. A., and Blobel, C. P. (2000). Cloning and characterization of ADAM28: evidence for autocatalytic pro- domain removal and for cell surface localization of mature ADAM28. *Biochem J* **348,** 21-27.

Huang, X., Huang, P., Robinson, M. K., Stern, M. J., and Jin, Y. (2003). UNC-71, a disintegrin and metalloprotease (ADAM) protein, regulates motor axon guidance and sex myoblast migration in C. elegans. *Development* **130,** 3147-61.

Hundhausen, C., Misztela, D., Berkhout, T. A., Broadway, N., Saftig, P., Reiss, K., Hartmann, D., Fahrenholz, F., Postina, R., Matthews, V., Kallen, K. J., Rose-John, S., and Ludwig, A. (2003). The disintegrin-like metalloproteinase ADAM10 is involved in constitutive cleavage of CX3CL1 (fractalkine) and regulates CX3CL1-mediated cell-cell adhesion. *Blood* **102,** 1186-95.

Kaji, K., and Kudo, A. (2004). The mechanism of sperm-oocyte fusion in mammals. *Reproduction* **127,** 423-9.

Kajita, M., Itoh, Y., Chiba, T., Mori, H., Okada, A., Kinoh, H., and Seiki, M. (2001). Membrane-type 1 matrix metalloproteinase cleaves CD44 and promotes cell migration. *J Cell Biol* **153,** 893-904.

Kawaguchi, N., Sundberg, C., Kveiborg, M., Moghadaszadeh, B., Asmar, M., Dietrich, N., Thodeti, C. K., Nielsen, F. C., Moller, P., Mercurio, A. M., Albrechtsen, R., and Wewer, U. M. (2003). ADAM12 induces actin cytoskeleton and extracellular matrix reorganization during early adipocyte differentiation by regulating {beta}1 integrin function. *J Cell Sci* **116,** 3893-3904.

Kojro, E., Gimpl, G., Lammich, S., Marz, W., and Fahrenholz, F. (2001). Low cholesterol stimulates the nonamyloidogenic pathway by its effect on the alpha -secretase ADAM 10. *Proc Natl Acad Sci U S A* **98,** 5815-20.

LaVoie, M. J., and Selkoe, D. J. (2003). The Notch ligands, Jagged and Delta, are sequentially processed by alpha-secretase and presenilin/gamma-secretase and release signaling fragments. *J Biol Chem* **278,** 34427-37.

Lee, M. H., Dodds, P., Verma, V., Maskos, K., Knaupcr, V., and Murphy, G. (2003). Tailoring tissue inhibitor of metalloproteinases-3 to overcome the weakening effects of the cysteine-rich domains of tumour necrosis factor-alpha converting enzyme. *Biochem J* **371,** 369-76.

Lemjabbar, H., and Basbaum, C. (2002). Platelet-activating factor receptor and ADAM10 mediate responses to Staphylococcus aureus in epithelial cells. *Nat Med* **8,** 41-6.

Lemjabbar, H., Li, D., Gallup, M., Sidhu, S., Drori, E., and Basbaum, C. (2003). Tobacco smoke-induced lung cell proliferation mediated by tumor necrosis factor alpha-converting enzyme and amphiregulin. *J Biol Chem* **278,** 26202-7.

Lind, D. L., Choudhry, S., Ung, N., Ziv, E., Avila, P. C., Salari, K., Coyle, N. E., Nazario, S., Rodriguez-Santana, J. R., Salas, J., Selman, M., Boushey, H. A., Weiss, S. T., Chapela, R., Ford, J. G., Rodriguez-Cintron, W., Silverman, E. K., Sheppard, D., Kwok, P. Y., and Burchard, E. G. (2003). ADAM33 is not associated with asthma in puerto rican or mexican populations. *Am J Respir Crit Care Med* **168,** 1312-6

Marambaud, P., Wen, P. H., Dutt, A., Shioi, J., Takashima, A., Siman, R., and Robakis, N. K. (2003). A CBP binding transcriptional repressor produced by the PS1/epsilon-cleavage of N-cadherin is inhibited by PS1 FAD mutations. *Cell* **114,** 635-45.

Marcinkiewicz, C. (2004). Functional characteristic of snake venom disintegrins: Potential therapeutic implication. *Curr Pharm Design,* in press.

Martin, J., Eynstone, L. V., Davies, M., Williams, J. D., and Steadman, R. (2002). The role of ADAM 15 in glomerular mesangial cell migration. *J Biol Chem* **277,** 33683-33689.

Maskos, K., Fernandez-Catalan, C., Huber, R., Bourenkov, G. P., Bartunik, H., Ellestad, G. A., Reddy, P., Wolfson, M. F., Rauch, C. T., Castner, B. J., Davis, R., Clarke, H. R. G., Peterson, M., Fitzner, J. N., Cerretti, D. P., March, C. J., Paxton, R. J., Black, R. A., and Bode, W. (1998). Crystal structure of the catalytic domain of human tumor necrosis factor a-converting enzyme. *Proc Nat'l Acad Sci USA* **95,** 3408-3412.

Matsui, T., Fujimura, Y., and Titani, K. (2000). Snake venom proteases affecting hemostasis and thrombosis. *Biochim Biophys Acta* **1477,** 146-56.

Matthews, V., Schuster, B., Schutze, S., Bussmeyer, I., Ludwig, A., Hundhausen, C., Sadowski, T., Saftig, P., Hartmann, D., Kallen, K. J., and Rose-John, S. (2003). Cellular cholesterol depletion triggers shedding of the human interleukin-6 receptor by ADAM10 and ADAM17 (TACE). *J Biol Chem* **278,** 38829-39.

Mechtersheimer, S., Gutwein, P., Agmon-Levin, N., Stoeck, A., Oleszewski, M., Riedle, S., Fogel, M., Lemmon, V., and Altevogt, P. (2001). Ectodomain shedding of L1 adhesion molecule promotes cell migration by autocrine binding to integrins. *J Cell Biol* **155,** 661-73.

Mohammed, F. F., Smookler, D. S., Taylor, S. E., Fingleton, B., Kassiri, Z., Sanchez, O. H., English, J. L., Matrisian, L. M., Au, B., Yeh, W. C., and Khokha, R. (2004). Abnormal TNF activity in Timp3(-/-) mice leads to chronic hepatic inflammation and failure of liver regeneration. *Nat Genet* **36,** 969-77.

Mori, S., Tanaka, M., Nanba, D., Nishiwaki, E., Ishiguro, H., Higashiyama, S., and Matsuura, N. (2003). PACSIN3 binds ADAM12/meltrin alpha and upregulates ectodomain shedding of heparin-binding EGF-like growth factor. *J Biol Chem* **278,** 46029-34

Moss, M. L., and Bartsch, J. W. (2004). Therapeutic benefits from targeting of ADAM family members. *Biochemistry* **43,** 7227-35.

Moss, M. L., Jin, S. L., Milla, M. E., Burkhart, W., Carter, H. L., Chen, W. J., Clay, W. C., Didsbury, J. R., Hassler, D., Hoffman, C. R., Kost, T. A., Lambert, M. H., Leesnitzer, M. A., McCauley, P., McGeehan, G., Mitchell, J., Moyer, M., Pahel, G., Rocque, W., Overton, L. K., Schoenen, F., Seaton, T., Su, J. L., Warner, J., Becherer, J. D., and et, a. (1997). Cloning of a disintegrin metalloproteinase that processes precursor tumour-necrosis factor-alpha. *Nature* **385,** 733-736.

Murai, T., Miyazaki, Y., Nishinakamura, H., Sugahara, K. N., Miyauchi, T., Sako, Y., Yanagida, T., and Miyasaka, M. (2004). Engagement of CD44 promotes Rac activation and CD44 cleavage during tumor cell migration. *J Biol Chem* **279,** 4541-50

Murakami, D., Okamoto, I., Nagano, O., Kawano, Y., Tomita, T., Iwatsubo, T., De Strooper, B., Yumoto, E., and Saya, H. (2003). Presenilin-dependent gamma-secretase activity mediates the intramembranous cleavage of CD44. *Oncogene* **22,** 1511-6.

Nagano, O., Murakami, D., Hartmann, D., De Strooper, B., Saftig, P., Iwatsubo, T., Nakajima, M., Shinohara, M., and Saya, H. (2004). Cell-matrix interaction via CD44 is independently regulated by different metalloproteinases activated in response to extracellular Ca2+ influx and PKC activation. *J Cell Biol* **165,** 893-902.

Nakamura, H., Suenaga, N., Taniwaki, K., Matsuki, H., Yonezawa, K., Fujii, M., Okada, Y., and Seiki, M. (2004a). Constitutive and induced CD44 shedding by ADAM-like proteases and membrane-type 1 matrix metalloproteinase. *Cancer Res* **64,** 876-82.

Nakamura, T., Abe, H., Hirata, A., and Shimoda, C. (2004b). ADAM family protein Mde10 is essential for development of spore envelopes in the fission yeast Schizosaccharomyces pombe. *Eukaryot Cell* **3,** 27-39.

Nanba, D., Mammoto, A., Hashimoto, K., and Higashiyama, S. (2003). Proteolytic release of the carboxy-terminal fragment of proHB-EGF causes nuclear export of PLZF. *J Cell Biol* **163,** 489-502.

Ni, C. Y., Murphy, M. P., Golde, T. E., and Carpenter, G. (2001). gamma -Secretase cleavage and nuclear localization of ErbB-4 receptor tyrosine kinase. *Science* **294,** 2179-81.

Novak, U. (2004). ADAM proteins in the brain. *J Clin Neurosci* **11,** 227-35.

Orth, P., Reichert, P., Wang, W., Prosise, W. W., Yarosh-Tomaine, T., Hammond, G., Ingram, R. N., Xiao, L., Mirza, U. A., Zou, J., Strickland, C., Taremi, S. S., Le, H. V., and Madison, V. (2004). Crystal structure of the catalytic domain of human ADAM33. *J Mol Biol* **335,** 129-37.

Overall, C. M. (2002). Molecular determinants of metalloproteinase substrate specificity: matrix metalloproteinase substrate binding domains, modules, and exosites. *Mol Biotechnol* **22,** 51-86.

Palecek, S. P., Loftus, J. C., Ginsberg, M. H., Lauffenburger, D. A., and Horwitz, A. F. (1997). Integrin-ligand binding properties govern cell migration speed through cell-substratum adhesiveness. *Nature* **385,** 537-40.

Pan, D., and Rubin, G. M. (1997). Kuzbanian controls proteolytic processing of notch and mediates lateral inhibition during Drosophila and vertebrate neurogenesis. *Cell* **90,** 271-280.

Peiretti, F., Deprez-Beauclair, P., Bonardo, B., Aubert, H., Juhan-Vague, I., and Nalbone, G. (2003). Identification of SAP97 as an intracellular binding partner of TACE. *J Cell Sci* **116,** 1949-57.

Peschon, J. J., Slack, J. L., Reddy, P., Stocking, K. L., Sunnarborg, B. J., Lee, D. C., Russell, W. E., Castner, B. J., Johnson, R. S., Fitzner, J. N., Boyce, R. W., Nelson, N., Kozlsky, C. J., Wolfson, M. F., Rauch, C. T., Cerretti, D. P., Paxton, R. J., March, C. J., and Black, R. A. (1998). An essential role for ectodomain shedding in mammalian development. *Science* **282,** 1281-1284.

Poghosyan, Z., Robbins, S. M., Houslay, M. D., Webster, A., Murphy, G., and Edwards, D. R. (2002). Phosphorylation-dependent interactions between ADAM15 cytoplasmic domain and Src family protein-tyrosine kinases. *J Biol Chem* **277,** 4999-5007.

Postina, R., Schroeder, A., Dewachter, I., Bohl, J., Schmitt, U., Kojro, E., Prinzen, C., Endres, K., Hiemke, C., Blessing, M., Flamez, P., Dequenne, A., Godaux, E., van Leuven, F., and Fahrenholz, F. (2004). A disintegrin-metalloproteinase prevents amyloid plaque formation and hippocampal defects in an Alzheimer disease mouse model. *J Clin Invest* **113,** 1456-64.

Prenzel, N., Zwick, E., Daub, H., Leserer, M., Abraham, R., Wallasch, C., and Ullrich, A. (1999). EGF receptor transactivation by G-protein-coupled receptors requires metalloproteinase cleavage of proHB-EGF. *Nature* **402,** 884-888.

Primakoff, P., Hyatt, H., and Tredick-Kline, J. (1987). Identification and purification of a sperm surface protein with a potential role in sperm-egg membrane fusion. *J Cell Biol* **104,** 141-149.

Reddy, P., Slack, J. L., Davis, R., Cerretti, D. P., Kozlosky, C. J., Blanton, R. A., Shows, D., Peschon, J. J., and Black, R. A. (2000). Functional analysis of the domain structure of tumor necrosis factor-alpha converting enzyme. *J Biol Chem* **275,** 14608-14.

Rooke, J., Pan, D., Xu, T., and Rubin, G. M. (1996). KUZ, a conserved metalloprotease-disintegrin protein with two roles in Drosophila neurogenesis. *Science* **273,** 1227-31.

Sahin, U., Weskamp, G., Kelly, K., Zhou, H. M., Higashiyama, S., Peschon, J., Hartmann, D., Saftig, P., and Blobel, C. P. (2004). Distinct roles for ADAM10 and ADAM17 in ectodomain shedding of six EGFR ligands. *J Cell Biol* **164,** 769-79.

Schafer, B., Gschwind, A., and Ullrich, A. (2004a). Multiple G-protein-coupled receptor signals converge on the epidermal growth factor receptor to promote migration and invasion. *Oncogene* **23,** 991-9.

Schafer, B., Marg, B., Gschwind, A., and Ullrich, A. (2004b). Distinct ADAM metalloproteinases regulate G protein coupled receptor-induced cell proliferation and survival. *J Biol Chem,* in press

Schlomann, U., Rathke-Hartlieb, S., Yamamoto, S., Jockusch, H., and Bartsch, J. W. (2000). Tumor necrosis factor alpha induces a metalloprotease-disintegrin, ADAM8 (CD 156): implications for neuron-glia interactions during neurodegeneration. *J Neurosci* **20,** 7964-7971.

Schlomann, U., Wildeboer, D., Webster, A., Antropova, O., Zeuschner, D., Knight, C. G., Docherty, A. J., Lambert, M., Skelton, L., Jockusch, H., and Bartsch, J. W. (2002). The metalloprotease disintegrin ADAM8. Processing by autocatalysis is required for proteolytic activity and cell adhesion. *J Biol Chem* **277,** 48210-9.

Seals, D. F., and Courtneidge, S. A. (2003). The ADAMs family of metalloproteases: multidomain proteins with multiple functions. *Genes Dev* **17,** 7-30.

Shao, M. X., Nakanaga, T., and Nadel, J. A. (2004). Cigarette smoke induces MUC5AC mucin overproduction via tumor necrosis factor-alpha-converting enzyme in human airway epithelial (NCI-H292) cells. *Am J Physiol Lung Cell Mol Physiol* **287,** L420-7.

Six, E., Ndiaye, D., Laabi, Y., Brou, C., Gupta-Rossi, N., Israel, A., and Logeat, F. (2003). The Notch ligand Delta1 is sequentially cleaved by an ADAM protease and gamma-secretase. *Proc Natl Acad Sci U S A* **100,** 7638-43.

Smith, K. M., Gaultier, A., Cousin, H., Alfandari, D., White, J. M., and DeSimone, D. (2002). The cysteine-rich domain regulates ADAM protease function in vivo. *J Cell Biol* **159,** 893-902.

Soejima, K., Matsumoto, M., Kokame, K., Yagi, H., Ishizashi, H., Maeda, H., Nozaki, C., Miyata, T., Fujimura, Y., and Nakagaki, T. (2003). ADAMTS-13 cysteine-rich/spacer domains are functionally essential for von Willebrand factor cleavage. *Blood* **102,** 3232-7.

Solomon, A., Rosenblum, G., Gonzales, P. E., Leonard, J. D., Mobashery, S., Milla, M. E., and Sagi, I. (2004). Pronounced diversity in electronic and chemical properties between the catalytic zinc sites of tumor necrosis factor-alpha-converting enzyme and matrix metalloproteinases despite their high structural similarity. *J Biol Chem* **279,** 31646-54.

Somerville, R. P., Longpre, J. M., Jungers, K. A., Engle, J. M., Ross, M., Evanko, S., Wight, T. N., Leduc, R., and Apte, S. S. (2003). Characterization of ADAMTS-9 and ADAMTS-20 as a distinct ADAMTS subfamily related to Caenorhabditis elegans GON-1. *J Biol Chem* **278,** 9503-13.

Tanaka, M., Nanba, D., Mori, S., Shiba, F., Ishiguro, H., Yoshino, K., Matsuura, N., and Higashiyama, S. (2004). ADAM-binding protein eve-1 is required for ectodomain shedding of EGF receptor ligands. *J Biol Chem,* in press

Thodeti, C. K., Albrechtsen, R., Grauslund, M., Asmar, M., Larsson, C., Takada, Y., Mercurio, A. M., Couchman, J. R., and Wewer, U. M. (2003). ADAM12/syndecan-4 signaling promotes beta 1 integrin-dependent cell spreading through protein kinase Calpha and RhoA. *J Biol Chem* **278,** 9576-84.

Tomczuk, M., Takahashi, Y., Huang, J., Murase, S., Mistretta, M., Klaffky, E., Sutherland, A., Bolling, L., Coonrod, S., Marcinkiewicz, C., Sheppard, D., Stepp, M. A., and White, J. M. (2003). Role of multiple beta1 integrins in cell adhesion to the disintegrin domains of ADAMs 2 and 3. *Exp Cell Res* **290,** 68-81.

Tomczuk, M. (2004) ADAMs in early mouse development. University of Virginia PhD thesis, pp.278.

Tschumperlin, D. J., Dai, G., Maly, I. V., Kikuchi, T., Laiho, L. H., McVittie, A. K., Haley, K. J., Lilly, C. M., So, P. T., Lauffenburger, D. A., Kamm, R. D., and Drazen, J. M. (2004). Mechanotransduction through growth-factor shedding into the extracellular space. *Nature* **429,** 83-86.

Van Eerdewegh, P., Little, R. D., Dupuis, J., Del Mastro, R. G., Falls, K., Simon, J., Torrey, D., Pandit, S., McKenny, J., Braunschweiger, K., Walsh, A., Liu, Z., Hayward, B., Folz, C., Manning, S. P., Bawa, A., Saracino, L., Thackston, M., Benchekroun, Y., Capparell, N., Wang, M., Adair, R., Feng, Y., Dubois, J., FitzGerald, M. G., Huang, H., Gibson, R., Allen, K. M., Pedan, A., Danzig, M. R., Umland, S. P., Egan, R. W., Cuss, F. M., Rorke, S., Clough, J. B., Holloway, J. W., Holgate, S. T., and Keith, T. P. (2002). Association of the ADAM33 gene with asthma and bronchial hyperresponsiveness. *Nature* **418,** 426-430.

Varney, T. R., Casademunt, E., Ho, H. N., Petty, C., Dolman, J., and Blumberg, D. D. (2002). A novel Dictyostelium gene encoding multiple repeats of adhesion inhibitor-like domains has effects on cell-cell and cell-substrate adhesion. *Dev Biol* **243,** 226-48.

Wakatsuki, S., Kurisaki, T., and Sehara-Fujisawa, A. (2004). Lipid rafts identified as locations of ectodomain shedding mediated by Meltrin beta/ADAM19. *J Neurochem* **89,** 119-23.

White, J. M. (2003). ADAMs: modulators of cell-cell and cell-matrix interactions. *Curr Opin Cell Biol* **15,** 598-606.

Wolfsberg, T. G., Bazan, J. J., Blobel, C. P., Myles, D. G., Primakoff, P., and White, J. M. (1993). The precursor region of a protein active in sperm-egg fusion contains a metalloprotease and a disintegrin domain: Structural, functional and evolutionary implications. *Proc. Natl. Acad. Sci. USA* **90,** 10783-10787.

Wolfe, M. S., and Kopan, R. (2004). Intramembrane proteolysis: Theme and variations. *Science* **305**, 1119-1123.

Wolfsberg, T. G., Straight, P. D., Gerena, R. L., Huovila, A.-P. J., Primakoff, P., Myles, D. G., and White, J. M. (1995). ADAM, a widely distributed and developmentally regulated gene family encoding membrane proteins with a disintegrin and metalloprotease domain. *Dev. Biol.* **169,** 378-383.

Yagami-Hiromasa, T., Sato, T., Kurisaki, T., Kamijo, K., Nabeshima, Y., and Fujisawa-Sehara, A. (1995). A metalloprotease-disintegrin participating in myoblast fusion. *Nature* **377,** 652-656.

Yan, Y., Shirakabe, K., and Werb, Z. (2002). The metalloprotease Kuzbanian (ADAM10) mediates the transactivation of EGF receptor by G protein-coupled receptors. *J Cell Biol* **158,** 221-6.

Zhao, L., Shey, M., Farnsworth, M., and Dailey, M. O. (2001). Regulation of membrane metalloproteolytic cleavage of L-selectin (CD62l) by the epidermal growth factor domain. *J Biol Chem* **276,** 30631-40.

Zheng, Y., Schlondorff, J., and Blobel, C. P. (2002). Evidence for regulation of the tumor necrosis factor alpha-convertase (TACE) by protein-tyrosine phosphatase PTPH1. *J Biol Chem* **277,** 42463-70.

Zhu, P., Sun, Y., Xu, R., Sang, Y., Zhao, J., Liu, G., Cai, L., Li, C., and Zhao, S. (2003). The interaction between ADAM 22 and 14-3-3zeta: regulation of cell adhesion and spreading. *Biochem Biophys Res Commun* **301,** 991-9.

Chapter 2

# STUDIES FROM ADAM KNOCKOUT MICE

Keisuke Hoiruchi and Carl P. Blobel
*Arthritis and Tissue Degeneration Program and Cell Biology Program, Hospital for Special Surgery at Weil Medical College of Cornell University, New York, NY, USA*

Abstract: ADAMs are membrane anchored glycoproteins that contain a disintegrin and metalloprotease domain. This chapter will focus on recent insights that have emerged from studies of "knockout" mice for ADAM proteases that are widely expressed or at least expressed in a variety of different cells and tissues (ADAMs 8, 9, 10, 12, 15, 17 and 19). These studies have shown that ADAM10 is important for signaling via the cell surface receptor Notch during development, while ADAM17 is critical for the development of the lung, epithelial structures and semilunar heart valves because of its role in the functional activation of ligands of the epidermal growth factor receptor. ADAM19 is essential for proper development of heart valves and the ventricular septum, although the underlying mechanism remains to be established. On the other hand, ADAMs 8, 9, 12 and 15 are dispensable for normal development and adult life in mice, at least under laboratory conditions. However, ADAM15 has a critical role in pathological neovascularization, making it a potential target for the design of inhibitors of angiogenesis. The availability of viable knockout mice for several widely expressed ADAM proteases sets the stage for a more comprehensive analysis of potential functions of these proteins in physiological and pathological processes. Furthermore, in light of the essential roles of ADAMs 10, 17 and 19 in development, it will be interesting to generate conditional knockout mice in order to evaluate the function of these proteases in adult animals.

Key words: ADAMs, knockout mice, metalloprotease-disintegrins, protein ectodomain shedding, Notch, EGFR, heart development, angiogenesis.

## 1. INTRODUCTION

ADAMs are a family of membrane anchored glycoproteins that have important functions in fertilization, neurogenesis, heart development, and in the ectodomain shedding of a number of membrane proteins, including the



*N. M. Hooper and U. Lendeckel (eds.), The ADAM Family of Proteases,* 29-64.

pro-inflammatory cytokine TNFα, and ligands of the epidermal growth factor receptor (EGFR) (for recent reviews, see Kheradmand and Werb 2002; Primakoff and Myles 2000; Schlondorff and Blobel 1999; Seals and Courtneidge 2003; White 2003). A typical ADAM has an N-terminal pro-domain, followed by a metalloprotease domain, disintegrin domain, cystein-rich region, EGF-repeat, transmembrane domain and cytoplasmic domain (Schlondorff and Blobel 1999; Wolfsberg *et al.* 1995a; Wolfsberg *et al.* 1995b). The first recognized ADAMs were the α and β subunit of the heterodimeric sperm surface protein fertilin, which is essential for fertilization in mice (Blobel *et al.* 1990; Blobel and White 1992; Cho *et al.* 1998; Primakoff *et al.* 1987; Wolfsberg *et al.* 1993). To date, over 30 ADAMs have been identified in a variety of species ranging from humans to mice, *Drosophila melanogaster, Caenorhabditis elegans* and *Schizosaccharomyces pombe* (a continuously updated list of ADAMs can be found at: www.people.virginia.edu/%7Ejw7g/Table_of_the_ADAMs.html). These ADAMs have been found in several different ways. ADAM10 (Kuzbanian) emerged from a genetic screen for novel genes with a role in the Notch signaling pathway or in neuronal extension in *Drosphila melanogaster* (Fambrough *et al.* 1996; Rooke *et al.* 1996), while a biochemical purification of the TNFα converting enzyme activity resulted in the identification of ADAM17/TACE (Black *et al.* 1997; Moss *et al.* 1997). The majority of ADAMs were found in PCR screens with degenerate primers that were designed to amplify cDNAs for disintegrin- or metalloprotease domains (see, for example Cai *et al.* 1998; Kratzschmar *et al.* 1996; Weskamp and Blobel 1994; Weskamp *et al.* 1996; Yagami-Hiromasa *et al.* 1995), while more recently additional ADAMs have been identified through the mouse and human genome sequencing project. Because only relatively minor gaps remain in the human and mouse genome sequences, it now appears likely that most, if not all, mouse and human ADAMs have been identified.

Predictions regarding the function of ADAMs that were not identified in a functional or genetic screen are mainly based on what is known about the role of their different protein domains in other ADAMs or in related proteins found in snake venom. The disintegrin domain of ADAMs, for example, is related to snake venom integrin ligands called disintegrins (Huang *et al.* 1987; Niewiarowski *et al.* 1994). Therefore "orphan" ADAMs, that is ADAMs of unknown function, might have roles in cell-cell or cell-matrix interactions (Blobel *et al.* 1992; White 2003). Furthermore, about half of the currently known ADAMs contain a catalytic site consensus sequence (HEXXH) in their metalloprotease domain (Jongeneel *et al.* 1989; Stocker *et al.* 1995), and are therefore predicted to be catalytically active. The remaining ADAMs lack this consensus sequence, presumably do not possess catalytic activity and are therefore not included in this book on ADAM proteases. Most ADAMs also contain putative cytoplasmic signaling motifs, such as proline-rich SH3-ligand domains and potential phosphorylation sites (Seals and Courtneidge 2003; Weskamp *et al.* 1996), raising the possibility that interactions with cytoplasmic proteins might regulate the function of ADAMs, or that ADAMs might have a role in intracellular signaling.

Following the identification of novel ADAM proteases, a number of different approaches were taken to elucidate their function, including purification and biochemical characterization of catalytically active enzymes (see, for example, Chesneau *et al.* 2003; Howard *et al.* 1996; Howard *et al.* 2001; Loechel *et al.* 1998; Roghani *et al.* 1999; Schlomann *et al.* 2002; Zou *et al.* 2004), evaluation of their potential role in cell adhesion and signaling (reviewed in White 2003), and studies of their expression patterns (see, for example Cai *et al.* 1998; Horiuchi *et al.* 2003; Kelly *et al.* 2004; Weskamp *et al.* 2002; Zhou *et al.* 2004). These approaches have provided much needed information about candidate substrates of catalytically active ADAMs, their possible role in cell adhesion, and about the cells and tissues in which they might have physiologically or pathologically relevant function. This has led to hypotheses about the roles of novel ADAMs which can be tested in a physiologically relevant model system. "Knockout" mice provide a particularly attractive model system for this purpose, and have the additional benefit that they can also deliver unexpected insights into the role of a given protein in development and disease. In this chapter, we summarize the findings that have emerged to date from studies of knockout mice for catalytically active ADAM proteases that are widely expressed or expressed in several different types of cells and tissues (ADAMs 8, 9, 10, 12, 15, 17 and 19, see Figure 1 for a dendrogram depicting the degree of sequence similarity among these 7 ADAMs and ADAM28).

## 2. ADAM8 (MS2, CD156a)

ADAM8 was originally identified in a macrophage cell line and named MS2 (Setoguchi *et al.* 1989). Expression of this gene was also found in central nervous system, bone cells, thymus, and lymphatic vessels. These observations and recent biochemical studies indicate possible roles of ADAM8 in inflammatory responses, neuropathology and bone metabolism. However, a gene targeting of ADAM8 revealed no apparent anomalies during development and in adult homeostasis. See also Chapter 3.

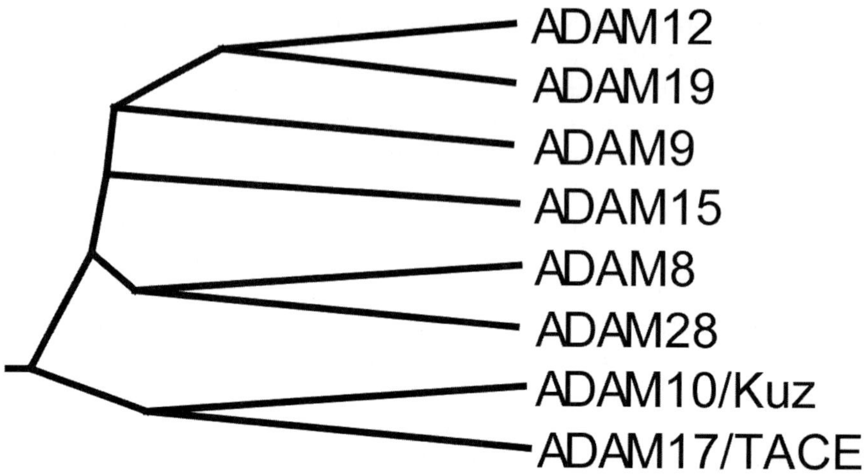


*Figure 1.* Dendrogram depicting the relative sequence relationship among catalytically active and widely expressed ADAMs in mice.

## 2.1 Expression pattern and putative functions of ADAM8 in immune system and neuropathology

ADAM8 was identified from a subtractive hybridization between cDNA from a macrophage cell line and a non-macrophage cell line (Setoguchi *et al.* 1989). Subsequent studies showed that ADAM8 is also expressed in human immune cells except for T-cells, and that the expression in macrophages is upregulated by macrophage stimulators (Yoshida *et al.* 1990; Yoshiyama *et al.* 1997). In adult mice, ADAM8 expression was also observed in the central nervous system and found to be induced by TNFα (Schlomann *et al.* 2000). At early developmental stages, prominent expression was found in extraembryonic tissues, while at later stages expression of ADAM8 was seen in several organs and tissues including the gonadal ridge, thymus, developing cartilage or bone, brain and spinal cord and in the mesenchyme in close proximity to developing blood vessels and lymphatic vessels (Kelly *et al.* 2004).

Recent biochemical studies showed that ADAM8 is catalytically active and is capable of processing myelin basic protein and peptides derived from the membrane-proximal region of several membrane bound molecules, including interleukin-1 receptor, kit ligand-1 (KL-1), amyloid precursor protein (APP), and CD23, a low affinity IgE receptor (Amour *et al.* 2002;

Fourie *et al.* 2003). Promoter studies have identified lipopolysaccharide and interferon-γ response elements in the 5' region of the ADAM8 gene (Kataoka *et al.* 1997).

Based on these observations, several studies aimed at uncovering a possible involvement of ADAM8 in inflammatory responses and neuropathological processes have been reported. Transgenic mice expressing the ectodomain of ADAM8 under the control of the α1-antitrypsin promoter showed a potential role of ADAM8 in leukocyte infiltration (Higuchi *et al.* 2002). Experiments in Wobbler mice, which have an accelerated course of neurodegeneration, showed an increase in ADAM8 expression in activated glial cells (astrocytes and activated microglia), suggesting that ADAM8 has a role in pathological neuron-glia interactions (Schlomann *et al.* 2000). It was also reported that ADAM8, but not ADAM17 or ADAM10, is responsible for the shedding of membrane bound form of CHL1, close homologue of L1, which is thought to have important roles in the nervous system (Naus *et al.* 2004). The shed form of this molecule is upregulated in the brain extracts of Wobbler mice, suggesting a possible involvement of ADAM8 in the pathology of neurodegeneration.

ADAM8 expression was also found in osteoclasts and shown to have a stimulatory effect on osteoclast differentiation (Choi *et al.* 2001); this was further supported by the observation that ADAM8 is found upregulated in the tissue surrounding loosened hip prosthesis in human patients (Mandelin *et al.* 2003). Finally, recent studies with a mouse model of asthma indicated a possible role of ADAM8 in the pathology of asthma (King *et al.* 2004).

## 2.2 *Adam8-/-* mice are viable and fertile with no apparent anomalies

ADAM8 gene-targeted mice were generated by replacing the coding sequence for the catalytic and disintegrin domain with a cassette containing the ß-galactosidase and neomycin genes (Kelly *et al.* 2004). Both male and female adult *Adam8-/-* mice are viable and fertile, producing litters at a frequency and size comparable to that of wild type controls. The appearance and behavior of *Adam8-/-* mice is indistinguishable from that of their wild type and heterozygous littermates. Close histopathological examination, especially of the tissues where ADAM8 expression is high, did not reveal any evident anomalies. Although the expression pattern of ADAM8 during embryogenesis indicated a possible role in lymphangiogenesis, immunostaining with Prox1, a marker for lymphatic endothelial cells did not reveal any defects in lymphatic development. Bone morphology was also studied to address potential defects in either osteoclast or osteoblast development, but there were no apparent differences between *Adam8-/-*

animals and wild type controls. Furthermore there were no significant changes in the differential blood counts, arguing against an essential and general function for ADAM8 in immune cell development. Thus, targeted inactivation of ADAM8 in mice did not lead to any overt defects in development, adult survival or fertility. These results demonstrate that ADAM8 is not essential for normal development and adult homeostasis. However, shedding of CHL1, which is related to the adhesion protein L1, is significantly decreased in brain extracts from *Adam8-/-* mice compared to wild type animals (Naus *et al.* 2004). Further studies, including functional challenges of the immune system, as well as studies of neurodegeneration or bone metabolism in *Adam8-/-* mice may help reveal the functions of this gene *in vivo*.

## 3. ADAM9 (MDC9, Meltrin γ)

ADAM9 was initially identified by PCR from mouse tissue and a mouse muscle cell line (Weskamp and Blobel 1994; Weskamp *et al.* 1996; Yagami-Hiromasa *et al.* 1995). ADAM9 is catalytically active (Roghani *et al.* 1999), and has been suggested to have a role in ectodomain shedding of heparin binding EGF-like growth factor (HB-EGF) and of the amyloid precursor protein (APP), a cell surface protein with a critical role in the pathogenesis of Alzheimer's diseases. However, cells derived from *Adam9-/-* mice have no detectable defects in processing HB-EGF or APP, arguing against a major role of ADAM9 in the cleavage of these molecules in mice. *Adam9-/-* mice arc viable and fertile, and do not display any evident pathological phenotypes. See also Chapter 4.

### 3.1 Expression pattern and putative functions of ADAM9 as HB-EGF and APP sheddase

ADAM9 was first identified in a search for novel members of the ADAM gene family by degenerative PCR (Weskamp and Blobel 1994; Yagami-Hiromasa *et al.* 1995). It is ubiquitously expressed in adult tissues including the heart, brain, placenta, lung, skeletal muscle, digestive system and reproductive system (Weskamp *et al.* 1996). ADAM9 is also widely expressed during development, with the most prominent expression in the developing mesenchyme, heart and brain.

Overexpression of ADAM9 in Vero-H cells reportedly results in increased shedding of HB-EGF, suggesting that ADAM9 might have an important role in this process (Izumi *et al.* 1998). However this activity was not observed in subsequent studies done by different groups (Hinkle *et al.* 2004; Prenzel *et al.* 1999; Sahin *et al.* 2004; Weskamp *et al.* 2002). ADAM9 is one of several ADAMs that has been suggested to function as an α-secretase for APP (also see Section 4.3 and 7.2) (Asai *et al.* 2003; Buxbaum *et al.* 1998; Koike *et al.* 1999; Lammich *et al.* 1999). The high expression of ADAM9 in the hippocampus in the brain would be consistent with this notion (Weskamp *et al.* 2002). Furthermore, ADAM9 has been shown to interact with integrins, including αvβ5 and α6β1 (Nath *et al.* 2000; Zhou *et al.* 2001).

Interestingly, several studies have indicated ADAM9 as a pro-oncogenic factor, although the actual effects of ADAM9 in oncogenesis and metastasis are still not clear. An upregulation of ADAM9 was found in pancreatic ductal adenocarcinoma, hepatocellular carcinoma and breast cancer (Grutzmann *et al.* 2003; Grutzmann *et al.* 2004; Le Pabic *et al.* 2003; O'Shea *et al.* 2003; Tannapfel *et al.* 2003), and overexpression of ADAM9 in non-small lung cancer cells was shown to correlate with brain metastasis in a mouse model (Shintani *et al.* 2004). These observations indicate a possible involvement of ADAM9 in the pathogenesis in these tumors.

## 3.2 Mice lacking ADAM9 have no evident major abnormalities

Mice lacking ADAM9 were generated by gene targeting within the second exon, and the absence of ADAM9 in fibroblasts from *Adam9-/-* embryos was confirmed by Western blot (Weskamp *et al.* 2002). Despite the ubiquitous expression pattern of ADAM9 during development and in adult tissues, *Adam9-/-* mice develop normally and are viable and fertile, and there was no significant difference in the mortality rate between *Adam9-/-* mice and wild type controls over the course of 2 years. Histopathological analysis of the organs from *Adam9-/-* mice did not reveal any apparent abnormalities, and differential blood counts and blood chemistry were all within normal range. Thus ADAM9 is not essential for normal development and homeostasis. The processing of HB-EGF in *Adam9-/-* fibroblasts was comparable to that of wild type controls. Furthermore hippocampal neurons isolated from *Adam9-/-* mice did not show detectable defects in the processing of APP at the metalloprotease-dependent α-secretase cleavage site, arguing against a critical role of ADAM9 as α-secretase. Thus ADAM9 is either not essential for the processing of APP or can be compensated by other sheddases such as ADAM10 and ADAM17. Further studies will be necessary to uncover possible functions of ADAM9 during development or in adult mice, and to address whether ADAM9 has a role in oncogenesis.

# 4. ADAM10 (Kuzbanian, MDAM)

ADAM10 was originally purified as an activity that efficiently degrades myelin basic protein in brain extracts. Peptide sequences obtained from the purified protein later allowed cloning of ADAM10 cDNA (referred to as MADM, mammalian disintegrin-metalloprotease) from a bovine cDNA library (Howard *et al.* 1996). Subsequently it became clear that ADAM10 is the mammalian homologue of *Drosophila* Kuzbanian (Kuz), which in turn has an essential role in Notch/Delta signaling. In addition, ADAM10 is considered one of several candidate α-secretases for APP. A targeted mutation of ADAM10 in mice revealed that ADAM10 is indeed essential for Notch/Delta signaling in mammals, but not for the α-secretase dependent processing of APP. See also Chapter 10.

## 4.1 Expression pattern

A study of the expression pattern of ADAM10 during embryogenesis was done in chicks. ADAM10 was found to be expressed ubiquitously in embyros, including in the dermatome, myotome of the somites, epidermis, gut endoderm, epithelial tissues of the kidney, liver, heart, and neural crest cells (Hall and Erickson 2003). Western blot analysis showed that ADAM10 is also highly expressed in a variety of adult tissues (Sahin *et al.* 2004). In adult brain, ADAM10 was found to be widely expressed throughout the brain, which is consistent with a putative role of this gene as an α-secretase for APP (Karkkainen *et al.* 2000).

## 4.2 Controversies over the functions of ADAM10 in Notch/Delta signaling

Studies in Drosophila were the first to uncover a critical role for ADAM10 in neural cell fate decisions (Rooke *et al.* 1996). Rooke *et al.* showed that loss-of-function mutations in the Kuz gene (the *Drosophila* orthologue of ADAM10) result in defects resembling those caused by mutations in Notch, a membrane anchored receptor with important roles in cell fate decisions during development. The mechanism underlying Notch/Delta dependent cell fate decisions is now thought to be conserved from worms to mammals and in a wide variety of cells (Artavanis-Tsakonas *et al.* 1995; Artavanis-Tsakonas *et al.* 1999; Greenwald 1998).

Studies on the processing of Notch over the past few years revealed that Notch undergoes three proteolytic cleavages, at distinct cleavage sites termed S1, S2 and S3. Several lines of evidence indicate that proteolysis of one of the extracellular cleavage sites, termed S2 site, is Kuz/ADAM10 dependent. Pan and Rubin demonstrated that Kuz functions upstream of Notch by showing that gain-of-function Notch is epistatic to coexpression of a dominant-negative form of Kuz (Pan and Rubin 1997). More recently, it has been shown that Kuz associates with Notch and that RNA interference against Kuz blocks S2 cleavage of Notch in flies (Lieber *et al.* 2002).

On the other hand, a study which aimed at identifying the protease responsible for S2 cleavage activity in mammalian cells in tissue culture found, quite unexpectedly, that ADAM17, but not ADAM10, is responsible for this activity (Brou *et al.* 2000). This group also demonstrated that differentiation of *Adam17-/-* myeloid precursor cells to macrophages, a process that is known to be inhibited by Notch signaling, could be blocked by rescuing these cells with ADAM17. Consistent with these observations, mammalian cells lacking ADAM10 still exhibit S2 Notch cleavage activity *in vitro* (Mumm *et al.* 2000). However, the finding that *Adam17-/-* mice do not show any apparent phenotype related to Notch or Notch ligand loss of function (see below), raises questions about how relevant ADAM17 is for Notch processing and activation during development *in vivo.* Thus much remains to be learned about the physiological roles of ADAM17 and ADAM10 in Notch/Delta signaling in mammals.

## 4.3 ADAM10 as a candidate for APP α-secretase

Intriguingly there is a strong resemblance between the biochemical processing of Notch and APP. Yet even though it appears clear that ADAMs have an important role in processing both proteins, the identity of the ADAM that is relevant for either process *in vivo* is not yet clearly defined. As in the case of Notch, APP is cleaved at several distinct sites by at least three different proteases, termed α-, β- and γ-secretase. Cleavage of APP by β- and γ-secreatase gives rise to the Aβ peptide, which is a major component of amyloid plaques and is thought to have a critical role in the pathogenesis of Alzheimer's disease (Selkoe and Schenk 2003). Cleavage of APP by α-secretase occurs between the β- and γ- cleavage site, thus preventing the production of Aβ peptides. Therefore the α-secretase is considered to be a protective factor against Alzheimer's disease.

Along with ADAM17 and ADAM9, ADAM10 is thought to be a potential α-secretase (Kojro *et al.* 2001; Lammich *et al.* 1999; Lopez-Perez *et al.* 2001). Overexpression of ADAM10 in HEK cells leads to an increased -secretase activity, and endogeneous α-secretase activity in these cells is inhibited by introducing a dominant negative form of ADAM10 (Lammich *et al.* 1999). ADAM10 is widely expressed in the brain, including neurons, whereas expression of ADAM17 is more or less limited to the endothelia and glia (Bernstein *et al.* 2003; Goddard *et al.* 2001; Karkkainen *et al.* 2000). The expression pattern in the brain is more consistent with a role for ADAM10 (and ADAM9) than ADAM17 as the APP α-secretase, although expression analysis is not a completely reliable indicator for where an enzyme might function (even low levels of an ADAM may be completely sufficient for its physiological activity). A recent study also showed that overexpression of ADAM10 in neurons alleviates amyloid plaque formation and hippocampal defects in an Alzheimer's disease mouse model (Postina *et al.* 2004). While this study suggests that increasing the activity of ADAM10 may be therapeutically desirable, it does not clarify the identity of the physiological α-secretase.

## 4.4 Other possible substrates of ADAM10

Apart from processing Notch/Delta and APP, ADAM10 has also been implicated in regulating neuronal repulsion through cleaving ephrinA2 (Hattori *et al.* 2000), and in the processing of various other cell surface molecules, including CD44 (Murai *et al.* 2004), type IV collagen (Millichip *et al.* 1998), L1 (Mechtersheimer *et al.* 2001), interleukin 6 receptor (Matthews *et al.* 2003), CXC chemokine ligand 16 (Abel *et al.* 2004; Gough *et al.* 2004), CX3C chemokine fractalkine (Hundhausen *et al.* 2003), neurotensin receptor-3 (Navarro *et al.* 2002), prion protein (Vincent *et al.* 2001), HB-EGF (Lemjabbar and Basbaum 2002; Schafer *et al.* 2004; Yan *et al.* 2002), betacellulin, EGF (Sahin *et al.* 2004), β-site amyloid precursor protein-cleaving enzyme 1 (Hussain *et al.* 2003). The physiological relevance of the processing of most of these candidate ADAM10 substrates during development *in vivo* remains to be determined.

## 4.5 Deletion mutation of ADAM10 leads to early embryonic lethality and multiple malformations which resemble those seen in the absence of Notch/Delta signaling

In light of the complexities surrounding the potential role of ADAM10, in the processing of Notch/Delta and APP, generation and analysis of *Adam10-/-* mice were expected to provide important new insights into these issues. Gene targeting of ADAM10 was performed by inserting a neomycin cassette into the second exon of the ADAM10 gene (Hartmann *et al.* 2002). *Adam10-/-* embryos were found to die around 9.5 dpc, with multiple malformations strikingly similar to that of a complex Notch deficiency as seen in presenilin1/presenilin2 or notch1/notch4 double mutant mice

(Herreman *et al.* 1999; Krebs *et al.* 2000). Consistent with these observations, the expression pattern of the genes involved in the Notch pathway, dll-1, one of the ligands of Notch, and hes-5, a transcription factor activated by Notch signaling, was severely disrupted. Theses findings further support the notion that, as in insects and worms, ADAM10, but not ADAM17, is essential for the Notch/Delta signaling during mammalian development.

It should be noted that the phenotype of *Adam10-/-* embryos is more severe than that of presenilin1/presenilin2 double knockout mice, where cleavage dependent Notch/Delta signaling is thought to be blocked (Herreman *et al.* 1999; Herreman *et al.* 2000). This strongly suggests that ADAM10 also has additional roles in processing of ligands or receptors other than Notch or Delta during embryogenesis.

As mentioned above, the role of ADAM10 in APP cleavage remains a somewhat controversial issue. Studies using hippocampal neurons derived from *Adam9-/-* mice revealed that absence of ADAM9 does not have a major impact on APP processing in these cells, while the absence of ADAM17 abolished regulated (PMA-stimulated), but not constitutive, α-secretase activity in mouse embryonic fibroblasts (Buxbaum *et al.* 1998; Weskamp *et al.* 2002). Due to the early embryonic lethality of *Adam10-/-* mice at a stage preceding neuronal development, APP processing can only be analyzed in immortalized cell lines derived from *Adam10-/-* embryos. The results might therefore not directly reflect the actual contribution of ADAM10 in APP cleavage in the brain. Nevertheless the data clearly showed that α-secretase activity was preserved in some *Adam10-/-* cell lines. It remains to be determined whether there is compensation by, or redundancy between, different ADAMs or other enzymes in the processing of APP (Asai *et al.* 2003) or whether different ADAMs have major roles in APP processing in distinct cells and tissues. Since further studies of the role of ADAM10 during development and in adult mice are hampered by the early embryonic lethality of *Adam10-/-* mice, it will be critical to generate mice carrying a conditional mutation in this gene to learn more about the function of ADAM10 during development and in adults, and in APP processing in the brain.

## 5. ADAM12 (Meltrin α)

ADAM12 was initially identified along with ADAM19 (meltrin β) and ADAM9 (meltrin γ) in myoblasts, where it was suggested to play a role in myoblast fusion during myoblast differentiation (Yagami-Hiromasa *et al.* 1995). A deletion mutation, however, did not lead to any overt defects, including in muscle development and regeneration, arguing against an essential role for ADAM12 in muscle fusion or differentiation during mouse development. See also Chapter 6.

## 5.1 Expression pattern

In the early stages of murine development, ADAM12 transcripts are expressed in the condensed mesenchymal cells which later develop to skeletal muscle, bones and visceral organs (Kurisaki *et al.* 1998). ADAM12 transcripts in the myotome appear at 10.5-11.5 dpc, after myotube formation takes place. Although ADAM12 expression persists in the tendinous region of the muscle in later stages of development, the most prominent expression is found in the developing bone, especially in the periostium and bone marrow (Kurisaki *et al.* 1998). ADAM12 was initially reported to have a very limited expression pattern in adult tissue, with high expression in bone (Yagami-Hiromasa *et al.* 1995). Due to its expression in bone and muscle cells, ADAM12 was thought to play a role in the fusion of myoblasts and osteoclasts. Later studies revealed that ADAM12 is also expressed quite ubiquitously in adult mice, and that osteoblasts are the major source of ADAM12 transcripts in the bone (Gilpin *et al.* 1998; Harris *et al.* 1997; Inoue *et al.* 1998).

## 5.2 Putative role of ADAM12 in myogenesis

During development myoblasts are known to fuse to form mature multinucleated myotubes. As mentioned above, ADAM12 was originally identified in a search for genes involved in muscle fusion by PCR with degenerate primers for conserved amino acids in the first recognized ADAMs, fertilin α and β, which were considered to be candidate fusion proteins at the time (we now know that fertilin α and β have an essential role in fertilization, but are not required for sperm-egg membrane fusion *in vivo*) (Primakoff and Myles 2002). Overexpression of a truncated form of ADAM12 lacking the pro- and metalloprotease domain in a myoblast-like cell line (C2 cells), significantly enhanced myoblast fusion *in vitro*, whereas C2 cells transfected with either full length ADAM12 or anti-sense RNA exhibited less membrane fusion. It was also shown that ADAM12 contains a sequence similar to the fusion peptides of paramyxoviruses (Yagami-Hiromasa *et al.* 1995).

A subsequent study demonstrated that an isoform of ADAM12 which lacks a transmembrane domain recruits host muscle cells into implanted human tumors in nude mice (Gilpin *et al.* 1998). Furthermore, the cytoplasmic domain of ADAM12 interacts with the muscle specific protein α-actinin-2 and this interaction is essential for ADAM12 to promote myoblast fusion (Galliano *et al.* 2000). In addition, introduction of ADAM12 in a mouse model of muscular dystrophy (mdx mice) significantly alleviates the muscle pathology (Kronqvist *et al.* 2002). Finally, the expression of ADAM12 is localized to muscle satellite cells and is upregulated in regenerating and denervated muscle (Borneman *et al.* 2000).

Despite these studies, however, it remains to be determined what the exact role of ADAM12 in muscle development and regeneration is. First, although the removal of the prodomain is known to be important for generation of mature and catalytically active ADAM12 (Cao *et al.* 2002; Loechel *et al.* 1998; Yagami-Hiromasa *et al.* 1995), there is little evidence for the existence of an ADAM12 which lacks its pro- and metalloprotease domain (the truncated form used in *in vitro* myotube formation assays (Yagami-Hiromasa *et al.* 1995)). Second, although ADAM12-S, a soluble isoform which contains the metalloprotease domain, has been implicated in recruitment of myogenic cells into tumors (Gilpin *et al.* 1998), transgenic mice overexpressing ADAM12-S under control of the muscle creatine kinase promoter resulted in induction of adipogenesis but had no overt defect in muscle differentiation (Kawaguchi *et al.* 2002).

A recent study by Cao *et al.* prompted a different view on this issue. This study showed that while inhibition of the expression of ADAM12 was accompanied by lower expression of markers for both quiescent and differentiating cells, overexpression of ADAM12 induced a quiescent cell-like phenotype and did not stimulate differentiation, indicating that ADAM12 has a role, as a negative regulator, in cell fate decision in myoblast differentiation (Cao *et al.* 2003). This observation offers some reconciliation with the early study in which stable transfection of C2 cells with the full-length ADAM12 led to inhibition of myoblast fusion (Yagami-Hiromasa *et al.* 1995), and may provide an explanation for the phenotype of ADAM12-S transgenic mice and ADAM12 null mice (discussed below). However it does not lend support to the notion that ADAM12 participates in the process of myoblast fusion or contributes to muscle regeneration in dystrophic mice model as previously shown. Finally, ADAM12 has also been reported to have a role in the formation of multinucleated osteoclasts, which are formed by cell-cell fusion of mononuclear precursors (Abe *et al.* 1999).

## 5.3 Putative substrates and interacting molecules

As with other members of this gene family, ADAM12 has been implicated as a protease of several molecules, including Insulin-like growth factor binding protein-3 and -5 (Loechel *et al.* 2000; Shi *et al.* 2000), oxytocinase (Ito *et al.* 2004) and HB-EGF (Asakura *et al.* 2002; Mori *et al.* 2003). However, considering the recent genetic and biochemical evidence pointing toward ADAM17 as the major sheddase for HB-EGF, the physiological relevance of ADAM12 in cleavage of HB-EGF remains to be determined (see Section 7.2)

Several molecules have also been shown to interact with the cyoplastic domain of ADAM12. These include Src (Kang *et al.* 2000), Grb2 (Suzuki *et al.* 2000), α-actinin-1 (Cao *et al.* 2001), Fish (Abram *et al.* 2003), phosphatidylinositol 3-kinase (Kang *et al.* 2001) and PACSIN3 (Mori *et al.* 2003). The disintegrin and cystein rich domain binds to integrin α9β1 and Syndecan-4 (Iba *et al.* 2000; Iba *et al.* 1999; Thodeti *et al.* 2003), respectively.

## 5.4 Removal of ADAM12 does not lead to overt anomalies in muscle

The initial identification of ADAM12 from muscle cells, and the functional studies with mutant forms of ADAM12 in these cells suggested that ADAM12 has an important role in myogenesis. To address this issue, *Adam12-/-* mice were generated in which the first exon of the ADAM12 gene was replaced with a neomycin cassette. Unexpectedly, *Adam12-/-* mice did not show any overt abnormalities, even though 30% less *Adam12-/-* mice were born from crosses of heterozygous parents than expected. Because the ratio of offspring from matings of *Adam12+/-* mice was Mendelian prior to birth, the missing 30% of mutant mice must have died several days after birth. The cause of this perinatal lethality remains to be determined (Kurisaki *et al.* 2003). Intriguingly, *Adam9/15/12-/-* triple knock out mice are viable and fertile and their survival rate was comparable to that of wild type controls (Sahin *et al.* 2004). One possible explanation for this discrepancy is that there may have been differences in the genetic background of mice used in these different studies.

Close histological analysis of *Adam12-/-* mice revealed subtle defects in the interscapular brown fat tissue, and in the neck and interscapular muscles of around 30% of the examined animals (Kurisaki *et al.* 2003). It remains to be determined why only the neck and interscapular region are affected by the loss of ADAM12, and whether these defects may be the cause for the 30% embryonic or perinatal lethality in *Adam12-/-* mice. Muscle regeneration in *Adam12-/-* mice was comparable to wild type controls, and when mice carrying the mdx mutation, which results in muscular dystrophy, were bred with *Adam12-/-* mice, no aggravation of muscle degeneration was seen in *Adam12-/-*mdx animals. Taken together, removal of ADAM12 did not lead to any major defects, including in muscle development and regeneration. On the other hand, phorbol ester-stimulated shedding of HB-EGF was significantly reduced in *Adam12-/-* fibroblasts (Kurisaki *et al.* 2003). This is in contrast with the results from Sahin *et al.* in which PMA-stimulated and constitutive shedding of HB-EGF was not strongly affected in *Adam9/15/12-/-* triple knockout fibroblasts. Nevertheless, *Adam12-/-* mice do not reveal any heart defects resembling those observed in HB-EGF null mice, while *Adam17-/-* mice have similar heart defects as *Hb-egf-/-* mice (see Secton 7.2), suggesting that ADAM12 is not essential for HB-EGF shedding and activation during heart development.

Given that ADAM12 may be involved in differentiation of self-renewing satellite cells *in vivo* (Cao *et al.* 2003), this provides a possible explanation for the lack of apparent muscle defects in *Adam12-/-* mice and ADAM12 transgenic mice. Based on the observations by Cao *et al.*, only high expression of ADAM12 prior to muscle differentiation can drive myoblasts toward a quiescent state. Hence either removal of ADAM12 or overexpression of ADAM12 in differentiated myofibers (with muscle specific creatine kinase promoter as done in previous studies) may not have a major impact on this event (Cao *et al.* 2003). Nevertheless, the lack of apparent muscle phenotype in *Adam12-/-* mice argues against an essential role for ADAM12 in myoblast fusion / differentiation (Yagami-Hiromasa *et al.* 1995). One possible explanation is that this result is due to compensation or redundancy with other ADAMs, even though neither *Adam9/12/15/17* nor *Adam9/12/15/19* quadruple mutant mice revealed apparent muscle defects during development (Sahin *et al.* 2004; Horiuchi *et al.*, manuscript in preparation). Further studies will be necessary to understand the role of ADAM12 in muscle differentiation and regeneration.

## 6. ADAM15 (MDC15, Metarginin)

Human ADAM15 was first discovered in a screen for novel ADAMs by PCR (Kratzschmar *et al.* 1996), and was named metargidin because it carries an RGD sequence in a similar position as snake venom disintegrins (Kratzschmar *et al.* 1996). However, the mouse and rat orthologues of ADAM15 lack an RGD sequence (Bosse *et al.* 2000; Lum *et al.* 1998), arguing against a conserved role of ADAM15 as ligands for RGD-binding integrins in these two rodents.

## 6.1 Expression pattern in vascular cells and putative function in angiogenesis

ADAM15 expression can be detected in most tissues in developing and adult mice and in specific regions of the brain and spinal cord (Bosse *et al.* 2000; Horiuchi *et al.* 2003; Lum *et al.* 1998). During development the highest levels of expression are observed in vascular cells, endocardial cells and in hypertrophic cells in developing bone (Horiuchi *et al.* 2003). The expression in developing vessels peaks at 12.5-14.5 dpc and subsequently subsides (unpublished observation). After birth, high expression of ADAM15 in the vasculature was seen in endothelial cells in the retina of animals subjected to a mouse model of retinopathy of prematurity. ADAM15 is also highly expressed in human umbilical vascular endothelial cells and vascular smooth muscle cells (Herren *et al.* 1997; Horiuchi *et al.* 2003), however it is not as highly expressed in normal vessels in adults (unpublished observation; Herren *et al.* 1997).

In accordance to its expression pattern in vasculature, several studies have indicated a possible involvement of ADAM15 in angiogenesis. Expression of ADAM15 has been shown to be upregulated in atherosclerotic lesions (Al-Fakhri *et al.* 2003; Herren *et al.* 1997). It has also been shown that ADAM15 colocalized with VE-cadherin, an endothelial cell specific cell-cell adhesion molecule, and the surface expression of ADAM15 was driven by this molecule (Ham *et al.* 2002).

Human ADAM15 contains an RGD sequence and has been shown to interact with $\alpha v\beta 3$ and $\alpha 5\beta 1$ integrins (Nath *et al.* 1999; Zhang *et al.* 1998), both of which are known to play crucial roles in angiogenesis (Eliceiri and Cheresh 2000; Hynes 2002), while mouse ADAM15 does not contain this sequence. It is therefore possible that human ADAM15 has different functions *in vivo* than mouse ADAM15 because it can interact with these two integrins. Interestingly, a recent study showed that the disintegrin domain of human ADAM15, which contains the RGD sequence, has a anti-oncogenic activity through suppressing tumor angiogenesis (Trochon-Joseph *et al.* 2004). Both mouse and human ADAM15 have been shown to interact with $\alpha 9\beta 1$ integrin in an RGD-independent manner, as is the case with other ADAMs, such as ADAM1, 2, 3, 9 and 12.

Although ADAM15 carries a catalytic site consensus sequence for metalloproteases, little is currently known about its catalytic activity. So far, ADAM15 has been implicated in processing CD23 (Fourie *et al.* 2003), type IV collagen, gelatin (Martin *et al.* 2002), TGF$\alpha$ and amphiregulin (Schafer *et al.* 2004) *in vitro*. However, ADAM15 null mice do not show any EGFR related developmental defects (described below) and fibroblasts from *Adam9/12/15-/-* embryos are still capable of processing TGF$\alpha$ and amphiregulin (Sahin *et al.* 2004). The biological relevance of these observations therefore remains to be clarified.

Several cytoplasmic proteins have also been shown to interact with the cytoplasmic domain of ADAM15. These include Src family protein kinases, the adaptor protein Grb2 (Poghosyan *et al.* 2002), SH3PX1 and endophilin I (Howard *et al.* 1999).

## 6.2 *Adam15-/-* mice are viable and fertile but have decreased pathological neovascularization in a mouse model for proliferative retinopathy

In order to generate *Adam15-/-* mice, a targeted mutation was introduced into the ADAM15 gene by replacing the exon carrying the initial methionine with a neomycin cassette. When heterozygous *Adam15+/-* mice of mixed genetic background (129/SvJ, C57Bl/6J) or of inbred genetic background (129/SvJ or C57Bl/6J) were mated, the genotype of the offspring followed a Mendelian distribution pattern, regardless of the background, and *Adam15-/-* mice were indistinguishable from control littermates in appearance and their behavior during routine handling. Matings of *Adam15-/-* mice resulted in viable and healthy litters with normal litter sizes. Histopathological analysis, especially in the tissues where ADAM15 expression is high, as well as clinical chemistry analysis of serum and differential blood count revealed no apparent anomalies. Furthermore, no significant difference in mortality or morbidity between *Adam15-/-* and wild type control mice was seen over two years. These observations show that ADAM15 is not essential for either development or adult homeostasis.

Because ADAM15 is highly expressed in developing blood vessels, *Adam15-/-* mice were subjected to a mouse model of retinopathy, in which neovascularization in the retina is induced by relative hypoxia. *Adam15-/-* mice showed remarkably fewer neovascularization tufts compared to wild type controls, indicating a possible involvement of ADAM15 in the pathology of proliferative retinopathy. Furthermore, growth of heterotopically injected tumors was decreased in *Adam15-/-* mice compared to wild type controls, which is consistent with a role for ADAM15 in neovascularization, even though there was no apparent decrease in the microvessel density in the tumors from *Adam15-/-* mice compared to that of wild type controls. This could be explained by defects in certain aspects of neovascularization, such as a delay in the initiation of new vascular sprouts or in the growth rate of new vessels. To further explore these possibilities, endothelial cells and aortae isolated from *Adam15-/-* and wild type control mice have been examined. However the cells and tissues from *Adam15-/-*

behaved identically to that of wild type controls in *in vitro* studies, including studies of proliferation, migration and tube formation of endothelial cells, and aortic explant sprouting assays (unpublished observation). Further studies are in progress to elucidate the mechanism underlying the role of ADAM15 in pathological neovascularization.

As described, α9β1 integrin can interact with several ADAMs including ADAM9, 12 and 15. Mice lacking α9β1 integrin develop a chylothorax after birth and die of respiratory failure (Huang *et al.* 2000). However single knockout mice for ADAMs 9, 12 or 15, or ADAM9/15 double mutant or even ADAM9/12/15 triple mutant mice are viable and fertile and do not display a chylothorax or respiratory failure (Sahin *et al.* 2004). The biological significance of the interaction between ADAMs and α9β1 integrin therefore remains to be determined.

## 7. ADAM17 (TACE)

ADAM17 can currently be considered the best-characterized ADAM besides ADAM10. Originally it was identified as an enzyme responsible for the cleavage of TNFα (Black *et al.* 1997; Moss *et al.* 1997), although further studies uncovered more diverse functions than initially expected. Gene targeting resulted in perinatal lethality with multiple defects in various organs which closely resemble those seen in mice lacking the EGFR, TGF or HB-EGF. See also Chapter 8.

### 7.1 Prominent role of ADAM17 as a multiple sheddase

ADAM17 is the first ADAM for which a role as a "sheddase" was clearly established. It was initially identified in a search for the TNFα converting enzyme (TACE), i.e. the enzyme that releases TNFα from cells (Black *et al.* 1997; Moss *et al.* 1997). ADAM17 is widely expressed with high levels of expression in the heart, placenta, testis, ovary, lung and spleen (Black *et al.* 1997; Sahin *et al.* 2004).

Apart from TNFα, a substantial number of other membrane proteins have been identified as substrates for ADAM17, including several EGFR ligands (HB-EGF, amphiregulin, TGF , epiregulin) (Sahin *et al.* 2004; Sunnarborg *et al.* 2002), p75 TNFR (Peschon *et al.* 1998a), IL6R (Matthews *et al.* 2003), APP (Buxbaum *et al.* 1998; Slack *et al.* 2001), MUC1 (Thathiah *et al.* 2003), growth hormone binding protein (Zhang *et al.* 2000), L-selectin (Condon *et al.* 2001; Peschon *et al.* 1998a), Fractalkine (CX3CL1) (Garton *et al.* 2001), collagen XVII (Franzke *et al.* 2004), prion protein (Vincent *et al.* 2001), CD40 (Contin *et al.* 2003), PAR1 (Contin *et al.* 2003), c-Kit (Cruz *et al.*

2004), VCAM-1 (Garton *et al.* 2001), p75 neurotrophin receptor (Weskamp *et al.* 2004) and ErbB4 (Rio *et al.* 2000). Not surprisingly, ADAM17 has been implicated in a variety of physiological and pathological processes, including tumor migration and proliferation (Borrell-Pages *et al.* 2003; Gschwind *et al.* 2003; Hart *et al.* 2004; Schafer *et al.* 2004), arthritis and inflammation (Newton *et al.* 2001; Ohta *et al.* 2001; Patel *et al.* 1998). Furthermore, it has also been shown that ADAM17 interacts with several other molecules, including α5β1 integrin (Bax *et al.* 2004), MAD2 (Nelson *et al.* 1999), PTPH1 (Zheng *et al.* 2002) and SAP97 (Peiretti *et al.* 2003). However, the physiological relevance of these interactions remains to be established.

## 7.2 Inactivation of ADAM17 leads to perinatal lethality with multiple defects resembling those seen in *Egfr-/-* *Tgfα-/-* and *Hb-egf-/-* mice

In order to inactivate *Adam17*, the exon carrying the $Zn^{2+}$ binding catalytic site was replaced with a neomycin cassette by homologous recombination (Black *et al.* 1997). T-cells derived from the mutant animals showed a significant reduction in TNF release and an increase in surface TNFα expression compared to the wild type controls. Initially, there was some concern about this gene targeting strategy because the resulting mutant form of ADAM17 lacking the catalytic site could conceivably have a dominant negative effect. However all data available to date suggests that this mutation leads to a loss of function instead of a dominant negative effect (see Schlondorff and Blobel 1999 for discussion).

Analysis of progeny derived from matings of *Adam17+/-* mice revealed that most *Adam17-/-* mice die between 17.5 dpc and the first day of birth. In light of the predicted role of ADAM17 in the processing of TNFα (at the time, TNFα was the only known substrate for ADAM17), this severe phenotype was quite unexpected because mice lacking TNFα or its receptors are overtly normal (Marino *et al.* 1997; Pasparakis *et al.* 1996; Peschon *et al.* 1998b). *Adam17-/-* mice have open eye lids resulting from a failure of eyelid fusion, lack a conjunctival sac, and have thinned corneas and several epidermal and hair defects. These defects in the eye, hair and skin are reminiscent of those seen in mice bearing a disruption of the gene for TGFα. Additional defects were observed in the epithelial maturation of multiple organs, including the intestine, lung, nonglandular stomach, thyroid, parathyroid and salivary gland, and in the spongiotrophoblast layer of the placenta. The epithelial defects are similar to those described in mice lacking the EGFR. Since all ligands of EGFR, including TGFα, HB-EGF, epiregulin, amphiregulin, EGF and betacellulin, are synthesized as membrane bound precursors and are subsequently released from the cell surface (Harris *et al.* 2003), this raised the possibility that ADAM17 might be responsible for the processing of one or more of these ligands in addition to TNFα.

Consistent with this notion, the processing of TGFα was indeed absent in the fibroblasts derived from *Adam17-/-* mice (Black *et al.* 1997). Moreover an anatomical analysis of *Adam17-/-* mice revealed multiple defects in the heart, especially in valvulogenesis, and in the lung, indicating that these defects in the major organs could be the cause of the perinatal lethality of *Adam17-/-* mice (Shi *et al.* 2003; Zhao *et al.* 2001). These findings are especially interesting in the light of the recent study showing very similar defects in the heart and lung in HB-EGF null mutant mice (Iwamoto *et al.* 2003; Jackson *et al.* 2003). These observations, along with emerging *in vitro* data, provide genetic and biochemical evidence that ADAM17 is also a major sheddase for HB-EGF. Although several ADAMs, including ADAM9, ADAM10 and ADAM12 have also been shown to take part in the processing of HB-EGF *in vitro* as described, the knockout mice for ADAMs 9 and 12 do not show a phenotype similar to that of HB-EGF (*Adam10-/-* mice die too early to allow an evaluation of heart valve development). Thus, even though it cannot be excluded that proteases other than ADAM17 might have a more prominent role in shedding HB-EGF under certain conditions (such as in pathological conditions e.g. cancer or wound healing), ADAM17 appears to be the physiologically relevant sheddase for HB-EGF during heart development. As for other EGFR ligands, ADAM17 has also been implicated in the cleavage of amphiregulin and epiregulin, further supporting the initial hypothesis that ADAM17 is important for activating several ligands of the EGFR (Gschwind *et al.* 2003; Hinkle *et al.* 2004; Sahin *et al.* 2004).

Due to the perinatal lethality of *Adam17-/-* mice, the role of ADAM17 during adult homeostasis as well as the physiological role it plays in regulating the function of other substrates cannot currently be addressed. Further elucidation of the role of ADAM17 in adult mice and in disease models will require the generation of conditional knockout for this multifunctional gene.

## 8. ADAM19 (Meltrin β, MADDAM)

ADAM19 is a widely expressed and catalytically active ADAM that was first identified in myoblasts along with ADAM12 and ADAM9 (Yagami-Hiromasa *et al.* 1995). While little is currently known about physiologically relevant substrates, targeted inactivation of ADAM19 has uncovered an essential role in cardiovascular morphogenesis. See also Chapter 9.

## 8.1 Expression pattern

ADAM19 is ubiquitously expressed in adult tissues, with most prominent expression in the heart, lung, cerebellum, placenta, lymph node, digestive system, leukocytes and in certain cells in the bone (Fritsche *et al.* 2000; Inoue *et al.* 1998; Wei *et al.* 2001). During development, expression of ADAM19 is first seen in the heart and tail bud at 8.0 dpc, and then in the cranial and dorsal root ganglia, and in the ventral horn of the spinal cord (Kurisaki *et al.* 1998). ADAM19 was also identified as a gene specifically expressed in dendritic cells but not in macrophages, indicating that ADAM19 may serve as a marker for the differentiation of dendritic cells from other cells of the myeloid lineage (Fritsche *et al.* 2000; Fritsche *et al.* 2003)

## 8.2 ADAM19 is implicated in the processing of TRANCE and neuregulin

Although ADAM19 carries a catalytic site consensus sequence for metalloproteases and possesses catalytic activity (Chesneau *et al.* 2003; Wei *et al.* 2001), little is currently known about its substrate profile. To date, only two membrane proteins have been identified as potential substrates of ADAM19, the TNF-family member TRANCE (OPGL, ODF, RANKL), a protein with important roles in osteoclast differentiation, dendritic cell survival and mammary development (Fata *et al.* 2000; Kong *et al.* 2000; Suda *et al.* 1999), and neuregulin, a ligand for ErbB receptors with important roles in development of the heart, nervous system and other organ systems (Chesneau *et al.* 2003; Falls 2003; Shirakabe *et al.* 2001; Wakatsuki *et al.* 2004). It should be noted, however, that there are conflicting results regarding the involvement of ADAM19 in processing neuregulin (discussed below).

## 8.3 *Adam19-/-* mice have multiple heart defects and most die shortly after birth

ADAM19 gene targeting was done by two independent groups with different strategies (Kurohara *et al.* 2004; Zhou *et al.* 2004). One was performed by using stem cells with a secretory gene trap insertion in ADAM19 (Mitchell *et al.* 2001; Zhou *et al.* 2004) and the other was done by removal of exons 10 to 12, which contain the catalytic sequence site (Kurohara *et al.* 2004). In both cases, the progeny of matings of heterozygous ADAM19 mutant mice resulted in a Mendelian distribution of the genotype at birth. However about 80% of *Adam19-/-* mice died in the first few days after birth. Histological analysis revealed very similar defects in heart development in both studies, including a membranous ventricular septum defect (VSD) and malformations of aortic, plumonic and tricuspid valves, but not the mitral valve. These heart defects are considered to be the most likely cause of the perinatal lethality in *Adam19-/-* mice. The penetrance of the VSD and aortic and pulmonic valve defects were complete, while the penetrance of the atrioventricular defects (ostium primum atrial septal and tricuspid valve defects) was only partial. This indicates that the proximal portion of the conotruncal endocardial cushion, i.e. the part of the endocardial cushion that gives rise to the ventricular septum and aortic and pulmonic valves, is most severely affected by the removal of ADAM19 activity. In addition to these multiple heart defects, Kurohara *et al.* also pointed out defects in the fasciculation of preganglionic neurons through the adrenal medulla and in muscle development, exemplified by a thinner diaphragm in the absence of ADAM19. The expression pattern of ADAM19 overlaps that of TRANCE in developing bone, and as described above, ADAM19 has been shown to cleave TRANCE *in vitro*. However, no evident major defect in bone development was found in newborn *Adam19-/-* mice (Zhou *et al.* 2004).

The two studies of *Adam19-/-* mice produced conflicting results with respect to the mechanism underlying the defect in heart development, and more specifically, the potential role of ADAM19 in processing of neuregulin. Kurohara *et al.*, showed that PMA-stimulated neuregulin shedding is abolished in *Adam19-/-* fibroblasts *in vitro* when they are *sparsely* plated, but not when they are densely plated. Based on this observation, they hypothesized that the cardiac defects (and also the defects in preganglionic neurons in the adrenal medulla) are caused by aberrant signaling between neuregulin and its ErbB receptors. However, Zhou *et al.* were unable to identify defects in stimulated or constitutive shedding of neuregulin β1 and β2 in *Adam19-/-* fibroblasts, regardless of the cell density. In addition, Zhou *et al.* demonstrated that overexpression of ADAM19 in Cos7 cells did not increase constitutive or stimulated processing of neuregulin β1 and β2, even though it did increase shedding of TRANCE, which was used to confirm that ADAM19 was active in Cos7 cells. The reason for this apparent discrepancy remains to be established, but it might be attributed to the differences in experimental design or the materials.

Quite intriguingly, there is a close similarity in the heart phenotype in *Adam17-/-* and *Adam19-/-* mice. A recent study suggests that the heart defects in *Adam17-/-* mice are caused by aberrant HB-EGF/EGFR signaling (see above, Jackson *et al.* 2003). As for ADAM19, there is no evidence that it participates in the processing and activation of HB-EGF (Zhou *et al.* 2004). Further studies on the roles of ADAMs in activation of ErbB receptors as well as other molecules with a role in heart development will be required to shed new light on the mechanism underlying the role of ADAM19 in heart development.

## 9. CONCLUSIONS

Evaluation of knockout mice for seven ADAM proteases has uncovered essential roles for ADAMs 10, 17, and 19 in mouse development, and has shown that ADAMs 8, 9, 12 and 15 are not essential for development and adult homeostasis, at least in the controlled and clean environment in which laboratory animals are kept (Table 1). Studies of *Adam10-/-* and *Adam17-/-* mice and cells derived from these animals have revealed that these ADAMs have key regulatory roles in some of the major signaling pathways in cells, including the EGFR and Notch/Delta pathways. Thus ADAM-dependent processing and ectodomain shedding of receptors, ligands and other proteins has emerged as a critical post-translational regulator of the function of at least some substrate proteins. Now that ADAMs have been implicated in a variety of shedding events and the physiological relevance of ectodomain shedding as a post-translational regulator of membrane proteins is well established for proteins such as Notch and some EGFR-ligands, it will be interesting to determine whether ADAM-dependent ectodomain shedding also regulates the function of other substrate proteins. Further studies of knockout mice for ADAMs that are not essential for development (ADAMs 8, 9, 12 and 15), including shedding experiments with cells derived from these animals, and gain of function studies in which these ADAMs that are overexpressed in cells or in transgenic mouse lines will be necessary to help understand what their substrates and functions might be. One important lesson from the published studies of ADAM knockout mice is that the expression pattern can provide important clues as to which tissue a given ADAM might have an important role in. ADAM19, for example, is highly expressed in developing heart valves, and is critical for their proper morphogenesis during development (Zhou *et al.* 2004), while ADAM15 is highly expressed in endothelial cells, and has an important role in pathological neovascularization (Horiuchi *et al.* 2003). Similar challenges that are designed based on the expression pattern of other ADAMs might also reveal unexpected functions of these proteins.

Once a function of an ADAM has been found, it will be important to determine the underlying molecular mechanism. In the case of ADAM15, for example, it will be interesting to test whether the decreased pathological neovascularization in *Adam15-/-* mice can be explained through a defect in shedding molecules with a role in angiogenesis and neovascularization, such as the receptors for the vascular endothelial growth factor (VEFGR1 and 2), and the angiopoietin receptor Tie 2.

*Table 1.* Ablation of ADAMs in mice and their resulting phenotypes

| Gene | Phenotype |
|---|---|
| ADAM8/MS2 | Fertile and viable with no overt phenotype |
| ADAM9/Meltrin γ | Fertile and viable with no overt phenotype |
| ADAM10/Kuz | Embryonic lethal at E9.5; multiple defects in several developing organs |
| ADAM12/Meltrin α | Fertile and viable with no overt phenotype (subtle defects in brown fat-tissue and muscle) |
| ADAM15/Metarginin | Fertile and viable with no overt phenotype (decreased neovascularization in proliferative retinopathy) |
| ADAM17/TACE | Perinatally lethal; Multiple defects in several organs |
| ADAM19/Meltrin β | Perinatally lethal; Multiple heart defects |

Finally, in the context of discussing the role of ADAMs in knockout mice it is interesting to note that ADAMs 10 and 17, which are essential for mouse development, have apparent orthologues in *Drosophila melanogaster*, whereas ADAMs 8, 9, 12, 15 and 19 do not. This suggests that the latter group of ADAMs evolved more recently. One possibility is that these ADAMs might therefore have functions in organ systems that have become more highly evolved and specialized in vertebrates, such as the immune system, cardiovascular system (as is the case for ADAMs 15 and 19) or the central nervous system. This further underscores the notion that specific challenges of knockout mice for ADAMs 8, 9, 12, and 15 might reveal roles for these proteins in cells or tissues that are not essential for development, but may instead have important functions in adult animals. It is also possible that the absence of a severe phenotype in mice lacking ADAMs 8, 9, 12 or 15 is due to functional redundancy with other ADAMs, or compensation by other ADAMs. However, mice lacking multiple ADAMs, such as *Adam9/12/15-/-* mice are viable and fertile, and appear normal and healthy (Sahin *et al.* 2004). Thus, to date there is no clear evidence for compensatory or redundant roles of ADAMs that are essential for development, although further studies will be necessary to address this issue more comprehensively. Taken together, the analysis of ADAM knockout mice has provided intriguing insights into the functions of some members of this protein family, and has established a critical role of ADAMs 10 and 17 as regulatory switches for the Notch and EGFR signaling pathways.

## REFERENCES


Abe, E., Mocharla, H., Yamate, T., Taguchi, Y., Manolagas, S. C., 1999, Meltrin-alpha, a fusion protein involved in multinucleated giant cell and osteoclast formation. *Calcif Tissue Int.* **64**: 508-515.

Abel, S., Hundhausen, C., Mentlein, R., Schulte, A., Berkhout, T. A., Broadway, N., Hartmann, D., Sedlacek, R., Dietrich, S., Muetze, B., Schuster, B., Kallen, K. J., Saftig, P., Rose-John, S., Ludwig, A., 2004, The transmembrane CXC-chemokine ligand 16 is induced by IFN-gamma and TNF-alpha and shed by the activity of the disintegrin-like metalloproteinase ADAM10. *J Immunol.* **172**: 6362-6372.

Abram, C. L., Seals, D. F., Pass, I., Salinsky, D., Maurer, L., Roth, T. M., Courtneidge, S. A., 2003, The adaptor protein fish associates with members of the ADAMs family and localizes to podosomes of Src-transformed cells. *J Biol Chem.* **278**: 16844-16851.

Al-Fakhri, N., Wilhelm, J., Hahn, M., Heidt, M., Hehrlein, F. W., Endisch, A. M., Hupp, T., Cherian, S. M., Bobryshev, Y. V., Lord, R. S., Katz, N., 2003, Increased expression of disintegrin-metalloproteinases ADAM-15 and ADAM-9 following upregulation of integrins alpha5beta1 and alphavbeta3 in atherosclerosis. *J Cell Biochem.* **89**: 808-823.

Amour, A., Knight, C. G., English, W. R., Webster, A., Slocombe, P. M., Knauper, V., Docherty, A. J., Becherer, J. D., Blobel, C. P., Murphy, G., 2002, The enzymatic activity of ADAM8 and ADAM9 is not regulated by TIMPs. *FEBS Lett.* **524**: 154-158.

Artavanis-Tsakonas, S., Matsuno, K., Fortini, M. E., 1995, Notch signaling. *Science.* **268**: 225-232.

Artavanis-Tsakonas, S., Rand, M. D., Lake, R. J., 1999, Notch signaling: cell fate control and signal integration in development. *Science.* **284**: 770-776.

Asai, M., Hattori, C., Szabo, B., Sasagawa, N., Maruyama, K., Tanuma, S., Ishiura, S., 2003, Putative function of ADAM9, ADAM10, and ADAM17 as APP alpha-secretase. *Biochem Biophys Res Commun.* **301**: 231-235.

Asakura, M., Kitakaze, M., Takashima, S., Liao, Y., Ishikura, F., Yoshinaka, T., Ohmoto, H., Node, K., Yoshino, K., Ishiguro, H., Asanuma, H., Sanada, S., Matsumura, Y., Takeda, H., Beppu, S., Tada, M., Hori, M., Higashiyama, S., 2002, Cardiac hypertrophy is inhibited by antagonism of ADAM12 processing of HB-EGF: metalloproteinase inhibitors as a new therapy. *Nat Med.* **8**: 35-40.

Bax, D. V., Messent, A. J., Tart, J., van Hoang, M., Kott, J., Maciewicz, R. A., Humphries, M. J., 2004, Integrin alpha5beta1 and ADAM-17 interact in vitro and co-localize in migrating HeLa cells. *J Biol Chem.* **279**: 22377-22386.

Bernstein, H. G., Bukowska, A., Krell, D., Bogerts, B., Ansorge, S., Lendeckel, U., 2003, Comparative localization of ADAMs 10 and 15 in human cerebral cortex normal aging, Alzheimer disease and Down syndrome. *J Neurocytol.* **32**: 153-160.

Black, R. A., Rauch, C. T., Kozlosky, C. J., Peschon, J. J., Slack, J. L., Wolfson, M. F., Castner, B. J., Stocking, K. L., Reddy, P., Srinivasan, S., Nelson, N., Boiani, N., Schooley, K. A., Gerhart, M., Davis, R., Fitzner, J. N., Johnson, R. S., Paxton, R. J., March, C. J., Cerretti, D. P., 1997, A metalloproteinase disintegrin that releases tumour-necrosis factor-alpha from cells. *Nature.* **385**: 729-733.

Blobel, C. P., Myles, D. G., Primakoff, P., White, J. M., 1990, Proteolytic processing of a protein involved in sperm-egg fusion correlates with acquisition of fertilization competence. *J Cell Biol.* **111**: 69-78.

Blobel, C. P. and White, J. M., 1992, Structure, function and evolutionary relationship of proteins containing a disintegrin domain. *Curr Opin Cell Biol.* **4**: 760-765.

Blobel, C. P., Wolfsberg, T. G., Turck, C. W., Myles, D. G., Primakoff, P., White, J. M., 1992, A potential fusion peptide and an integrin ligand domain in a protein active in sperm-egg fusion. *Nature.* **356**: 248-252.

Borneman, A., Kuschel, R., Fujisawa-Sehara, A., 2000, Analysis for transcript expression of meltrin alpha in normal, regenerating, and denervated rat muscle. *J Muscle Res Cell Motil.* **21**: 475-480.

Borrell-Pages, M., Rojo, F., Albanell, J., Baselga, J., Arribas, J., 2003, TACE is required for the activation of the EGFR by TGF-alpha in tumors. *Embo J.* **22**: 1114-1124.

Bosse, F., Petzold, G., Greiner-Petter, R., Pippirs, U., Gillen, C., Muller, H. W., 2000, Cellular localization of the disintegrin CRII-7/rMDC15 mRNA in rat PNS and CNS and regulated expression in postnatal development and after nerve injury. *Glia.* **32**: 313-327.

Brou, C., Logeat, F., Gupta, N., Bessia, C., LeBail, O., Doedens, J. R., Cumano, A., Roux, P., Black, R. A., Israel, A., 2000, A novel proteolytic cleavage involved in Notch signaling: the role of the disintegrin-metalloprotease TACE. *Mol Cell.* **5**: 207-216.

Buxbaum, J. D., Liu, K. N., Luo, Y., Slack, J. L., Stocking, K. L., Peschon, J. J., Johnson, R. S., Castner, B. J., Cerretti, D. P., Black, R. A., 1998, Evidence that tumor necrosis factor alpha converting enzyme is involved in regulated alpha-secretase cleavage of the Alzheimer amyloid protein precursor. *J Biol Chem.* **273**: 27765-27767.

Cai, H., Kratzschmar, J., Alfandari, D., Hunnicutt, G., Blobel, C. P., 1998, Neural crest-specific and general expression of distinct metalloprotease-disintegrins in early Xenopus laevis development. *Dev Biol.* **204**: 508-524.

Cao, Y., Kang, Q., Zhao, Z., Zolkiewska, A., 2002, Intracellular processing of metalloprotease disintegrin ADAM12. *J Biol Chem.* **277**: 26403-26411.

Cao, Y., Kang, Q., Zolkiewska, A., 2001, Metalloprotease-disintegrin ADAM 12 interacts with alpha-actinin-1. *Biochem J.* **357**: 353-361.

Cao, Y., Zhao, Z., Gruszczynska-Biegala, J., Zolkiewska, A., 2003, Role of metalloprotease disintegrin ADAM12 in determination of quiescent reserve cells during myogenic differentiation in vitro. *Mol Cell Biol.* **23**: 6725-6738.

Chesneau, V., Becherer, J. D., Zheng, Y., Erdjument-Bromage, H., Tempst, P., Blobel, C. P., 2003, Catalytic properties of ADAM19. *J Biol Chem.* **278**: 22331-22340.

Cho, C., Bunch, D. O., Faure, J. E., Goulding, E. H., Eddy, E. M., Primakoff, P., Myles, D. G., 1998, Fertilization defects in sperm from mice lacking fertilin beta. *Science.* **281**: 1857-1859.

Choi, S. J., Han, J. H., Roodman, G. D., 2001, ADAM8: a novel osteoclast stimulating factor. *J Bone Miner Res.* **16**: 814-822.

Condon, T. P., Flournoy, S., Sawyer, G. J., Baker, B. F., Kishimoto, T. K., Bennett, C. F., 2001, ADAM17 but not ADAM10 mediates tumor necrosis factor-alpha and L-selectin shedding from leukocyte membranes. *Antisense Nucleic Acid Drug Dev.* **11**: 107-116.

Contin, C., Pitard, V., Itai, T., Nagata, S., Moreau, J. F., Dechanet-Merville, J., 2003, Membrane-anchored CD40 is processed by the tumor necrosis factor-alpha-converting enzyme. Implications for CD40 signaling. *J Biol Chem.* **278**: 32801-32809.

Cruz, A. C., Frank, B. T., Edwards, S. T., Dazin, P. F., Peschon, J. J., Fang, K. C., 2004, Tumor necrosis factor-alpha-converting enzyme controls surface expression of c-Kit and survival of embryonic stem cell-derived mast cells. *J Biol Chem.* **279**: 5612-5620.

Eliceiri, B. P. and Cheresh, D. A., 2000, Role of alpha v integrins during angiogenesis. *Cancer J.* **6**: S245-249.

Falls, D. L., 2003, Neuregulins: functions, forms, and signaling strategies. *Exp Cell Res.* **284**: 14-30.

Fambrough, D., Pan, D., Rubin, G. M., Goodman, C. S., 1996, The cell surface metalloprotease/disintegrin Kuzbanian is required for axonal extension in Drosophila. *Proc Natl Acad Sci U S A.* **93**: 13233-13238.

Fata, J. E., Kong, Y. Y., Li, J., Sasaki, T., Irie-Sasaki, J., Moorehead, R. A., Elliott, R., Scully, S., Voura, E. B., Lacey, D. L., Boyle, W. J., Khokha, R., Penninger, J. M., 2000, The osteoclast differentiation factor osteoprotegerin-ligand is essential for mammary gland development. *Cell.* **103**: 41-50.

Fourie, A. M., Coles, F., Moreno, V., Karlsson, L., 2003, Catalytic activity of ADAM8, ADAM15, and MDC-L (ADAM28) on synthetic peptide substrates and in ectodomain cleavage of CD23. *J Biol Chem.* **278**: 30469-30477.

Franzke, C. W., Tasanen, K., Borradori, L., Huotari, V., Bruckner-Tuderman, L., 2004, Shedding of Collagen XVII/BP180: STRUCTURAL MOTIFS INFLUENCE CLEAVAGE FROM CELL SURFACE. *J Biol Chem.* **279**: 24521-24529.

Fritsche, J., Moser, M., Faust, S., Peuker, A., Buttner, R., Andreesen, R., Kreutz, M., 2000, Molecular cloning and characterization of a human metalloprotease disintegrin--a novel marker for dendritic cell differentiation. *Blood.* **96**: 732-739.

Fritsche, J., Muller, A., Hausmann, M., Rogler, G., Andreesen, R., Kreutz, M., 2003, Inverse regulation of the ADAM-family members, decysin and MADDAM/ADAM19 during monocyte differentiation. *Immunology.* **110**: 450-457.

Galliano, M. F., Huet, C., Frygelius, J., Polgren, A., Wewer, U. M., Engvall, E., 2000, Binding of ADAM12, a marker of skeletal muscle regeneration, to the muscle-specific actin-binding protein, alpha -actinin-2, is required for myoblast fusion. *J Biol Chem.* **275**: 13933-13939.

Garton, K. J., Gough, P. J., Blobel, C. P., Murphy, G., Greaves, D. R., Dempsey, P. J., Raines, E. W., 2001, Tumor necrosis factor-alpha-converting enzyme (ADAM17) mediates the cleavage and shedding of fractalkine (CX3CL1). *J Biol Chem.* **276**: 37993-38001.

Gilpin, B. J., Loechel, F., Mattei, M. G., Engvall, E., Albrechtsen, R., Wewer, U. M., 1998, A novel, secreted form of human ADAM 12 (meltrin alpha) provokes myogenesis in vivo. *J Biol Chem.* **273**: 157-166.

Goddard, D. R., Bunning, R. A., Woodroofe, M. N., 2001, Astrocyte and endothelial cell expression of ADAM 17 (TACE) in adult human CNS. *Glia.* **34**: 267-271.

Gough, P. J., Garton, K. J., Wille, P. T., Rychlewski, M., Dempsey, P. J., Raines, E. W., 2004, A disintegrin and metalloproteinase 10-mediated cleavage and shedding regulates the cell surface expression of CXC chemokine ligand 16. *J Immunol.* **172**: 3678-3685.

Greenwald, I., 1998, LIN-12/Notch signaling: lessons from worms and flies. *Genes Dev.* **12**: 1751-1762.

Grutzmann, R., Foerder, M., Alldinger, I., Staub, E., Brummendorf, T., Ropcke, S., Li, X., Kristiansen, G., Jesnowski, R., Sipos, B., Lohr, M., Luttges, J., Ockert, D., Kloppel, G., Saeger, H. D., Pilarsky, C., 2003, Gene expression profiles of microdissected pancreatic ductal adenocarcinoma. *Virchows Arch.* **443**: 508-517.

Grutzmann, R., Luttges, J., Sipos, B., Ammerpohl, O., Dobrowolski, F., Alldinger, I., Kersting, S., Ockert, D., Koch, R., Kalthoff, H., Schackert, H. K., Saeger, H. D., Kloppel, G., Pilarsky, C., 2004, ADAM9 expression in pancreatic cancer is associated with tumour type and is a prognostic factor in ductal adenocarcinoma. *Br J Cancer.* **90**: 1053-1058.

Gschwind, A., Hart, S., Fischer, O. M., Ullrich, A., 2003, TACE cleavage of proamphiregulin regulates GPCR-induced proliferation and motility of cancer cells. *Embo J.* **22**: 2411-2421.

Hall, R. J. and Erickson, C. A., 2003, ADAM 10: an active metalloprotease expressed during avian epithelial morphogenesis. *Dev Biol.* **256**: 146-159.

Ham, C., Levkau, B., Raines, E. W., Herren, B., 2002, ADAM15 is an adherens junction molecule whose surface expression can be driven by VE-cadherin. *Exp Cell Res.* **279**: 239-247.

Harris, H. A., Murrills, R. J., Komm, B. S., 1997, Expression of meltrin-alpha mRNA is not restricted to fusagenic cells. *J Cell Biochem.* **67**: 136-142.

Harris, R. C., Chung, E., Coffey, R. J., 2003, EGF receptor ligands. *Exp Cell Res.* **284**: 2-13.

Hart, S., Fischer, O. M., Ullrich, A., 2004, Cannabinoids induce cancer cell proliferation via tumor necrosis factor alpha-converting enzyme (TACE/ADAM17)-mediated transactivation of the epidermal growth factor receptor. *Cancer Res.* **64**: 1943-1950.

Hartmann, D., de Strooper, B., Serneels, L., Craessaerts, K., Herreman, A., Annaert, W., Umans, L., Lubke, T., Lena Illert, A., von Figura, K., Saftig, P., 2002, The disintegrin/metalloprotease ADAM 10 is essential for Notch signalling but not for alpha-secretase activity in fibroblasts. *Hum Mol Genet.* **11**: 2615-2624.

Hattori, M., Osterfield, M., Flanagan, J. G., 2000, Regulated cleavage of a contact-mediated axon repellent. *Science.* **289**: 1360-1365.

Herreman, A., Hartmann, D., Annaert, W., Saftig, P., Craessaerts, K., Serneels, L., Umans, L., Schrijvers, V., Checler, F., Vanderstichele, H., Baekelandt, V., Dressel, R., Cupers, P., Huylebroeck, D., Zwijsen, A., Van Leuven, F., De Strooper, B., 1999, Presenilin 2 deficiency causes a mild pulmonary phenotype and no changes in amyloid precursor protein processing but enhances the embryonic lethal phenotype of presenilin 1 deficiency. *Proc Natl Acad Sci U S A.* **96**: 11872-11877.

Herreman, A., Serneels, L., Annaert, W., Collen, D., Schoonjans, L., De Strooper, B., 2000, Total inactivation of gamma-secretase activity in presenilin-deficient embryonic stem cells. *Nat Cell Biol.* **2**: 461-462.

Herren, B., Raines, E. W., Ross, R., 1997, Expression of a disintegrin-like protein in cultured human vascular cells and in vivo. *Faseb J.* **11**: 173-180.

Higuchi, Y., Yasui, A., Matsuura, K., Yamamoto, S., 2002, CD156 transgenic mice. Different responses between inflammatory types. *Pathobiology.* **70**: 47-54.

Hinkle, C. L., Sunnarborg, S. W., Loiselle, D., Parker, C. E., Stevenson, M., Russell, W. E., Lee, D. C., 2004, Selective roles for tumor necrosis factor alpha-converting enzyme/ADAM17 in the shedding of the epidermal growth factor receptor ligand family: the juxtamembrane stalk determines cleavage efficiency. *J Biol Chem.* **279**: 24179-24188.

Horiuchi, K., Weskamp, G., Lum, L., Hammes, H. P., Cai, H., Brodie, T. A., Ludwig, T., Chiusaroli, R., Baron, R., Preissner, K. T., Manova, K., Blobel, C. P., 2003, Potential role for ADAM15 in pathological neovascularization in mice. *Mol Cell Biol.* **23**: 5614-5624.

Howard, L., Lu, X., Mitchell, S., Griffiths, S., Glynn, P., 1996, Molecular cloning of MADM: a catalytically active mammalian disintegrin-metalloprotease expressed in various cell types. *Biochem J.* **317**: 45-50.

Howard, L., Nelson, K. K., Maciewicz, R. A., Blobel, C. P., 1999, Interaction of the metalloprotease disintegrins MDC9 and MDC15 with two SH3 domain-containing proteins, endophilin I and SH3PX1. *J Biol Chem.* **274**: 31693-31699.

Howard, L., Zheng, Y., Horrocks, M., Maciewicz, R. A., Blobel, C., 2001, Catalytic activity of ADAM28. *FEBS Lett.* **498**: 82-86.

Huang, T. F., Holt, J. C., Lukasiewicz, H., Niewiarowski, S., 1987, Trigramin. A low molecular weight peptide inhibiting fibrinogen interaction with platelet receptors expressed on glycoprotein IIb-IIIa complex. *J Biol Chem.* **262**: 16157-16163.

Huang, X. Z., Wu, J. F., Ferrando, R., Lee, J. H., Wang, Y. L., Farese, R. V., Jr., Sheppard, D., 2000, Fatal bilateral chylothorax in mice lacking the integrin alpha9beta1. *Mol Cell Biol.* **20**: 5208-5215.

Hundhausen, C., Misztela, D., Berkhout, T. A., Broadway, N., Saftig, P., Reiss, K., Hartmann, D., Fahrenholz, F., Postina, R., Matthews, V., Kallen, K. J., Rose-John, S., Ludwig, A., 2003, The disintegrin-like metalloproteinase ADAM10 is involved in constitutive cleavage of CX3CL1 (fractalkine) and regulates CX3CL1-mediated cell-cell adhesion. *Blood.* **102**: 1186-1195.

Hussain, I., Hawkins, J., Shikotra, A., Riddell, D. R., Faller, A., Dingwall, C., 2003, Characterization of the ectodomain shedding of the beta-site amyloid precursor protein-cleaving enzyme 1 (BACE1). *J Biol Chem.* **278**: 36264-36268.

Hynes, R. O., 2002, A reevaluation of integrins as regulators of angiogenesis. *Nat Med.* **8**: 918-921.

Iba, K., Albrechtsen, R., Gilpin, B., Frohlich, C., Loechel, F., Zolkiewska, A., Ishiguro, K., Kojima, T., Liu, W., Langford, J. K., Sanderson, R. D., Brakebusch, C., Fassler, R., Wewer, U. M., 2000, The cysteine-rich domain of human ADAM 12 supports cell adhesion through syndecans and triggers signaling events that lead to beta1 integrin-dependent cell spreading. *J Cell Biol.* **149**: 1143-1156.

Iba, K., Albrechtsen, R., Gilpin, B. J., Loechel, F., Wewer, U. M., 1999, Cysteine-rich domain of human ADAM 12 (meltrin alpha) supports tumor cell adhesion. *Am J Pathol.* **154**: 1489-1501.

Inoue, D., Reid, M., Lum, L., Kratzschmar, J., Weskamp, G., Myung, Y. M., Baron, R., Blobel, C. P., 1998, Cloning and initial characterization of mouse meltrin beta and analysis of the expression of four metalloprotease-disintegrins in bone cells. *J Biol Chem.* **273**: 4180-4187.

Ito, N., Nomura, S., Iwase, A., Ito, T., Kikkawa, F., Tsujimoto, M., Ishiura, S., Mizutani, S., 2004, ADAMs, a disintegrin and metalloproteinases, mediate shedding of oxytocinase. *Biochem Biophys Res Commun.* **314**: 1008-1013.

Iwamoto, R., Yamazaki, S., Asakura, M., Takashima, S., Hasuwa, H., Miyado, K., Adachi, S., Kitakaze, M., Hashimoto, K., Raab, G., Nanba, D., Higashiyama, S., Hori, M., Klagsbrun, M., Mekada, E., 2003, Heparin-binding EGF-like growth factor and ErbB signaling is essential for heart function. *Proc Natl Acad Sci U S A.* **100**: 3221-3226.

Izumi, Y., Hirata, M., Hasuwa, H., Iwamoto, R., Umata, T., Miyado, K., Tamai, Y., Kurisaki, T., Sehara-Fujisawa, A., Ohno, S., Mekada, E., 1998, A metalloprotease-disintegrin, MDC9/meltrin-gamma/ADAM9 and PKCdelta are involved in TPA-induced ectodomain shedding of membrane-anchored heparin-binding EGF-like growth factor. *Embo J.* **17**: 7260-7272.

Jackson, L. F., Qiu, T. H., Sunnarborg, S. W., Chang, A., Zhang, C., Patterson, C., Lee, D. C., 2003, Defective valvulogenesis in HB-EGF and TACE-null mice is associated with aberrant BMP signaling. *Embo J.* **22**: 2704-2716.

Jongeneel, C. V., Bouvier, J., Bairoch, A., 1989, A unique signature identifies a family of zinc-dependent metallopeptidases. *FEBS Lett.* **242**: 211-214.

Kang, Q., Cao, Y., Zolkiewska, A., 2000, Metalloprotease-disintegrin ADAM 12 binds to the SH3 domain of Src and activates Src tyrosine kinase in C2C12 cells. *Biochem J.* **352**: 883-892.

Kang, Q., Cao, Y., Zolkiewska, A., 2001, Direct interaction between the cytoplasmic tail of ADAM 12 and the Src homology 3 domain of p85alpha activates phosphatidylinositol 3-kinase in C2C12 cells. *J Biol Chem.* **276**: 24466-24472.

Karkkainen, I., Rybnikova, E., Pelto-Huikko, M., Huovila, A. P., 2000, Metalloprotease-disintegrin (ADAM) genes are widely and differentially expressed in the adult CNS. *Mol Cell Neurosci.* **15**: 547-560.

Kataoka, M., Yoshiyama, K., Matsuura, K., Hijiya, N., Higuchi, Y., Yamamoto, S., 1997, Structure of the murine CD156 gene, characterization of its promoter, and chromosomal location. *J Biol Chem.* **272**: 18209-18215.

Kawaguchi, N., Xu, X., Tajima, R., Kronqvist, P., Sundberg, C., Loechel, F., Albrechtsen, R., Wewer, U. M., 2002, ADAM 12 protease induces adipogenesis in transgenic mice. *Am J Pathol.* **160**: 1895-1903.

Kelly, K., Hutchinson, G., Klewe-Nebenius, D., Smith, A., Bartsch, J. W., Horiuchi, K., Manova, K., Docherty, A. J., Blobel, C. P., 2004, Metalloprotease-disintegrin ADAM8: expression pattern analysis and targeted deletion in mice. *Dev Dyn.* (in press).

Kheradmand, F. and Werb, Z., 2002, Shedding light on sheddases: role in growth and development. *Bioessays.* **24**: 8-12.

King, N. E., Zimmermann, N., Pope, S. M., Fulkerson, P. C., Nikolaidis, N. M., Mishra, A., Witte, D. P., Rothenberg, M. E., 2004, Expression and regulation of a disintegrin and metalloproteinase (ADAM)8 in experimental asthma. *Am J Respir Cell Mol Biol.* **31**: 257-265.

Koike, H., Tomioka, S., Sorimachi, H., Saido, T. C., Maruyama, K., Okuyama, A., Fujisawa-Sehara, A., Ohno, S., Suzuki, K., Ishiura, S., 1999, Membrane-anchored metalloprotease MDC9 has an alpha-secretase activity responsible for processing the amyloid precursor protein. *Biochem J.* **343**: 371-375.

Kojro, E., Gimpl, G., Lammich, S., Marz, W., Fahrenholz, F., 2001, Low cholesterol stimulates the nonamyloidogenic pathway by its effect on the alpha -secretase ADAM 10. *Proc Natl Acad Sci U S A.* **98**: 5815-5820.

Kong, Y. Y., Boyle, W. J., Penninger, J. M., 2000, Osteoprotegerin ligand: a regulator of immune responses and bone physiology. *Immunol Today.* **21**: 495-502.

Kratzschmar, J., Lum, L., Blobel, C. P., 1996, Metargidin, a membrane-anchored metalloprotease-disintegrin protein with an RGD integrin binding sequence. *J Biol Chem.* **271**: 4593-4596.

Krebs, L. T., Xue, Y., Norton, C. R., Shutter, J. R., Maguire, M., Sundberg, J. P., Gallahan, D., Closson, V., Kitajewski, J., Callahan, R., Smith, G. H., Stark, K. L., Gridley, T., 2000, Notch signaling is essential for vascular morphogenesis in mice. *Genes Dev.* **14**: 1343-1352.

Kronqvist, P., Kawaguchi, N., Albrechtsen, R., Xu, X., Schroder, H. D., Moghadaszadeh, B., Nielsen, F. C., Frohlich, C., Engvall, E., Wewer, U. M., 2002, ADAM12 alleviates the skeletal muscle pathology in mdx dystrophic mice. *Am J Pathol.* **161**: 1535-1540.

Kurisaki, T., Masuda, A., Osumi, N., Nabeshima, Y., Fujisawa-Sehara, A., 1998, Spatially- and temporally-restricted expression of meltrin alpha (ADAM12) and beta (ADAM19) in mouse embryo. *Mech Dev.* **73**: 211-215.

Kurisaki, T., Masuda, A., Sudo, K., Sakagami, J., Higashiyama, S., Matsuda, Y., Nagabukuro, A., Tsuji, A., Nabeshima, Y., Asano, M., Iwakura, Y., Sehara-Fujisawa, A., 2003, Phenotypic analysis of Meltrin alpha (ADAM12)-deficient mice: involvement of Meltrin alpha in adipogenesis and myogenesis. *Mol Cell Biol.* **23**: 55-61.

Kurohara, K., Komatsu, K., Kurisaki, T., Masuda, A., Irie, N., Asano, M., Sudo, K., Nabeshima, Y., Iwakura, Y., Sehara-Fujisawa, A., 2004, Essential roles of Meltrin beta (ADAM19) in heart development. *Dev Biol.* **267**: 14-28.

Lammich, S., Kojro, E., Postina, R., Gilbert, S., Pfeiffer, R., Jasionowski, M., Haass, C., Fahrenholz, F., 1999, Constitutive and regulated alpha-secretase cleavage of Alzheimer's amyloid precursor protein by a disintegrin metalloprotease. *Proc Natl Acad Sci U S A.* **96**: 3922-3927.

Le Pabic, H., Bonnier, D., Wewer, U. M., Coutand, A., Musso, O., Baffet, G., Clement, B., Theret, N., 2003, ADAM12 in human liver cancers: TGF-beta-regulated expression in stellate cells is associated with matrix remodeling. *Hepatology.* **37**: 1056-1066.

Lemjabbar, H. and Basbaum, C., 2002, Platelet-activating factor receptor and ADAM10 mediate responses to Staphylococcus aureus in epithelial cells. *Nat Med.* **8**: 41-46.

Lieber, T., Kidd, S., Young, M. W., 2002, kuzbanian-mediated cleavage of Drosophila Notch. *Genes Dev.* **16**: 209-221.

Loechel, F., Fox, J. W., Murphy, G., Albrechtsen, R., Wewer, U. M., 2000, ADAM 12-S cleaves IGFBP-3 and IGFBP-5 and is inhibited by TIMP-3. *Biochem Biophys Res Commun.* **278**: 511-515.

Loechel, F., Gilpin, B. J., Engvall, E., Albrechtsen, R., Wewer, U. M., 1998, Human ADAM 12 (meltrin alpha) is an active metalloprotease. *J Biol Chem.* **273**: 16993-16997.

Lopez-Perez, E., Zhang, Y., Frank, S. J., Creemers, J., Seidah, N., Checler, F., 2001, Constitutive alpha-secretase cleavage of the beta-amyloid precursor protein in the furin-deficient LoVo cell line: involvement of the pro-hormone convertase 7 and the disintegrin metalloprotease ADAM10. *J Neurochem.* **76**: 1532-1539.

Lum, L., Reid, M. S., Blobel, C. P., 1998, Intracellular maturation of the mouse metalloprotease disintegrin MDC15. *J Biol Chem.* **273**: 26236-26247.

Mandelin, J., Li, T. F., Hukkanen, M. V., Liljestrom, M., Chen, Z. K., Santavirta, S., Kitti, U., Konttinen, Y. T., 2003, Increased expression of a novel osteoclast-stimulating factor, ADAM8, in interface tissue around loosened hip prostheses. *J Rheumatol.* **30**: 2033-2038.

Marino, M. W., Dunn, A., Grail, D., Inglese, M., Noguchi, Y., Richards, E., Jungbluth, A., Wada, H., Moore, M., Williamson, B., Basu, S., Old, L. J., 1997, Characterization of tumor necrosis factor-deficient mice. *Proc Natl Acad Sci U S A.* **94**: 8093-8098.

Martin, J., Eynstone, L. V., Davies, M., Williams, J. D., Steadman, R., 2002, The role of ADAM 15 in glomerular mesangial cell migration. *J Biol Chem.* **277**: 33683-33689.

Matthews, V., Schuster, B., Schutze, S., Bussmeyer, I., Ludwig, A., Hundhausen, C., Sadowski, T., Saftig, P., Hartmann, D., Kallen, K. J., Rose-John, S., 2003, Cellular cholesterol depletion triggers shedding of the human interleukin-6 receptor by ADAM10 and ADAM17 (TACE). *J Biol Chem.* **278**: 38829-38839.

Mechtersheimer, S., Gutwein, P., Agmon-Levin, N., Stoeck, A., Oleszewski, M., Riedle, S., Postina, R., Fahrenholz, F., Fogel, M., Lemmon, V., Altevogt, P., 2001, Ectodomain shedding of L1 adhesion molecule promotes cell migration by autocrine binding to integrins. *J Cell Biol.* **155**: 661-673.

Millichip, M. I., Dallas, D. J., Wu, E., Dale, S., McKie, N., 1998, The metallo-disintegrin ADAM10 (MADM) from bovine kidney has type IV collagenase activity in vitro. *Biochem Biophys Res Commun.* **245**: 594-598.

Mitchell, K. J., Pinson, K. I., Kelly, O. G., Brennan, J., Zupicich, J., Scherz, P., Leighton, P. A., Goodrich, L. V., Lu, X., Avery, B. J., Tate, P., Dill, K., Pangilinan, E., Wakenight, P., Tessier-Lavigne, M., Skarnes, W. C., 2001, Functional analysis of secreted and transmembrane proteins critical to mouse development. *Nat Genet.* **28**: 241-249.

Mori, S., Tanaka, M., Nanba, D., Nishiwaki, E., Ishiguro, H., Higashiyama, S., Matsuura, N., 2003, PACSIN3 binds ADAM12/meltrin alpha and up-regulates ectodomain shedding of heparin-binding epidermal growth factor-like growth factor. *J Biol Chem.* **278**: 46029-46034.

Moss, M. L., Jin, S. L., Milla, M. E., Bickett, D. M., Burkhart, W., Carter, H. L., Chen, W. J., Clay, W. C., Didsbury, J. R., Hassler, D., Hoffman, C. R., Kost, T. A., Lambert, M. H., Leesnitzer, M. A., McCauley, P., McGeehan, G., Mitchell, J., Moyer, M., Pahel, G., Rocque, W., Overton, L. K., Schoenen, F., Seaton, T., Su, J. L., Becherer, J. D., *et al.*., 1997, Cloning of a disintegrin metalloproteinase that processes precursor tumour-necrosis factor-alpha. *Nature.* **385**: 733-736.

Mumm, J. S., Schroeter, E. H., Saxena, M. T., Griesemer, A., Tian, X., Pan, D. J., Ray, W. J., Kopan, R., 2000, A ligand-induced extracellular cleavage regulates gamma-secretase-like proteolytic activation of Notch1. *Mol Cell.* **5**: 197-206.

Murai, T., Miyazaki, Y., Nishinakamura, H., Sugahara, K. N., Miyauchi, T., Sako, Y., Yanagida, T., Miyasaka, M., 2004, Engagement of CD44 promotes Rac activation and CD44 cleavage during tumor cell migration. *J Biol Chem.* **279**: 4541-4550.

Nath, D., Slocombe, P. M., Stephens, P. E., Warn, A., Hutchinson, G. R., Yamada, K. M., Docherty, A. J., Murphy, G., 1999, Interaction of metargidin (ADAM-15) with alphavbeta3 and alpha5beta1 integrins on different haemopoietic cells. *J Cell Sci.* **112**: 579-587.

Nath, D., Slocombe, P. M., Webster, A., Stephens, P. E., Docherty, A. J., Murphy, G., 2000, Meltrin gamma(ADAM-9) mediates cellular adhesion through alpha(6)beta(1 )integrin, leading to a marked induction of fibroblast cell motility. *J Cell Sci.* **113**: 2319-2328.

Naus, S., Richter, M., Wildeboer, D., Schachner, M., Moss, M., Bartsch, J. W., 2004, Ectodomain shedding of the neural recognition molecule CHL1 by the metalloprotease-disintegrin ADAM8 promotes neurite outgrowth and suppresses neuronal cell death. *J Biol Chem.*

Navarro, V., Vincent, J. P., Mazella, J., 2002, Shedding of the luminal domain of the neurotensin receptor-3/sortilin in the HT29 cell line. *Biochem Biophys Res Commun.* **298**: 760-764.

Nelson, K. K., Schlondorff, J., Blobel, C. P., 1999, Evidence for an interaction of the metalloprotease-disintegrin tumour necrosis factor alpha convertase (TACE) with mitotic arrest deficient 2 (MAD2), and of the metalloprotease-disintegrin MDC9 with a novel MAD2-related protein, MAD2beta. *Biochem J.* **343**: 673-680.

Newton, R. C., Solomon, K. A., Covington, M. B., Decicco, C. P., Haley, P. J., Friedman, S. M., Vaddi, K., 2001, Biology of TACE inhibition. *Ann Rheum Dis.* **60**: iii25-32.

Niewiarowski, S., McLane, M. A., Kloczewiak, M., Stewart, G. J., 1994, Disintegrins and other naturally occurring antagonists of platelet fibrinogen receptors. *Semin Hematol.* **31**: 289-300.

Ohta, S., Harigai, M., Tanaka, M., Kawaguchi, Y., Sugiura, T., Takagi, K., Fukasawa, C., Hara, M., Kamatani, N., 2001, Tumor necrosis factor-alpha (TNF-alpha) converting enzyme contributes to production of TNF-alpha in synovial tissues from patients with rheumatoid arthritis. *J Rheumatol.* **28**: 1756-1763.

O'Shea, C., McKie, N., Buggy, Y., Duggan, C., Hill, A. D., McDermott, E., O'Higgins, N., Duffy, M. J., 2003, Expression of ADAM-9 mRNA and protein in human breast cancer. *Int J Cancer.* **105**: 754-761.

Pan, D. and Rubin, G. M., 1997, Kuzbanian controls proteolytic processing of Notch and mediates lateral inhibition during Drosophila and vertebrate neurogenesis. *Cell.* **90**: 271-280.

Pasparakis, M., Alexopoulou, L., Episkopou, V., Kollias, G., 1996, Immune and inflammatory responses in TNF alpha-deficient mice: a critical requirement for TNF alpha in the formation of primary B cell follicles, follicular dendritic cell networks and germinal centers, and in the maturation of the humoral immune response. *J Exp Med.* **184**: 1397-1411.

Patel, I. R., Attur, M. G., Patel, R. N., Stuchin, S. A., Abagyan, R. A., Abramson, S. B., Amin, A. R., 1998, TNF-alpha convertase enzyme from human arthritis-affected cartilage: isolation of cDNA by differential display, expression of the active enzyme, and regulation of TNF-alpha. *J Immunol.* **160**: 4570-4579.

Peiretti, F., Deprez-Beauclair, P., Bonardo, B., Aubert, H., Juhan-Vague, I., Nalbone, G., 2003, Identification of SAP97 as an intracellular binding partner of TACE. *J Cell Sci.* **116**: 1949-1957.

Peschon, J. J., Slack, J. L., Reddy, P., Stocking, K. L., Sunnarborg, S. W., Lee, D. C., Russell, W. E., Castner, B. J., Johnson, R. S., Fitzner, J. N., Boyce, R. W., Nelson, N., Kozlosky, C. J., Wolfson, M. F., Rauch, C. T., Cerretti, D. P., Paxton, R. J., March, C. J., Black, R. A., 1998a, An essential role for ectodomain shedding in mammalian development. *Science.* **282**: 1281-1284.

Peschon, J. J., Torrance, D. S., Stocking, K. L., Glaccum, M. B., Otten, C., Willis, C. R., Charrier, K., Morrissey, P. J., Ware, C. B., Mohler, K. M., 1998b, TNF receptor-deficient mice reveal divergent roles for p55 and p75 in several models of inflammation. *J Immunol.* **160**: 943-952.

Poghosyan, Z., Robbins, S. M., Houslay, M. D., Webster, A., Murphy, G., Edwards, D. R., 2002, Phosphorylation-dependent interactions between ADAM15 cytoplasmic domain and Src family protein-tyrosine kinases. *J Biol Chem.* **277**: 4999-5007.

Postina, R., Schroeder, A., Dewachter, I., Bohl, J., Schmitt, U., Kojro, E., Prinzen, C., Endres, K., Hiemke, C., Blessing, M., Flamez, P., Dequenne, A., Godaux, E., Van Leuven, F., Fahrenholz, F., 2004, A disintegrin-metalloproteinase prevents amyloid plaque formation and hippocampal defects in an Alzheimer disease mouse model. *J Clin Invest.* **113**: 1456-1464.

Prenzel, N., Zwick, E., Daub, H., Leserer, M., Abraham, R., Wallasch, C., Ullrich, A., 1999, EGF receptor transactivation by G-protein-coupled receptors requires metalloproteinase cleavage of proHB-EGF. *Nature.* **402**: 884-888.

Primakoff, P., Hyatt, H., Tredick-Kline, J., 1987, Identification and purification of a sperm surface protein with a potential role in sperm-egg membrane fusion. *J Cell Biol.* **104**: 141-149.

Primakoff, P. and Myles, D. G., 2000, The ADAM gene family: surface proteins with adhesion and protease activity. *Trends Genet.* **16**: 83-87.

Primakoff, P. and Myles, D. G., 2002, Penetration, adhesion, and fusion in mammalian sperm-egg interaction. *Science.* **296**: 2183-2185.

Rio, C., Buxbaum, J. D., Peschon, J. J., Corfas, G., 2000, Tumor necrosis factor-alpha-converting enzyme is required for cleavage of erbB4/HER4. *J Biol Chem.* **275**: 10379-10387.

Roghani, M., Becherer, J. D., Moss, M. L., Atherton, R. E., Erdjument-Bromage, H., Arribas, J., Blackburn, R. K., Weskamp, G., Tempst, P., Blobel, C. P., 1999, Metalloprotease-disintegrin MDC9: intracellular maturation and catalytic activity. *J Biol Chem.* **274**: 3531-3540.

Rooke, J., Pan, D., Xu, T., Rubin, G. M., 1996, KUZ, a conserved metalloprotease-disintegrin protein with two roles in Drosophila neurogenesis. *Science.* **273**: 1227-1231.

Sahin, U., Weskamp, G., Kelly, K., Zhou, H. M., Higashiyama, S., Peschon, J., Hartmann, D., Saftig, P., Blobel, C. P., 2004, Distinct roles for ADAM10 and ADAM17 in ectodomain shedding of six EGFR ligands. *J Cell Biol.* **164**: 769-779.

Schafer, B., Gschwind, A., Ullrich, A., 2004, Multiple G-protein-coupled receptor signals converge on the epidermal growth factor receptor to promote migration and invasion. *Oncogene.* **23**: 991-999.

Schlomann, U., Rathke-Hartlieb, S., Yamamoto, S., Jockusch, H., Bartsch, J. W., 2000, Tumor necrosis factor alpha induces a metalloprotease-disintegrin, ADAM8 (CD 156): implications for neuron-glia interactions during neurodegeneration. *J Neurosci.* **20**: 7964-7971.

Schlomann, U., Wildeboer, D., Webster, A., Antropova, O., Zeuschner, D., Knight, C. G., Docherty, A. J., Lambert, M., Skelton, L., Jockusch, H., Bartsch, J. W., 2002, The metalloprotease disintegrin ADAM8. Processing by autocatalysis is required for proteolytic activity and cell adhesion. *J Biol Chem.* **277**: 48210-48219.

Schlondorff, J. and Blobel, C. P., 1999, Metalloprotease-disintegrins: modular proteins capable of promoting cell-cell interactions and triggering signals by protein-ectodomain shedding. *J Cell Sci.* **112**: 3603-3617.

Seals, D. F. and Courtneidge, S. A., 2003, The ADAMs family of metalloproteases: multidomain proteins with multiple functions. *Genes Dev.* **17**: 7-30.

Selkoe, D. J. and Schenk, D., 2003, Alzheimer's disease: molecular understanding predicts amyloid-based therapeutics. *Annu Rev Pharmacol Toxicol.* **43**: 545-584.

Setoguchi, M., Nasu, N., Yoshida, S., Higuchi, Y., Akizuki, S., Yamamoto, S., 1989, Mouse and human CD14 (myeloid cell-specific leucine-rich glycoprotein) primary structure deduced from cDNA clones. *Biochim Biophys Acta.* **1008**: 213-222.

Shi, W., Chen, H., Sun, J., Buckley, S., Zhao, J., Anderson, K. D., Williams, R. G., Warburton, D., 2003, TACE is required for fetal murine cardiac development and modeling. *Dev Biol.* **261**: 371-380.

Shi, Z., Xu, W., Loechel, F., Wewer, U. M., Murphy, L. J., 2000, ADAM 12, a disintegrin metalloprotease, interacts with insulin-like growth factor-binding protein-3. *J Biol Chem.* **275**: 18574-18580.

Shintani, Y., Higashiyama, S., Ohta, M., Hirabayashi, H., Yamamoto, S., Yoshimasu, T., Matsuda, H., Matsuura, N., 2004, Overexpression of ADAM9 in Non-Small Cell Lung Cancer Correlates with Brain Metastasis. *Cancer Res.* **64**: 4190-4196.

Shirakabe, K., Wakatsuki, S., Kurisaki, T., Fujisawa-Sehara, A., 2001, Roles of Meltrin beta /ADAM19 in the processing of neuregulin. *J Biol Chem.* **276**: 9352-9358.

Slack, B. E., Ma, L. K., Seah, C. C., 2001, Constitutive shedding of the amyloid precursor protein ectodomain is up-regulated by tumour necrosis factor-alpha converting enzyme. *Biochem J.* **357**: 787-794.

Stocker, W., Grams, F., Baumann, U., Reinemer, P., Gomis-Ruth, F. X., McKay, D. B., Bode, W., 1995, The metzincins--topological and sequential relations between the astacins, adamalysins, serralysins, and matrixins (collagenases) define a superfamily of zinc-peptidases. *Protein Sci.* **4**: 823-840.

Suda, T., Takahashi, N., Udagawa, N., Jimi, E., Gillespie, M. T., Martin, T. J., 1999, Modulation of osteoclast differentiation and function by the new members of the tumor necrosis factor receptor and ligand families. *Endocr Rev.* **20**: 345-357.

Sunnarborg, S. W., Hinkle, C. L., Stevenson, M., Russell, W. E., Raska, C. S., Peschon, J. J., Castner, B. J., Gerhart, M. J., Paxton, R. J., Black, R. A., Lee, D. C., 2002, Tumor necrosis factor-alpha converting enzyme (TACE) regulates epidermal growth factor receptor ligand availability. *J Biol Chem.* **277**: 12838-12845.

Suzuki, A., Kadota, N., Hara, T., Nakagami, Y., Izumi, T., Takenawa, T., Sabe, H., Endo, T., 2000, Meltrin alpha cytoplasmic domain interacts with SH3 domains of Src and Grb2 and is phosphorylated by v-Src. *Oncogene*. **19**: 5842-5850.

Tannapfel, A., Anhalt, K., Hausermann, P., Sommerer, F., Benicke, M., Uhlmann, D., Witzigmann, H., Hauss, J., Wittekind, C., 2003, Identification of novel proteins associated with hepatocellular carcinomas using protein microarrays. *J Pathol*. **201**: 238-249.

Thathiah, A., Blobel, C. P., Carson, D. D., 2003, Tumor necrosis factor-alpha converting enzyme/ADAM 17 mediates MUC1 shedding. *J Biol Chem*. **278**: 3386-3394.

Thodeti, C. K., Albrechtsen, R., Grauslund, M., Asmar, M., Larsson, C., Takada, Y., Mercurio, A. M., Couchman, J. R., Wewer, U. M., 2003, ADAM12/syndecan-4 signaling promotes beta 1 integrin-dependent cell spreading through protein kinase Calpha and RhoA. *J Biol Chem*. **278**: 9576-9584.

Trochon-Joseph, V., Martel-Renoir, D., Mir, L. M., Thomaidis, A., Opolon, P., Connault, E., Li, H., Grenet, C., Fauvel-Lafeve, F., Soria, J., Legrand, C., Soria, C., Perricaudet, M., Lu, H., 2004, Evidence of antiangiogenic and antimetastatic activities of the recombinant disintegrin domain of metargidin. *Cancer Res*. **64**: 2062-2069.

Vincent, B., Paitel, E., Saftig, P., Frobert, Y., Hartmann, D., De Strooper, B., Grassi, J., Lopez-Perez, E., Checler, F., 2001, The disintegrins ADAM10 and TACE contribute to the constitutive and phorbol ester-regulated normal cleavage of the cellular prion protein. *J Biol Chem*. **276**: 37743-37746.

Wakatsuki, S., Kurisaki, T., Sehara-Fujisawa, A., 2004, Lipid rafts identified as locations of ectodomain shedding mediated by Meltrin beta/ADAM19. *J Neurochem*. **89**: 119-123.

Wei, P., Zhao, Y. G., Zhuang, L., Ruben, S., Sang, Q. X., 2001, Expression and enzymatic activity of human disintegrin and metalloproteinase ADAM19/meltrin beta. *Biochem Biophys Res Commun*. **280**: 744-755.

Weskamp, G. and Blobel, C. P., 1994, A family of cellular proteins related to snake venom disintegrins. *Proc Natl Acad Sci U S A*. **91**: 2748-2751.

Weskamp, G., Cai, H., Brodie, T. A., Higashyama, S., Manova, K., Ludwig, T., Blobel, C. P., 2002, Mice lacking the metalloprotease-disintegrin MDC9 (ADAM9) have no evident major abnormalities during development or adult life. *Mol Cell Biol*. **22**: 1537-1544.

Weskamp, G., Kratzschmar, J., Reid, M. S., Blobel, C. P., 1996, MDC9, a widely expressed cellular disintegrin containing cytoplasmic SH3 ligand domains. *J Cell Biol*. **132**: 717-726.

Weskamp, G., Schlondorff, J., Lum, L., Becherer, J. D., Kim, T. W., Saftig, P., Hartmann, D., Murphy, G., Blobel, C. P., 2004, Evidence for a critical role of the tumor necrosis factor alpha convertase (TACE) in ectodomain shedding of the p75 neurotrophin receptor (p75NTR). *J Biol Chem*. **279**: 4241-4249.

White, J. M., 2003, ADAMs: modulators of cell-cell and cell-matrix interactions. *Curr Opin Cell Biol*. **15**: 598-606.

Wolfsberg, T. G., Bazan, J. F., Blobel, C. P., Myles, D. G., Primakoff, P., White, J. M., 1993, The precursor region of a protein active in sperm-egg fusion contains a metalloprotease and a disintegrin domain: structural, functional, and evolutionary implications. *Proc Natl Acad Sci U S A*. **90**: 10783-10787.

Wolfsberg, T. G., Primakoff, P., Myles, D. G., White, J. M., 1995a, ADAM, a novel family of membrane proteins containing A Disintegrin And Metalloprotease domain: multipotential functions in cell-cell and cell-matrix interactions. *J Cell Biol*. **131**: 275-278.

Wolfsberg, T. G., Straight, P. D., Gerena, R. L., Huovila, A. P., Primakoff, P., Myles, D. G., White, J. M., 1995b, ADAM, a widely distributed and developmentally regulated gene family encoding membrane proteins with a disintegrin and metalloprotease domain. *Dev Biol*. **169**: 378-383.

Yagami-Hiromasa, T., Sato, T., Kurisaki, T., Kamijo, K., Nabeshima, Y., Fujisawa-Sehara, A., 1995, A metalloprotease-disintegrin participating in myoblast fusion. *Nature.* **377**: 652-656.

Yan, Y., Shirakabe, K., Werb, Z., 2002, The metalloprotease Kuzbanian (ADAM10) mediates the transactivation of EGF receptor by G protein-coupled receptors. *J Cell Biol.* **158**: 221-226.

Yoshida, S., Setoguchi, M., Higuchi, Y., Akizuki, S., Yamamoto, S., 1990, Molecular cloning of cDNA encoding MS2 antigen, a novel cell surface antigen strongly expressed in murine monocytic lineage. *Int Immunol.* **2**: 585-591.

Yoshiyama, K., Higuchi, Y., Kataoka, M., Matsuura, K., Yamamoto, S., 1997, CD156 (human ADAM8): expression, primary amino acid sequence, and gene location. *Genomics.* **41**: 56-62.

Zhang, X. P., Kamata, T., Yokoyama, K., Puzon-McLaughlin, W., Takada, Y., 1998, Specific interaction of the recombinant disintegrin-like domain of MDC-15 (metargidin, ADAM-15) with integrin alphavbeta3. *J Biol Chem.* **273**: 7345-7350.

Zhang, Y., Jiang, J., Black, R. A., Baumann, G., Frank, S. J., 2000, Tumor necrosis factor-alpha converting enzyme (TACE) is a growth hormone binding protein (GHBP) sheddase: the metalloprotease TACE/ADAM-17 is critical for (PMA-induced) GH receptor proteolysis and GHBP generation. *Endocrinology.* **141**: 4342-4348.

Zhao, J., Chen, H., Peschon, J. J., Shi, W., Zhang, Y., Frank, S. J., Warburton, D., 2001, Pulmonary hypoplasia in mice lacking tumor necrosis factor-alpha converting enzyme indicates an indispensable role for cell surface protein shedding during embryonic lung branching morphogenesis. *Dev Biol.* **232**: 204-218.

Zheng, Y., Schlondorff, J., Blobel, C. P., 2002, Evidence for regulation of the tumor necrosis factor alpha-convertase (TACE) by protein-tyrosine phosphatase PTPH1. *J Biol Chem.* **277**: 42463-42470.

Zhou, H. M., Weskamp, G., Chesneau, V., Sahin, U., Vortkamp, A., Horiuchi, K., Chiusaroli, R., Hahn, R., Wilkes, D., Fisher, P., Baron, R., Manova, K., Basson, C. T., Hempstead, B., Blobel, C. P., 2004, Essential role for ADAM19 in cardiovascular morphogenesis. *Mol Cell Biol.* **24**: 96-104.

Zhou, M., Graham, R., Russell, G., Croucher, P. I., 2001, MDC-9 (ADAM-9/Meltrin gamma) functions as an adhesion molecule by binding the alpha(v)beta(5) integrin. *Biochem Biophys Res Commun.* **280**: 574-580.

Zou, J., Zhu, F., Liu, J., Wang, W., Zhang, R., Garlisi, C. G., Liu, Y. H., Wang, S., Shah, H., Wan, Y., Umland, S. P., 2004, Catalytic Activity of Human ADAM33. *J Biol Chem.* **279**: 9818-9830.

# Chapter 3

# ADAM8/MS2/CD156a

## *A metalloprotease-disintegrin involved in immune responses*

Jörg W. Bartsch, Silvia Naus, Andrea Rittger, Uwe Schlomann, Dirk Wildeboer

*Developmental Biology and Molecular Pathology, W7, Bielefeld University, 33615 Bielefeld, Germany*

Abstract: From all ADAM family members known, interesting features of some members of this the family are is their distinct expression patterns. ADAM8 is such an example, as it was identified originally in monocytes and is expressed in many specialised cell types, among them macrophages, B-cells, granulocytes, follicle cells, glandular epithelial cells, osteoclasts, oligodendrocytes, microglia, neurons and astrocytes. ADAM8 is activated by autocatalytic prodomain removal and the substrates like the Close Homologue of L1 (CHL1) and CD23 identified so far are either involved in cell adhesion or immune responses. In turn, ADAM8 expression in some cell types such as macrophages, astrocytes and microglia is regulated by inflammatory mediators including tumor necrosis factor-α, lipopolysaccharides (LPS) and prostaglandins. Whereas embryonal development in ADAM8 deficient mice appears normal, its upregulation under inflammatory conditions like that seen in chronic neurodegeneration, after administration of LPS and in allergic asthma, seems to reflect a specific function of ADAM8 in cytokine response. From recent experiments it can be concluded that the ADAM8 induction by inflammatory cytokines serves protective functions, e.g. by shedding of receptors mediating inflammatory responses or by degrading immune mediators directly.

Key words: ADAM8, Autocatalytic prodomain removal, Inflammation, CD23, Close Homologue of L1, Tumor Ntumor ecrosis factor-α, Neurodegeneration, Asthma, Aasthma, llergy.



*N. M. Hooper and U. Lendeckel (eds.), The ADAM Family of Proteases,* 65-73.

# 1. INTRODUCTION

In 1990, some years before ADAMs were assembled in a large gene family (Wolfsberg, 1995). ADAM8 was identified as MS2 on the basis of its expression in mouse macrophages and macrophage cell lines (Yoshida *et al.*1990). The 3.2 kb corresponding cDNA encodes a protein with 826 amino acids and all structural features of an ADAM. Later, human ADAM8 cDNA was isolated from THP-1 cells and showed 65.5% homology to mouse ADAM8. The human ADAM8 gene was mapped to human chromosome 10q26.3 and to mouse chromosome 7 (Yoshiyama *et al.* 1997). Despite its early discovery, the biochemical features of ADAM8 have been revealed recently and functional data are still accumulating stating that ADAM8 plays an important role in diseases with chronic inflammation.

# 2. BIOCHEMISTRY OF ADAM8

## 2.1 Protease activation by autocatalytic prodomain removal

Many active ADAM proteases like ADAM9, ADAM10, TACE, and ADAM19 undergo proteolytic prodomain removal by furinlike proteases or prohormone convertases. The respective recognition motifs RXR/KR are located in the hinge region between the pro (PD) and the metalloprotease (MP) domain (Roghani *et al.* 1999; Anders *et al.* 2001; Clarke *et al.* 1998; Kang *et al.* 2002). Multiple sequence alignmentsSequence alignment of this processing region (Figure 1) revealed that ADAM8 as well as ADAM28 do not contain these consensus motifs, raising the question by which mechanism prodomain removal occurs.

Both ADAMs show the highest homology to hemorrhagic snake venom proteases which are activated by autocatalysis, raising the question whether ADAM8 and ADAM28 are processed in the same manner. Indeed, direct evidence for autocatalytic prodomain removal in ADAM8 came from two types of experiments:

1) When $Glu^{330}$ in the HEXXH consensus motif was exchanged by a glutamine (EQ-ADAM8), prodomain removal was completely abolished (Schlomann *et al.* 2002). A similar experiment performed with ADAM28 (glutamate to alanine exchange) gave an identical result (Howard *et al.* 2000).

2) Cotransfection of inactive EQ-ADAM8 with a construct encoding the ADAM8 ectodomain led to processing of EQ-ADAM8, demonstrating that

active ADAM8 can process EQ-ADAM8 in *trans* during transport through the Trans-Golgi Network. This type of autocatalysis has some restrictions: when the entire MP domain of ADAM8 was expressed, processing of EQ-ADAM8 was not observed, allowing for the conclusion that the presence of the disintegrin domain influences autocatalytic prodomain cleavage, e.g. by mediating the interaction of at least two molecules of ADAM8. Full length mouse or human cDNA of ADAM8 expressed in COS-7 cells resulted in ADAM8 protein represented by 3 bands of defined lengths, corresponding to the proform of ADAM8 (120 kD), the processed (90 kD) and a remnant form (60 kD) which is also detected in tissue extracts from mouse lung, spleen, and kidney.

| | | | | |
|---|---|---|---|---|
| mADAM9 | (NP_031430) | QGDEEEHPS—VTQLL | **RRRR** | AVLPQTRYVELFIVVD |
| mADAM10 | (NP_031425) | RAHPEKHAASSGPELL | **RKKR** | TTLAERNTCQLYIQTD |
| mADAM17 | (NP_033745) | KGLIDREPSEEFVR**RV** | **KRRA** | EPNPLKNTCKLLVVAD |
| mADAM19 | (NP_033746) | WALQFTHQT--KKQP**R** | **RMKR** | EDLHSMKYVELYLVAD |
| Consensus | | | **RXKR** | |
| | | | **RXRR** | |
| mADAM8 | (NP_031428) | RALEIYRAQ--PRN-W | **LIPR** | E----TRYVELYVVAD |
| mADAM28 | (AFF71993) | VALPATRLI--KLNDG | **MVQE** | P----KKYIEYYVVLD |

*Figure 1:* Sequence alignments of putative furin cleavage sites in mouse ADAMs 9,10,17, and 19, using ClustalX. The proprotein convertase sequences are shown in bold letters and consensus sequences are underlined. For ADAMs 8 and 28, no such consensus site can be delineated in the respective positions.

## 2.2 Proteolytic activity of ADAM8

Due to the perfect metzincin consensus sequence present in ADAM8, a protease activity could be expected. Experiments with soluble forms of ADAM8 either transfected in HEK293 cells (Choi *et al.* 2001), in COS-7 (Schlomann *et al.* 2002) or NS0 cells (Amour et al., 2002) clearly indicated protease activity, which was only dependent on the presence of the MP domain, but did not require further domains. Protease activity was initially monitored by cleavage of Myelin Basic Protein (MBP), and the cleavage was inhibited by metalloprotease inhibitors 1,10-phenanthroline and BB-94 ($IC_{50}$ value: 51 nM) and weakly (5μM) by a hydroxamate inhibitor (Amour *et al.* 2002). In contrast, no inhibition was obtained by using recombinant TIMP (Tissue Inhibitor of Metalloproteinases) even when high concentrations of TIMPs up to 500 nM were used (Amour *et al.* 2002; Schlomann *et al.* 2002). By demonstration of autocatalysis, a first substrate of ADAM8 was identified. For autocatalytic prodomain removal, cleavage occurs within a stretch of 10 amino acids, and the precise cleavage site is dependent on the construct used for expression of soluble ADAM8 protease (Figure 2). The ADAM8 cleavage site in myelin basic protein (MBP) is within the sequence TTHYGSLP↓QKAQGQ, similar to the cleavage sites observed for ADAM10 and ADAM28 (Howard *et al.* 2001). In further studies from Amour *et al.*(2002), Fourie *et al.*(2003) and Naus *et al.*(2004), substrates were identified on the basis of *in vitro* cleavage experiments or by peptides representing the juxtamembraneous regions of several membrane proteins, candidates for physiologically relevant ADAM8 cleavage either by their cellular localisations or their functions (Table 1).

*Table 1:* ADAM8-mediated cleavage of peptides based on proteolytically sensitive sequences of shed proteins. Known cleavage sites are indicated by vertical arrows. Peptide sequences labelled by an asterisk are derived from N-terminal sequencing of respective proteins.

| TM protein | Peptide (↓ cleavage site) | Reference |
|---|---|---|
| TNF-α | SPLA↓QA↓VRSSSRK | Amour et al., 2002 |
| APP | YEVHH↓QKLVFF | Amour et al., 2002 |
| IL-1 rec II | TVKEAS↓STFSWG | Amour et al., 2002 |
| KL | LPPVAA↓SSLR | Amour et al., 2002 |
| A8PD* | AQPR↓NWLI | Schlomann et al., 2002 |
| MBP | GSLP↓QKAQ | Schlomann et al., 2002 |
| CHL1*(FN V) | EVLVP↓GAEHI | Naus et al., 2004 |
| CHL1*(FN II) | VSWKP↓QGAPE | Naus et al., 2004 |
| CD23 | GDQMA↓QKSQS | unpublished |
| TNFRI | CMKLCLPPPL | unpublished |

## 2.3 Function of the ADAM8 disintegrin domain

In the original work published by Yamamoto and colleagues, the function of the ADAM8 disintegrin (DI) domain was attributed to leukocyte adhesion (Kataoka *et al.*1997). Functional evidence for the ADAM8 DI domain came from an autoimmunity model in the rat (Schluesener 1998). In this animal multiple sclerosis model, a MBP peptide was used to induce severe generalised autoimmune symptoms in the brain (GANS). When rats were pretreated with the recombinant disintegrin domain of ADAM8, the autoimmune symptoms were dramatically reduced, indicating that either the DI domain competes with endogenous ADAM8 for cell-cell interactions or has an indirect effect e.g. by modulatinginfluencing ADAM8 protease activity. Experimental evidence for the ability of ADAM8 DI domain to mediate cell adhesion was given when the disintegrin/cysteine-rich domain was expressed in *E. coli*. Recombinant DI protein coated as a substrate on cell culture dishes was able to mediate adhesion specifically to cells expressing ADAM8, arguing for disintegrin dependent cell adhesion in a homophilic manner (Schlomann *et al.* 2002).

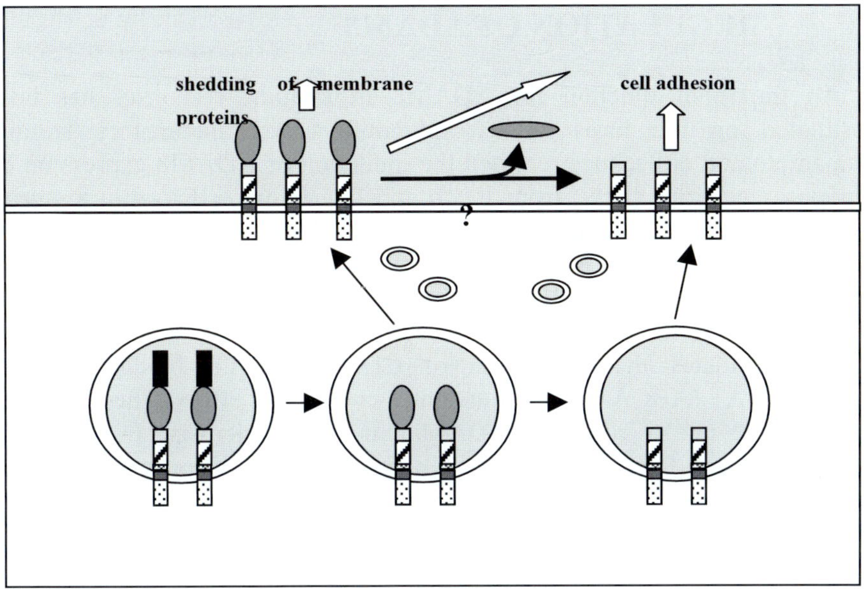


**TGN**

*Figure 2:* Model for ADAM8 processing and function. After passage through the medial Golgi, pro-ADAM8 is processed by autocatalysis. In the processed form, ADAM8 is transported to the cell surface where it acts as a shedding enzyme. A major proportion is further processed by MP removal leading to an ADAM8 protein capable of cell adhesion. Whether MP removal occurs intracellular or extracellular, is still unknown (indicated by the question mark), but MP domain has never been detected in supernatants from cells transfected with full length ADAM8. Prodomain, black bars; MP domain, ovals; DI domain, gray squares, Cys-rich domain, hatched boxes. TGN, Trans-Golgi Network. From: Schlomann et al. (2002).

In addition to homophilic cell adhesion, interaction of the ADAM8 DI domain with integrins is very likely, although not proven. As the DI domains of ADAM8 and ADAM28 are highly homologous, it can be speculated that similar to ADAM28, the DI domain forms an extended surface in the integrin binding loop, which in the case of ADAM28 can interact with integrin $\alpha 4\beta 1$ (Bridges *et al.* 2003). This interaction has particular importance for the migration of lymphocytes. For ADAM8 DI domain this interaction remains to be determined. The current hypothesis states that the interaction of ADAM8 with integrins occurs indirectly e.g. via adaptor molecules like CD9 or other tetraspanin-like proteins (reviewed in Moss and Bartsch, 2004).).

## 3. REGULATION OF ADAM8

An important function for ADAM8 in immune responses has been postulated on the basis of its inducibility by inflammatory stimuli. Yamamoto and colleagues reported the induction of ADAM8 expression by Lipopolysaccharide (LPS) from *E.coli* and Interferon-γ (IFN-γ) in a murine macrophage cell line (aHINSB3; Kataoka *et al.* 1997). In another murine macrophage cell line RAW264.7, induction of ADAM8 mRNA was observed after administration of prostaglandins effecting PPARγ, the peroxisome proliferator-activated receptor (Hodgkinson and Ye, 2003). PPARγ regulates the expression of CD36 and ATP-binding cassette transporter A1 (ABCA1), which mediates cholesterol efflux. These findings could point out a role for ADAM8 in arteriosclerosis. In the same macrophage cell line, shedding of the tumor necrosis factor receptor type I (TNRI, p55) is dependent on treatment with LPS which induces ADAM8 in these cells (Wildeboer, Naus, Rittger, Bartsch, unpublished observations).

A role in leukocyte infiltration has been demonstrated for ADAM8 in transgenic mice overexpressing the ADAM8 ectodomain under control of the α1-antitrypsin promoter. In these mice the expression of L-selectin in neutrophils (PMN) from peripheral blood leukocytes (PBL) was more strongly down regulated than in non-transgenic mice, suggesting that L-selectin in PMN transgenic mice was shed by sCD156. In addition, these experiments suggest that sCD156 may activate the endothelial cells and lead to an upregulation of E-selectin thereby regulating leukocyte infiltration directly or indirectly (Higuchi *et al.* 2002).

Furthermore, ADAM8 was also found in other tissues such as the central nervous system. In an experimental model for neurodegeneration, the Wobbler mouse, a model for Amyotrophic Lateral Sclerosis (ALS), ADAM8 mRNA was found to be prominently upregulated in degenerating neurons and in glia cells. The inducer for enhanced transcription is the Tumor Necrosis Factor α (TNF-α), a cytokine with wide ranging effects in the organism, in particular under inflammatory conditions (Schlomann *et al.* 2000). In lungs of mice undergoing asthma induction with different allergens like ovalbumin and *Aspergillus fumigatus*, ADAM8 is strongly induced (King *et al.* 2004) suggesting that ADAM8 plays a role in pathogenesis of allergic asthma. Under these conditions, the induction of ADAM8 was increased by Interleukins 4 and 13 (IL-4, IL-13). The pathways leading to increased expression of ADAM8 in allergic asthma were critically

dependent on the IL-4 receptor α-chain and on a Signal Transducer and Activator of Transcription (STAT) protein family member, STAT6.
The cytoplasmic region of ADAM8 contains putative proline rich SH3 binding sequences, suggesting that it may be capable of intracellular signaling (Yoshida *et al.* 1990; Yoshiyama *et al.* 1997). So far, no evidence for an intracellular regulation of ADAM8 e.g. by the phorbol ester TPA has been given. This is in contrast to ADAM10 and TACE which can be activated by TPA leading to enhanced shedding of membrane proteins.

## 4. ADAM8 EXPRESSION IN THE ORGANISM

A systematic analysis of ADAM8 expression in the mouse came from ADAM8 deficient mice in which the exon 8 (MP domain) of the ADAM8 gene was exchanged by a LacZ-neo cassette. In early embryonal development, ADAM8 is expressed by maternal cells in the decidua and by trophoblast derivatives of the embryo. At later stages of embryonal development, ADAM8 is expressed in the gonadal ridge, thymus, developing cartilage or bone, in mesenchymal tissue in close proximity to developing blood vessels and lymphatic vessels. Also in the developing brain and spinal cord, ADAM8 expression was detected. Despite this prominent expression in embryonal development, no major defects or abnormalities were evident (Kelly *et al.* in press).

In adult tissues, ADAM8 expression occurs in lymphatic organs thymus and spleen as well as in lung, epithelia of salivary glands, and in tubular epithelial cells of the kidney. In lungs, expression is restricted to epithelial cells of distinct respiratory bronchioles, and under experimental induction of allergic asthma, ADAM8 expression is upregulated in these cells (King *et al.* in press). It is noteworthy, that no expression was observed in smooth muscle layers underlying the bronchioles. In the spleen, expression of ADAM8 is weak and restricted to cells in the nodules. After subcutaneous injection of LPS in heterozygous ADAM8 deficient mice, upregulation of ADAM8 in these cells was observed (Rittger and Bartsch, unpublished observations). In the brain, ADAM8 is detectable in neurons and oligodendrocytes, although very weakly under normal conditions. Under inflammatory stimuli, like exogenous LPS injected subcutaneously or endogenous TNF-α, ADAM8 expression levels wereare enhanced, caused by additional expression in astrocytes and microglia, and particularly in degenerating neurons.

# 5. CONCLUSION

From all observations made so far itIt is quite obvious that ADAM8 does not play an essential role in normal development, as demonstrated in ADAM8 deficient mice. In contrast, at least under inflammatory conditions ADAM8 is upregulated and is supposed to play a role in immunomodulation. In this respect, substrates modulating the immune response like CD23, an allergic mediator, or TNFRI are of particular interest, creating a challenge for future research in unraveling the specific function of ectodomain shedding of immune-relevant membrane proteins by ADAM8. In addition, modulating ADAM8 expression may influence inflammation processes, and eventually pinpoint ADAM8 as a therapeutic target.

# REFERENCES

Amour, A., Knight, C.G., English, W.R., Webster, A., Slocombe, P.M., Knauper, V., Docherty, A.J.P., Becherer, J.D., Blobel, C.P., and Murphy G., 2002, The enzymatic activity of ADAM8 and ADAM9 is not regulated by TIMPs. *FEBS Lett* **524:** 154-158.

Anders, A., Gilbert, S., Garten, W., Postina, R., Fahrenholz, F., 2001, Regulation of the alpha-secretase ADAM10 by its prodomain and proprotein convertases. *FASEB J.* **15:** 1837-1839.

Bridges, L.C., Hanson, K.R., Tani, P.H., Mather, T., Bowditch, R.D., 2003, Integrin alpha4beta1-dependent adhesion to ADAM28 (MDC-L) requires an extended surface of disintegrin domain. *Biochemistry* **87:** 3734-3741.

Choi, S.J., Han, J.H., Roodman, G.D., 2001, A novel osteoclast stimulating factor. *J. Bone Miner Res.* **16:** 814-822.

Clarke, H.R., Wolfson, M.F., Rauch, C.T., Castner, B.J., Huang, C.P., Gerhart, M.J., Johnson, R.S., Cerretti, D.P., Paxton, R.J., Price, V.L., Black R.A., 1998, Expression and purification of correctly processed, active human TACE catalytic domain in Saccharomyces cerevisae. *Protein Expr. Purif.* **13:** 104-110.

Fourie, A.M., Coles, F., Moreno, V., and Karlsson, L., 2003, Catalytic activity of ADAM8, ADAM15 and MDC-L (ADAM28) on synthetic peptide substrates, and in ectodomain cleavage of CD23. *J. Biol.Chem.* **278:** 30469-30477.

Higuchi, Y., Yasui, A., Matsuura, K., Yamamoto, S., 2002, CD156 transgenic mice. Different responses between inflammatory types. *Pathobiology* **70:** 47-54.

Hodgkinson, C.P., Ye, S., 2003, Microarray analysis of peroxisome proliferator-activated receptor-gamma induced changes in gene expression in macrophages. *Biochem. Biophys. Res. Commun.* **308:** 505-510.

Howard, L., Maciewicz, R.A., Blobel, C.P., 2000, Cloning and characterization of ADAM28: evidence for autocatalytic pro-domain removal and for cell surface localization of mature ADAM28. *Biochem. J.* **348:** 21-27.

Howard, L., Zheng, Y., Horrocks, M., Maciewicz, R.A., Blobel, C., 2001, Catalytic activity of ADAM28. *FEBS Lett.* **498:** 82-86.

Kang, T., Zhao, Y.G., Pei, D., Sucic, J.F., Sang, Q.X., 2002, Intracellular activation of human adamalysin 19/disintegrin and metalloproteinase 19 by furin occurs via one of the two consecutive recognition sites. *J. Biol. Chem.* **277:** 25583-25591.

Kataoka, M., Yoshiyama, K., Matsuura, K., Hijiya, N., Higuchi, Y., Yamamoto, S., 1997, Structure of the murine CD156 gene, characterization of its promoter, and chromosomal location. *J. Biol. Chem.* **272:** 18209-18215.

Kelly, K., Hutchinson, G., Klewe-Nebenius, D., Smith, A., Bartsch, J.W., Rittger, A., Horiuchi, K., Sahin, U., Manova, K., Docherty, A.J.P., Blobel, C.P., 2004, Metalloprotease-disintegrin ADAM8: Expression analysis and targeted deletion in mice. *Dev. Dynamics*, in press.

King, N.E., Zimmermann, N., Pope, S.M., Fulkerson, P.C., Nikolaidis, N.M., Mishra, A., Witte, D.P., Rothenberg, M.E., 2004, Expression and regulation of a disintegrin and metalloproteinase (ADAM)8 in experimental asthma. *Am. J. Respir. Cell. Mol. Biol.* **31:** 257-265.

Moss, M.L., and Bartsch, J.W., 2004, Therapeutic benefits form targeting of ADAM family members. *Biochemistry* **43:** 7227-7235.

Naus, S., Richter, M., Wildeboer D., Schachner M., Moss, M., and Bartsch J.W., 2004, Ectodomain shedding of the neural recognition molecule CHL1 by the metalloprotease-disintegrin ADAM8 promotes neurite outgrowth and suppresses neuronal cell death. *J. Biol. Chem.* **279:** 16040-16049.

Roghani, M., Becherer, J.D., Moss, M.L., Atherton, R.E., Erdjument-Bromage, H., Arribas, J., Blackburn, R.K., Weskamp, G., Tempst, P., Blobel, C.P., 1999, Metalloprotease-disintegrin MDC9: intracellular maturation and catalytic activity. *J. Biol. Chem.* **274:** 3531-3540.

Schlomann, U., Rathke-Hartlieb, S., Yamamoto, S., Jockusch, H., Bartsch, J.W., 2000, Tumor necrosis factor alpha induces a metalloprotease-disintegrin, ADAM8 (CD 156): implications for neuron-glia interactions during neurodegeneration. *J. Neurosci.* **20:** 7964-7971.

Schlomann, U., Wildeboer, D., Webster, A., Antropova, O., Zeuschner, D., Knight, C.G., Docherty, A.J., Lambert, M., Skelton, L., Jockusch, H., Bartsch, J.W., 2002, The metalloprotease disintegrin ADAM8. Processing by autocatalysis is required for proteolytic activity and cell adhesion. *J. Biol. Chem.* **277:** 48210-48219.

Schluesener, H.J., 1998, The disintegrin domain of ADAM8 enhances protection against rat experimental autoimmune encephalomyelitis, neuritis and uveitis by a polyvalent autoantigen vaccine. *J. Neuroimmunol.* **87:** 197-202.

Wolfsberg, T.G., Straight, P.D., Gerena, R.L., Huovila, A.P., Primakoff, P., Myles, D.G., and White, J.M.,1995, ADAM, a widely distributed and developmentally regulated gene family encoding membrane proteins with a disintegrin and metalloprotease domain. *Dev. Biol.* **69:** 378-383.

Yoshida, S., Setoguchi, M., Higuchi, Y., Akizuki, S., Yamamoto, S., 1990, Molecular cloning of cDNA encoding MS2 antigen, a novel cell surface antigen strongly expressed in murine monocytic lineage. *Int. Immunol.* **2:** 585-91.

Yoshiyama, K., Higuchi, Y., Kataoka, M., Matsuura, K., Yamamoto, S., 1997, CD156 (human ADAM8): expression, primary amino acid sequence, and gene location. *Genomics* **41:** 56-62.

# Chapter 4

# ADAM9

Shoichi Ishiura
*Department of Life Sciences, Graduate School of Arts and Sciences, The University of Tokyo, Japan*

Abstract: ADAM9 (meltrin γ, MDC9) was first described by Yagami *et al.* (1995) as one of fertilin-like myoblast-expressed gene products. Although only meltrin α (ADAM12) is required for myotube formation, meltrins β (ADAM19) and γ are involved in various cellular processes. The sequence of the predicted ADAM9 contains several domains like other ADAMs: a signal peptide, a zinc-dependent metalloprotease domain, a disintegrin-like domain, a cysteine-rich domain, an epidermal growth factor (EGF)-like domain, a transmembrane domain, and a C-terminal cytoplasmic tail. ADAM9 is predicted to be an active metalloprotease due to a catalytic site consensus sequence HEXXH in its metalloprotease domain.

Key words: ADAM9, metalloprotease, disintegrin

## 1. MOLECULAR ASPECTS OF ADAM9

Mouse and human ADAM9 were cloned and shown to be ubiquitously expressed, especially at the cell surface. ADAM9 mRNA is highly expressed in developing mesenchyme, heart and brain (Weskamp *et al.* 2002). The apparent molecular mass of active ADAM9 was shown to be 84 kDa. Full-length human ADAM9 consists of an open reading frame of 2460 nucleotides encoding 819 amino acid residues. A novel short form of ADAM9 was cloned from a human cDNA library (Hotoda *et al.* 2002) (Fig. 1). It has a 106 bp deletion in the EGF-like domain, with a TGA stop codon, and consequently lacks the transmembrane and cytoplasmic domains. It was shown to be secreted and retains shedding activity.



*N. M. Hooper and U. Lendeckel (eds.), The ADAM Family of Proteases,* 75-83.

TPA. A dominant-negative form of ADAM9 suppressed the TPA-induced shedding of the ectodomain of proHB-EGF (Izumi *et al.* 1998). A direct interaction between PKCδ and ADAM9 was shown to be involved in the "regulated" shedding of proHB-EGF.

We independently demonstrated that ADAM9 has another *in vivo* substrate (Koike *et al.* 1999). When ADAM9 was coexpressed in COS cells with the Alzheimer's amyloid precursor protein (APP), APP was digested at the α–secretory site in ADAM9 expressing cells (Fig. 2). When an artificial α–secretory site mutant (the HHQK of Aβ13-16 was replaced with LHHAA to confer greater resistance to α–secretory cleavage) was also coexpressed with ADAM9, secreted APP was not increased in conditioned medium. These results suggest that ADAM9 has an α–secretase-like activity. The activity was activated by phorbol ester. Interestingly, inhibition of ADAM9 by a hydroxamate-based inhibitor SI-27 enhanced β–secretase cleavage. Increasing the concentration of SI-27 had reciprocal effects on α– and β–secretory cleavages.

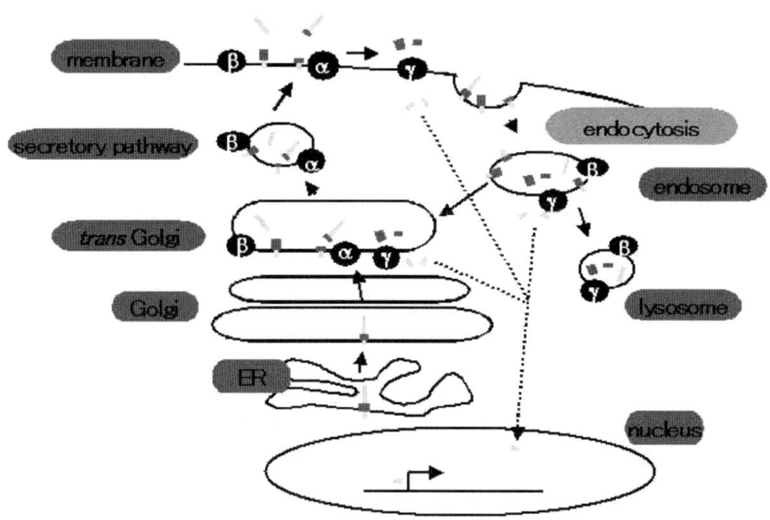


*Figure 2.* Ectodomain shedding of APP in the secretory and endocytic pathways

Several studies have suggested that not only ADAM9 but also ADAM10 and ADAM17 are candidate α–secretases. In a comparative study (Asai *et al.* 2003) these ADAMs were expressed in COS cells and both "constitutive" and "regulated or TPA-stimulated" α–secretase activities were determined. The results indicated that ADAM9, ADAM10 and ADAM17 all catalyze α–secretory cleavage. Suppression of ADAMs in human glioblastoma A172 cells, which contain large amounts of endogenous ADAMs, by lipofection of the double-stranded RNA indicates that ADAM9, ADAM10 and ADAM17 all act as α–secretases in A172 cells. These results indicate that endogenous APP α–secretase is composed of several ADAMs. Recent results suggest that ADAM19 but not ADAM12 also has α–secretase activity (Hotoda *et al.* unpublished).

Overexpression of ADAM9 or ADAM12 in CHO cells increased placental leucine aminopeptidase (P-LAP) release (Ito *et al.* 2004). Since P-LAP is responsible for oxytocin degradation during pregnancy, ectodomain shedding of P-LAP and the release of its soluble form is a key phenomenon for oxytocin balance. Immunohistochemical observation suggests that ADAM12 is highly expressed in syncytiotrophoblasts, while ADAM9 expression is low. These results indicate that ADAM12 may be involved in P-LAP shedding in human placenta.

## 3. MATRIX DEGRADATION AND CELL-CELL CONTACT

ADAMs and MMPs are assumed to be involved in extracellular matrix remodeling. Flannery *et al.* (1999) described the altered expression of ADAM9 in cultured human chondrocytes upon stimulation by the cellular environment. ADAM9 mRNA levels were decreased in chondrocytes exposed to IL-1 or retinoic acid. Purified metalloprotease domain of ADAM9, which was expressed in *Pichia pastoris*, digested fibronectin as well as gelatin and β-casein (Schwettmann and Tschesche 2001).

Integrin ligands (RGD) have been identified for only two ADAMs, ADAM2 and ADAM15. Nath *et al.* (2000) showed that ADAM9 specifically bound to integrin α6β1, which is a receptor for laminins, via its disintegrin domain, although ADAM9 lacks an RGD motif. When Wehi 164 or HT 1080 fibroblasts were plated on ADAM9, there was a dramatic 8–10 fold increase in their motility compared to plating on laminin. This motility was reduced by a p160 ROCK kinase inhibitor, suggesting an important role of ADAM9 in regulating cell motility. Another authors showed that ADAM9 binds the αvβ5 integrin in divalent cation-dependent manner (Zhou *et al.* 2001). This also suggests that ADAM9 can function as an adhesion molecule.

Up-regulation of integrin $\alpha5\beta1$ and $\alpha v\beta3$ by incubating human atrial and venous vascular smooth muscle cells with PDGF up to 3 days increased ADAM9 mRNA. Increased expression of ADAM9 was also observed in atherosclerotic arteries (Al-Fakhri *et al.* 2003).

## 4. ROLE OF ADAM9 IN BLOOD CELLS AND OTHER CELL TYPES

Since ADAM9 has a potential fusion peptide, its role in multinucleation processes (monocyte fusion) was investigated (Namba *et al.* 2001). Expression of ADAM9 was induced in blood monocytes when stimulated with anti-CD98 antibody or RANKL+M-CSF. SI-27, a hydroxamate-based metalloprotease inhibitor, suppressed multinucleated giant cell formation.

Expression of ADAM9 in lens was demonstrated by Lim *et al.* (2002). Significant reduction of ADAM9 expression was observed in lens epithelial cells from patients with anterior polar cataracts. It may serve as a marker for such symptoms.

ADAM9 was also shown to be IGFBP5-degrading protease in osteoblasts (Mohan *et al.* 2002). They purified IGFBP5-degrading protease from human U2 osteosarcoma cells by several chromatographic procedures and determined the N-terminal sequence. It was identical to that of $\alpha2$ macroglobulin ($\alpha$2M). Since $\alpha$2M is not a protease, they identified ADAM9 as an IGFBP5-degrading protease in the purified chromatographic fraction. ADAM9 cleaved IGFBP5 and bound to $\alpha$2M.

ADAM9 is expressed in liver cells. The up-regulation of ADAM9 and ADAM12 was shown to be correlated with an increase in MMP-2, and associated with tumor aggressiveness and progression (Le Pabic *et al.* 2003; Tannapfel *et al.* 2003). Overexpression of ADAM9 mRNA was observed in breast carcinomas (72/110, 66%) and fibroadenomas (21/38, 55%) (O'Sea *et al.* 2003). Overexpressed ADAM9 was also present in 70% of pancreatic ductal adenocarcinomas (Grutzmann *et al.* 2003, 2004).

To determine the relationship between metastasis and ADAM9 expression, sublines of an EBC-1 lung cancer cell line that were highly metastatic to either brain or bone were examined (Shintani *et al.* 2004). ADAM9 mRNA levels were significantly higher in brain-metastatic sublines. Stable transfection of ADAM9 in EBC-1 cells resulted in increased invasive capacity, increased adhesion to brain tissue, and increased expression of integrin $\alpha3\beta1$. These results suggest a prognostic role of ADAM9 in several cancers.

## 5. MODEL ANIMALS OF ADAM9 DEFICIENCY

Mice lacking ADAM9 were created by Weskamp *et al.* (2002). Despite the ubiquitous expression of ADAM9, ADAM9(–/–) mice developed normally, were viable and fertile, and did not show any abnormality. "Constitutive" and "regulated" shedding of HB-EGF and APP was observed in embryonic fibroblasts isolated from ADAM9 null mice. These data suggest a redundant role of ADAM9 or the presence of some compensatory mechanism in such ectodomain shedding.

## 6. CONCLUSION

ADAMs were first thought to be involved in cell-cell interaction or cell-matrix adhesion. Although ADAM9 is an active protease and is widely expressed in mammalian tissues, some specific functions such as involvement in metastasis and ectodomain shedding of important membrane proteins have been revealed. Further extensive characterization of ADAM9 is necessary for elucidating its physiological roles and especially for potential therapeutic implications.

## REFERENCES

Al-Fakhri N., Wilhelm, J., Hahn, M., Heidt, M., Hehrlein, F.W., Endisch, A.M., Hupp, T., Cherian, S.M., Bobryshev, Y.V., Lord, R.S., Katz, N., 2003, Increased expression of disintegrin-metalloproteinases ADAM-15 and ADAM-9 following upregulation of integrins alpha5beta1 and alphavbeta3 in atherosclerosis. *J Cell Biochem.* **89**:808-823.

Amour, A., Knight, C.G., English, W.R., Webster, A., Slocombe, P.M., Knauper, V., Docherty, A.J., Becherer, J.D., Blobel, C.P., Murphy, G., 2002, The enzymatic activity of ADAM8 and ADAM9 is not regulated by TIMPs. *FEBS Lett.* **524**:154-158.

Asai, M., Hattori, C., Szabo, B., Sasagawa, N., Maruyama, K., Tanuma, S., Ishiura, S., 2003, Putative function of ADAM9, ADAM10, and ADAM17 as APP alpha-secretase. *Biochem Biophys Res Commun.* **301**:231-235.

Cao, Y., Kang, Q., Zhao, Z., Zolkiewska, A., 2002, Intracellular processing of metalloprotease disintegrin ADAM12. *J Biol Chem.* **277**:26403-26411.

Flannery, C.R., Little, C.B., Caterson, B., Hughes, C.E., 1999, Effects of culture conditions and exposure to catabolic stimulators (IL-1 and retinoic acid) on the expression of matrix metalloproteases (MMPs) and disintegrin metalloproteases (ADAMs) by articular cartilage chondrocyte. *Matrix Biol* **18**: 225-237.

Grutzmann, R., Foerder, M., Alldinger, I., Staub, E., Brummendorf, T,, Ropcke, S., Li, X., Kristiansen, G., Jesnowski, R., Sipos, B., Lohr, M., Luttges, J., Ockert, D., Kloppel, G., Saeger, H.D., Pilarsky, C., 2003, Gene expression profiles of microdissected pancreatic ductal adenocarcinoma. *Virchows Arch.* **443**:508-17.

Hotoda, N., Koike, H., Sasagawa, N., Ishiura, S., 2002, A secreted form of human ADAM9 has an alpha-secretase activity for APP. *Biochem Biophys Res Commun.* **293**:800-805.

Grutzmann, R., Luttges, J., Sipos, B., Ammerpohl, O., Dobrowolski, F., Alldinger, I., Kersting, S., Ockert, D., Koch, R., Kalthoff, H., Schackert, H.K., Saeger, H.D., Kloppel, G., Pilarsky, C., 2004, ADAM9 expression in pancreatic cancer is associated with tumour type and is a prognostic factor in ductal adenocarcinoma. *Br J Cancer.* 90:1053-1058.

Howard, L., Nelson, K.K., Maciewicz, R.A., Blobel, C.P., 1999, Interaction of the metalloprotease disintegrins MDC9 and MDC15 with two SH3 domain-containing proteins, endophilin I and SH3PX1. *J Biol Chem.* **274**:31693-31699.

Inoue, D., Reid, M., Lum, L., Kratzschmar J., Weskamp, G., Myung, Y.M., Baron, R., Blobel C.P., 1998, Cloning and initial characterization of mouse meltrin beta and analysis of the expression of four metalloprotease-disintegrins in bone cells. *J Biol Chem* **273**: 4180-4187.

Ito, N., Nomura, S., Iwase, A., Ito, T., Kikkawa, F., Tsujimoto, M., Ishiura, S., Mizutani, S., 2004, ADAMs, a disintegrin and metalloproteinases, mediate shedding of oxytocinase. *Biochem Biophys Res Commun.* **314**:1008-1013.

Izumi, Y., Hirata, M., Hasuwa, H., Iwamoto, R., Umata, T., Miyado, K.,Tamai, Y., Kurisaki, T., Sehara-Fujisawa, A., Ohno, S., Mekada, M., 1998, A metalloprotease-disintegrin, MDC9/meltrin-gamma/ADAM9 and PKCdelta are involvedin TPA-induced ectodomain shedding of membrane-anchored heparin-binding EGF-like growth factor. *EMBO J* **17**: 7260-7272

Koike, H., Tomioka, S., Sorimachi, H., Saido, T.C., Maruyama, K., Okuyama, A., Fujisawa-Sehara, A., Ohno, S., Suzuki, K., Ishiura, S., 1999, Membrane-anchored metalloprotease MDC9 has an alpha-secretase activity responsible for processing the amyloid precursor protein. *Biochem J* **343**: 371-375.

Le Pabic, H., Bonnier, D., Wewer, U.M., Coutand, A., Musso, O., Baffet, G., Clement, B., Theret, N., 2003, ADAM12 in human liver cancers: TGF-beta-regulated expression in stellate cells is associated with matrix remodeling. *Hepatology.* **37**:1056-1066.

Lim, J.M., Lee, J.H., Wee, W.R., Joo, C.K., 2002, Downregulated expression of ADAM9 in anterior polar cataracts. *J Cataract Refract Surg.* **28**:697-702.

Mahimkar, R.M., Baricos, W.H., Visaya, O., Pollock, A.S., Lovett, D.H., 2000, Identification, cellular distribution and potential function of the metalloprotease-disintegrin MDC9 in the kidney. *J Am Soc Nephrol.* **11**:595-603.

Mohan, S., Thompson, G.R., Amaar, Y.G., Hathaway, G., Tschesche, H., Baylink, D.J., 2002, ADAM-9 is an insulin-like growth factor binding protein-5 protease produced and secreted by human osteoblasts. *Biochemistry.* **41**:15394-15403.

Namba, K., Nishio, M., Mori, K., Miyamoto, N., Tsurudome, M., Ito, M., Kawano, M., Uchida, A., Ito, Y., 2001, Involvement of ADAM9 in multinucleated giant cell formation of blood monocytes. *Cell Immunol.* **213**:104-113.

Nath, D., Slocombe, P.M., Webster, A., Stephens, P.E., Docherty, A.J., Murphy, G., 2000, Meltrin gamma(ADAM-9) mediates cellular adhesion through alpha(6)beta(1)integrin, leading to a marked induction of fibroblast cell motility. *J Cell Sci.* **113**:2319-2328.

Nelson, K.K., Schlondorff, J., Blobel, C.P., 1999, Evidence for an interaction of the metalloprotease-disintegrin tumour necrosis factor alpha convertase (TACE) with mitotic arrest deficient 2 (MAD2), and of the metalloprotease-disintegrin MDC9 with a novel MAD2-related protein, MAD2beta. *Biochem J* **343**:673-680.

Olson, G.E., Winfrey, V.P., Matrisian, P.E., NagDas, S.K., Hoffman, L.H., 1998, Blastocyst-dependent upregulation of metalloprotease/disintegrin MDC9 expression in rabbit endometrium. *Cell Tissue Res* **293**: 489-498.

O'Shea, C., McKie, N., Buggy, Y., Duggan, C., Hill, A.D., McDermott, E., O'Higgins, N., Duffy, M.J.,2003, Expression of ADAM-9 mRNA and protein in human breast cancer. Int J Cancer. **105**:754-761.

Roghani, M., Becherer, J.D., Moss, M.L., Atherton, R.E., Erdjument-Bromage, H., Arribas, J., Blackburn, R.K., Weskamp, G., Tempst, P., Blobel, C.P., 1999, Metalloprotease-disintegrin MDC9: intracellular maturation and catalytic avtivity. *J Biol Chem* **274**: 3531-3540.

Schwettmann, L., Tschesche, H., 2001, Cloning and expression in Pichia pastoris of metalloprotease domain of ADAM 9 catalytically active against fibronectin. *Protein Expr Purif.* **21**:65-70.

Shintani, Y., Higashiyama, S., Ohta, M., Hirabayashi, H., Yamamoto, S., Yoshimasu, T., Matsuda, H., Matsuura, N., 2004, Overexpression of ADAM9 in non-small cell lung cancer correlates with brain metastasis. *Cancer Res.* **64**:4190-4196.

Tannapfel, A., Anhalt, K., Hausermann, P., Sommerer, F., Benicke, M., Uhlmann, D., Witzigmann, H., Hauss, J., Wittekind, C., 2003, Identification of novel proteins associated with hepatocellular carcinomas using protein microarrays. *J Pathol.* **201**:238-249.

Weskamp, G., Kratzschmar, J., Reid, M.S., Blobel, C.P., 1996, MDC9, a widely expressed cellular disintegrin containing cytoplasmic SH3 ligand domains. *J Cell Biol* **132**: 717-726.

Weskamp, G., Cai, H., Brodie, T.A., Higashyama, S., Manova, K., Ludwig, T., Blobel, C.P., 2002, Mice lacking the metalloprotease-disintegrin MDC9 (ADAM9) have no evident major abnormalities during development or adult life. *Mol Cell Biol.* **22**:1537-1544.

Yagami-Hiromasa, T., Sato, T., Kurisaki, T., Kamijo, K., Nabeshima, Y., Fujisawa-Sehara, A., 1995, A metalloprotease-disintegrin participating in myoblast fusion. *Nature* **377**: 575-576.

Zhou, M., Graham, R., Russell, G., Croucher, P.I., 2001, MDC-9 (ADAM-9/Meltrin gamma) functions as an adhesion molecule by binding the alpha(v)beta(5) integrin. *Biochem Biophys Res Commun.* **280**:574-580.

# Chapter 5

# ADAM10

## *A major membrane protein ectodomain sheddase involved in regulated intramembrane proteolysis*

Paul Saftig[1] and Dieter Hartmann[2]
*[1]Biochemical Institute, Christians-Albrecht-Universität Kiel, Olshausenstr. 40, 24018 Kiel, Germany [2] Vlaams Interuniversitair Instituut voor Biotechnologie, VIB 4; and Department for Human Genetics, Katholieke Universiteit Leuven, Campus Gasthuisberg, Herestraat 49, B-3000 Leuven, Belgium*

Abstract: Our understanding of ADAM10 function has undergone a striking evolution during the past 10 years. After its independent identification in the bovine brain (then termed MADM) and invertebrates (sup-17 in *Caenorhabditis elegans* and kuzbanian in *Drosophila melanogaster*) it was first characterized as a key player in the Notch pathway in both flies and nematodes and as a candidate Amyloid Precursor Protein α-secretase in mammals. From that point on, the list of ADAM10 substrates expanded rapidly, and at present includes adhesion molecules (cadherins, CD44, L1), guideposts of cell migration and axon navigation (robo receptors, ephrins) and more recently key signaling factors of the immune system. An especially fascinating aspect of ADAM10 function is the high proportion of substrates that are further processed by I-Clips like γ-secretase in regulated intramembrane proteolysis (RIP), as if ADAM10 were the "RIPping ADAM". However, important putative functions of this protein still prove to be surprisingly tough to address. First, for several substrates like APP there appears to exist a considerable overlap and mutual compensation with other metalloproteases, which makes it difficult to define the in vivo relevance of α-secretase activity identified by overexpression in vitro, the more so as knockout models have so far failed to confirm this function. Second, essential results in ADAM10 research like Notch site 2 cleavage have been obtained in invertebrate models. Even if the underlying assumption that ADAM10 is the orthologue of sup-17 and kuzbanian has so far worked surprisingly well, there is some evidence that important functions of sup-17 or kuzbanian may be spread over several of the more numerous mammalian ADAMs. Finally, the structural model of ADAM10 clearly predicts additional functions as a disintegrin or a cell fusion protein.



*N. M. Hooper and U. Lendeckel (eds.), The ADAM Family of Proteases,* 85-121.

Key words: ADAM10, sup-17, kuzbanian, ectodomain shedding, regulated intramembrane proteolysis, Notch, APP, EGFR ligands, cadherins, chemokines

# 1. INTRODUCTION

Of the 33 disintegrin metalloproteases of the ADAM family that have been identified so far in the mammalian genome, only a few have attracted as much interest by scientists as ADAM10. This is mainly due to a number of both biologically and medically important substrates like the amyloid precursor protein (APP), but also to its key involvement in the regulated intramembrane proteolysis ('RIPping') of cell surface proteins. The early discovery of invertebrate orthologues like kuzbanian in *Drosophila* or sup-17 and the related mig-17 in *C. elegans* opened the road to a very fruitful genetic screening, to which we owe the identification of key functions like in the Notch pathway and in axonal navigation.

The picture of ADAM10 as reflected in the research literature has over the past years undergone a significant shift. Whereas the original "image" was dominated by its assumed role in the non-amyloidogenic α-secretase cleavage of APP in mammals and Notch site2 processing in invertebrate models, more recent work has led to the identification of a widespread set of substrates, ranging from adhesion molecules like CD44 or cadherins to a surprisingly large set of molecules with functions in immune defense control (Fig.1). It is thereby intriguing that many substrates of ADAM10 after ectodomain shedding undergo further regulated intramembrane cleavage resulting in the generation of a nucleus-targeted signal, as if ADAM10 were the 'RIPping sheddase' of the ADAM family. Finally, the concept of ADAM10 as a constitutive sheddase, which is regularly supplemented by ADAM17 as the PMA/PKCε inducible sheddase, has undergone considerable revision by the demonstration of a far-reaching inducability of ADAM10 as well and the constitutive shedding of some EGF-R ligands by ADAM17 (Sahin *et al.* 2004).

It is thereby not without irony that one of its assumed prime roles, APP α-secretase cleavage, now appears to be only weakly founded, with key evidence being solely based on overexpression experiments (Buxbaum *et al.* 1998, Koike *et al.* 1999, Lammich *et al.* 1999) and being strikingly contradicted by knockout data (Hartmann *et al.* 2002). It is likewise intriguing that other supposed functions of ADAM10 as deduced from its domain structure, i.e. in cell adhesion and cell fusion, linked to its disintegrin and cysteine-rich domains, have as yet not been identified.

Still, knockout experiments in all species investigated so far clearly show that ADAM10 and its orthologues are absolutely essential proteins, causing the most devastating phenotypes of all ADAMs that have so far been inactivated. Even nematodes without functional sup-17 barely survive into late embryogenesis (Wen *et al.* 1997), and both flies and mice die with severe malformations during early development (Pan *et al*, 1997, Sotillos *et al.* 1997, Hartmann *et al.* 2002).

*Figure 1.* Schematic view of ADAM10 mediated shedding of several type I membrane proteins.

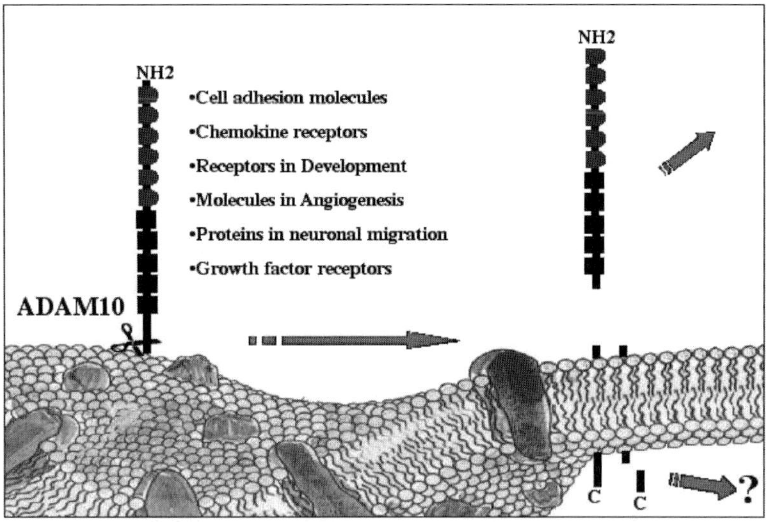


In the present chapter, we will summarize recent evidence from both mammalian and invertebrate *in vitro* and *in vivo* models on the currently established spectrum of ADAM10 functions. However, special caution should be applied when drawing together data from the different models reviewed here. First, the definition of mammalian ADAM10, nematode sup-17 and fly kuzbanian as orthologues now appears less straightforward than in the beginning, as the number of disintegrin metalloproteases in invertebrates is (still) rather limited as compared to the 33 mammalian proteins, of which even additional splice variants exist. Second, *in vitro* data based upon overexpression experiments or using dominant negative ADAM 10 variants in standard model cell lines have proven to be highly problematic when compared to data from knockout models, as ADAMs tend to broadly overlap with respect to substrates that they are able to cleave under *in vitro* conditions. Likewise, both knockout and RNA interference data have to be used with care, as the ADAM expression profile and thus the compensation capacity for the loss of one varies greatly between different cells and even may change due to immortalization or transformation of cells derived from knockout animals.

## 2. DISCOVERY OF ADAM10

The term "ADAM" was coined in a now classical paper by Wolfsberg *et al* in 1995 (Wolfsberg *et al.* 1995a, b) based upon common structural features of the first cloned testicular ADAMs with snake venom disintegrin proteases and the identification in the genome of guinea pigs and mice of four additional genes for proteins with similar structural features. Within this timeline, the cloning of ADAM10 (then still called MADM for **MA**mmalian **D**isintegrin **M**etalloprotease) by the group of Paul Glynn in 1996 (Howard *et al.* 1996) make it one of the longest known members of this family.

ADAM10/MADM was originally isolated in 1989 from bovine brain myelin preparations by the same group (Chantry *et al.* 1989) due to its ability to degrade myelin basic protein (MBP) in *in-vitro* preparations. For the sake of clarity, it should be pointed out that MBP as an intracellular cytosolic protein is most probably not a physiologic substrate of ADAM10, but becomes only accessible during tissue extraction (Chantry *et al.* 1989, Howard *et al.* 1996).

MADM / ADAM10 received its original designation due to its structural and sequence similarity to snake venom disintegrin metalloproteases, most notably jararhagin (Howard *et al.* 1996). Initial screening already indicated a low-level expression in numerous adult bovine tissues as well as several mammalian cell lines. The authors also noted another curious fact, the absence of several residues within the cysteine-rich and disintegrin domain conserved among other then known reprolysins, and hypothesized an early evolutionary divergence of this gene / enzyme from other metalloproteases.

Invertebrate orthologues of ADAM10 were likewise identified early on, first in a mutagenesis screen in *C. elegans* (sup-17) aimed at the identification of *sup*pressors of the phenotype caused by the constituively active Notch / lin-12(d) mutation (Tax *et al.* 1997, Wen *et al*, 1997), and then in *Drosophila* screening for neurodevelopmental genes, where kuzbanian surfaced both as a neurogenic gene and a key factor in axon development (Rooke *et al.* 1996, Pan and Rubin 1997, Sotillos *et al.* 1997). The human gene was mapped to 15q21.3-q23 in 1997 by Yamazaki *et al* (Yamazaki *et al.* 1997a) and the murine gene to chromosome 9 (Yamazaki *et al.* 1997b).

# 3. ESSENTIALS OF ADAM10 FUNCTION

## 3.1 Cell Biology of ADAM10

ADAM10 like the other ADAM proteins is a membrane-anchored protease that regulates cell behavior by proteolytically modifying the cell surface. ADAMs are characterized by a well defined domain structure, consisting of a N-terminal prodomain followed by a metalloproteinase domain, a disintegrin domain, a cysteine-rich domain, an EGF-like domain, a transmembrane and a cytoplasmic domain (Fig. 2) (Seals and Courtneidge 2003).

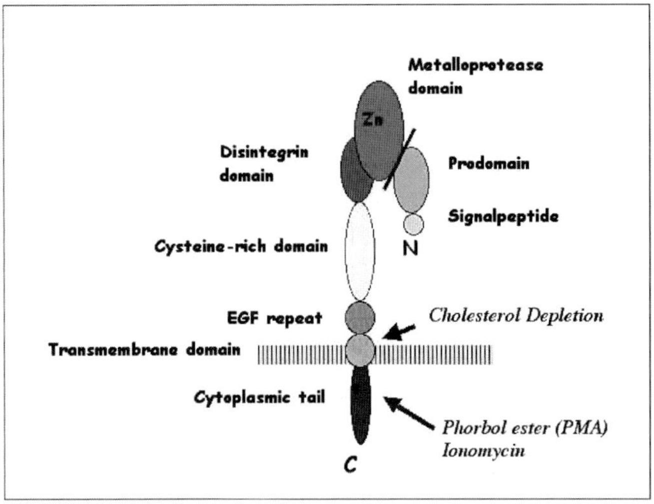


*Figure 2.* Postulated domain structure and sites of activation of ADAM10. Cholesterol depletion acts (probably) in the membrane plane. ADAM10 appears to be a non-raft enzyme, and one effect of cholesterol depletion may be to make substrates available to it by reducing raft domains.

ADAM10 like the other ADAMs is synthesized in the rough endoplasmic reticulum and is matured in a late Golgi compartment. The N-terminus of ADAM10 contains a signal sequence that directs it into the secretory pathway. Maturation of ADAM10 involves the removal of the prodomain from the ADAM 10 precursor protein, which may in part be due to the prohormone convertase PC7 (Endres *et al.* 2003). If the potential ADAM10 proprotein convertase recognition sequence is mutated no maturation and activation of ADAM10 is observed. Anders and co-workers suggested that the prodomain of ADAM10 acts as an intramolecular chaperone facilitating the correct protein folding and maintaining the proteinase in a latent state before getting activated by prohormone convertase cleavages (Anders *et al.* 2001).

By confocal microscopy ADAM10 was localized predominantly in the Golgi apparatus. After cell surface iodination ADAM10 could be immunoprecipitated from the lysate of breast tumor cells suggesting that the active and processed protease is located at the cell surface. Also subcellular localization analyses revealed a minor part of the mature ADAM10 in ER/plasma membrane-enriched fractions (Gutwein *et al.* 2003) suggesting the plasma membrane as the location of ADAM10 activity.

There is a remarkable conservation within the catalytic site among the various proteolytically active ADAM family members as well as some distinguishing structural features that may determine substrate specificity as well as the specificity for various proteinase inhibitors (Seals and Courtneidge 2003). Selective inhibitors for distinct metalloproteinases are not only of great use for the experimental analysis of ADAMs, but may have clinical value as well. In addition, natural tissue inhibitors of metalloproteinases (TIMPs) display differential activities for selected metalloproteinases of the ADAM family. For example, TIMP1 blocks ADAM10 but not ADAM17, TIMP-3 acts on ADAM17 and with lower effectivity on ADAM10, and TIMP2 affects neither protease, but e.g. MT1-MMP. However, the lack of effective methods for systemic delivery of these protein inhibitors has limited their utility in animal experiments and as therapeutic drugs in clinical trials.

Like other membrane-anchored disintegrin metalloproteases, ADAM10 contains a candidate "adhesive" domain downstream of its metalloprotease domains. The mechanism by which membrane-anchored cell surface proteases utilize these putative adhesive domains to regulate protease function *in vivo* is not well understood, and experimental evidence as summarized below is only available for other members of the ADAM family.

In a recent study Smith and co-workers analysed the relative contributions of downstream extracellular domains (disintegrin, cysteine rich, and EGF-like repeat) of the ADAM13 metalloprotease during *Xenopus laevis* development. When expressed in embryos, ADAM13 induces hyperplasia of the cement gland, whereas ADAM10 does not. Using chimeric constructs the metalloprotease domain of ADAM10 can substitute for that of ADAM13. However specificity for cement gland expansion requires a downstream extracellular domain of ADAM13. Analysis of the chimeras indicated an essential role for the cysteine-rich domain and a supporting role for the disintegrin domain (Smith *et al.* 2002). Cao and co-workers recently showed that the region extending from the disintegrin to the transmembrane domain of ADAM12 and containing cell adhesion activity as well as the cytoplasmic domain of ADAM12 are required for ADAM12-mediated cell cycle arrest of reserve cells during myoblast differentiation, while the metalloprotease domain is not essential (Cao *et al.* 2003). Interestingly, in a subpopulation of ADAM12 deficient mice an impaired formation of the neck and interscapular muscles was observed (Kurisaki *et al.* 2003) underlining the essential role of this protease in muscle development. An essential role of the disintegrin domain has been also shown for ADAM2 and 15, which bind to the $\alpha 9\beta 1$ integrin of oocytes. The ADAM2 and 15 disintegrin domains also effectively blocked binding of sperm to oocytes suggesting that ADAM integrin association is involved in sperm-egg interaction during fertilization (Eto *et al.* 2000).

Although not yet directly demonstrated, the disintegrin and EGF-like repeat domain of ADAM10 may be also of modulatory function for the proteolytic activity of this protease. The cytoplasmic tail of ADAM10 contains binding site motifs for SH3 domain-containing proteins that have been postulated to be involved in the inside-out regulation of metalloproteinase activity, the outside-in regulation of cell signaling and the control of maturation and subcellular localization (Seals and Courtneidge 2003).

## 3.2. Expression and regulation of ADAM10

Howard and co-workers found that ADAM10 is expressed in a wide variety of mammalian cell lines (Howard *et al.* 1996). Using *in situ* hybridization a widespread expression of ADAM10 mRNA was found among CNS neurons, while ADAM17 showed a more restricted pattern in neurons, but is present in brain endothelia and glia (Karkkainen *et al.* 2000). Using immunohistochemical techniques ADAM 10 was found in neurons of the perinatal cortex. During aging the intraneuronal staining intensity increased. Interestingly, in Alzheimer Disease brains ADAM 10 immunoreactivity was associated with diffuse and neuritic plaques. ADAM 10 immunoreactivity appeared to be decreased in neurons in Alzheimer Disease brains in comparison to controls (Bernstein *et al.* 2003) supporting the discussed role of ADAM 10 in preventing A$\beta$ generation.

In specialized tissues like bone, ADAM10 was found in periosteal cells, osteoblasts, and osteocytes at locations of active bone formation. Osteoclasts did not express ADAM10 (Dallas *et al.* 1999). Using RT-PCR and Northern blotting the expression of ADAM10 was also observed in cells derived from a variety of haematological malignancies including leukaemia, lymphoma and myeloma.

In general, ADAM - mediated shedding of diverse membrane proteins occurs both constitutively and in response to a wide variety of stimuli. These include serum factors, peptide growth factors, changes in the intracellular calcium concentration, osmotic and mechanical stress and protein kinase C activation (Fig.2). It is therefore likely that the activity of ADAM10 and also of other ADAMs is regulated by multiple and still not yet defined mechanisms (Pandiella and Massague 1991).

# 4. PUTATIVE SUBSTRATES OF ADAM10

## 4.1 Notch receptors

Notch receptors, originally discovered in *Drosophila*, are large (~ 300 kDa) heterodimeric receptor proteins of a type1 transmembrane architecture. Together with their ligands delta and serrate (Drosophila) or the orthologous delta-like and jagged families in mammals they play a key role in cell fate selection of differentiating stem cells during development and regeneration. The most drastic example so far has been found in *Drosophila* embryos during the specification of neural vs. epidermal precursor cells from the common ectodermal lineage. Loss of Notch function here results in a complete conversion of the ectoderm into neural precursors and embryonic lethality.

Notch receptors upon ligand binding are activated by a classical sequence of regulated intramembrane proteolysis, involving first an ectodomain shedding by a metalloprotease-like activity and then intramembrane cleavage by $\gamma$-secretase.

During cloning and initial analysis of kuzbanian, the *Drosophila* orthologue of ADAM10 (Rooke *et al.* 1996) a key role in partitioning neural vs ectodermal cells was identified, whereby kuzbanian (like the Notch receptor itself) was necessary for cells to receive a Notch signal and to inhibit neural fate selection. Further evidence along this line was provided by Sotillos *et al.* (1997), who showed by the use of kuzbanian genetic mosaics that this metalloprotease was required in cells for Notch signalling in a number of salient Notch-dependent developmental events, like neurogenesis, wing margin formation and wing vein patterning. Finally, it was shown that kuzbanian was not mandatory for signaling from a constitutively active Notch receptor, pinpointing the role of kuzbanian – mediated proteolysis to receptor *activation*. These results were paralleled by those of Pan and Rubin (Pan and Rubin 1997), who demonstrated that a dominant-negative kuzbanian could perturb cell fate selection even in *Xenopus*, indicating that metalloprotease cleavage of Notch is an evolutionary conserved mechanism (see also the review by Nye (Nye 1997). These data were further extended by Lieber *et al.* (2002) using *Drosophila* cells in vitro, where they confirmed a direct role for kuzbanian in Notch S2 cleavage.

In *C. elegans*, a similar relationship was found between its Notch orthologue lin-12 and the sup-17 gene (Tax *et al.* 1997, Wen *et al.* 1997) that likewise encodes for a metalloprotease highly similar to fly kuzbanian and vertebrate ADAM10. Sup-17 acts cell-autonomously to facilitate lin-12 signalling by acting on the lin-12 ectodomain, in line with the S2 cleavage mechanism proposed by Lieber *et al.* in *Drosophila*.

In striking contrast to these invertebrate data, it has proven to be surprisingly difficult to identify the Notch S2 protease in mammalian models. The initial analysis was dominated by two conflicting studies, both appearing in 2000 in the same volume of Molecular Cell. Mumm and coworkers (Mumm *et al.* 2000) showed that ligand binding to Notch facilitates shedding of the Notch receptor ectodomain by ADAM10, generating a transmembrane stub of Notch termed NEXT (**N**otch **EX**tracellular **T**runcation) which is then further cleaved by γ-secretase to release the signal transducing NICD (**N**otch **I**ntra**C**ellular **D**omain) fragment.

Brou *et al.* basically described the same mechanism of sequential proteolysis to activate Notch signal transduction after ligand binding, but identified ADAM17 / TACE as the responsible Notch S2 protease, using ADAM17 – deficient monocytic precursor cells derived from the bone marrow (Brou *et al.* 2000).

As in both cases (similar to the findings in *Drosophila* and *C. elegans*) Notch S2 cleavage was seen as essential for signal transduction, deficiency for the S2 protease should cause a Notch deficiency - like phenotype (as was shown for *Drosophila* and *C. elegans*, see below for details). It is thus striking that already before these two papers were published, Peschon *et al.* (1998) published the ADAM17 knockout mouse and reported basically a perinatally lethal phenotype related to a loss of TGF-α function, but without any traces of the much more severe loss of Notch signaling. Further supporting the opposing concept of ADAM10 being the key Notch S2 sheddase, such a 'Notch phenotype' was observed in the ADAM10 deficient mouse (Hartmann *et al.* 2002). However, with a view to reconcile the knockout mouse findings with the conflicting data pointing towards a role of also ADAM17 in Notch S2 cleavage, one can as yet not fully exclude that ADAM17 might play such a role in specific cell types (such as bone marrow stem cells), which could not be analyzed in ADAM10 knockout mice due to their early lethality (see also Boissy *et al.* 2003).

## 4.2 Notch ligands

Similar to their receptors, Notch ligands of both the delta/delta-like and the serrate/jagged families are type1 transmembrane proteins, and the original concept of Notch signaling therefore described an interaction limited to immediately neighbouring cells. It has thereby remained enigmatic how such a mechanism could result in two distinct populations of cells defined by either ligand or receptor expression/function when starting from an uniform population. Only recently evidence is accumulating that the Notch/Notch ligand system indeed creates *reciprocal* signals, modifying gene expression in both ligand- and receptor – bearing cells.

The first observation leading into this extended concept was published by Qi *et al.* (1999), again in the *Drosophila* system. They described the release of the delta ectodomain by kuzbanian - mediated proteolysis, which was then able to bind to Notch receptors, where it could act as an agonist. This *in-vitro* study basically transformed the concept of Notch signaling by adding a kind of 'long-range option' to it, based upon paracrine diffusion of Notch ligands.

However, about 3 years later the same group (Mishra-Gorur *et al.* 2002) published a more extensive study, in which they described several variants of the shed delta ectodomain, all of which appeared to be inactive, indicating that delta shedding by kuzbanian would be a mechanism to downregulate delta signaling on that cell, rendering it into a 'receptor – only' cell and thus contribute to the creation of asymmetry in an initially homogenous cell population.

This concept of Notch ligands contributing directly to reciprocal signaling significantly moved forward by the demonstration that in mammalian cells the delta orthologue delta-like 1 (dll-1) not only undergoes ectodomain shedding by ADAM10 as well as other metalloproteases, but also intramembrane cleavage by $\gamma$-secretase, whereby the dll intracellular domain is then partly transferred to the nucleus (Six *et al.* 2003), there probably engaging in transcription regulation. This RIPping of dll is increased upon coculture of dll-expressing cells with Notch – expressing cells, indicating that kuzbanian / ADAM10 plays an important role on both sides of a true bi-directional signaling pathway (Bland *et al.* 2003). Similar findings were reported by La Voie and Selkoe (LaVoie and Selkoe, 2003) for both dll and jagged, again noting stimulation by contact to Notch receptors and a nuclear translocation of both ligands intracellular domains, but focusing predominantly on an "ADAM17-like activity" as the responsible metalloprotease. With respect to signal transduction, jagged-ICD signaling appears to involve binding to AP1 and is counteracted by co-expression of the Notch ICD, pointing towards a direct antagonism of Notch and Notch ligand ICDs. It is striking thereby, that even at the level of Notch ligand processing, a similar discrepancy exists in the literature regarding ADAM10 or ADAM17 and their respective orthologues as the involved metalloproteases.

When summarizing the role of ADAM10 vs. other ADAM proteases in the Notch/Notch ligand signaling pathway, it appears clear from all available knockout data in both mice and invertebrates that ADAM10 is the key protease at least during development. Likewise, an even partial compensation by other ADAMs is highly unlikely, as the ADAM10 knockout mouse described further below virtually phenocopies a double knockout of both presenilins (Herreman *et al.* 1999), which completely interrupts Notch signaling (Herreman *et al.* 2000). A relevant contribution of e.g. ADAM17 may thus only come into play within specific cell types and/or at a later stage of development, as detailed above for bone marrow stem cells. A key step to clarify this issue would be the generation of conditional knockout mice for both ADAMs that would allow characterizing the impact on Notch signaling in a 'tissue-by-tissue' approach during different stages of development (for instance looking at hematopoiesis in ADAM17 – deficient mice).

At the moment, the situation is much less clear for Notch ligand shedding, which still awaits a more systematic analysis in ADAM10 and ADAM17 deficient cells. Likewise, it is an exiting question as yet to what degree a hypothetical isolated shedding defect of the Notch ligands would be able to fully or partially phenocopy a defective Notch receptor S2 cleavage.

## 4.3 Amyloid Precursor Protein (APP) family

Amyloid precursor protein (APP) is an ubiquitously expressed typeI transmembrane protein with putative functions in cell motility (Sabo *et al.* 2001, 2003) and intracellular (axonal) vesicular transport (Kamal *et al.* 2001, Sisodia 2002). Similar to Notch, APP is subject to a processing scheme related to RIPping, with the notable exception that the underlying physiologically relevant regulatory mechanisms (ligand binding or regulation of shedding activity via PKC) are not well understood (Kimberly *et al.* 2001). Two additional proteins, termed APLP1 (**A**myloid **P**recursor **L**ike **P**rotein 1) and APLP2 have been identified in the mammalian genome. AP(L)P orthologues appear to be highly conserved during evolution and have been identified in Drosophila (APL) and C. elegans (apl-1).

The *enzyme activities* responsible for APP cleavage have early on been designated as α- and γ-secretase, performing ectodomain shedding and intramembrane cleavage, respectively (Selkoe 2001). This releases a large ectodomain fragment termed APPsα, and then the transmembrane 'stub' is split into a small soluble p3 fragment (extracellular) and AICD, the **A**PP **I**ntra**C**ellular **D**omain that similar to the Notch NICD appears to serve nuclear signalling functions ( Fig. 3).

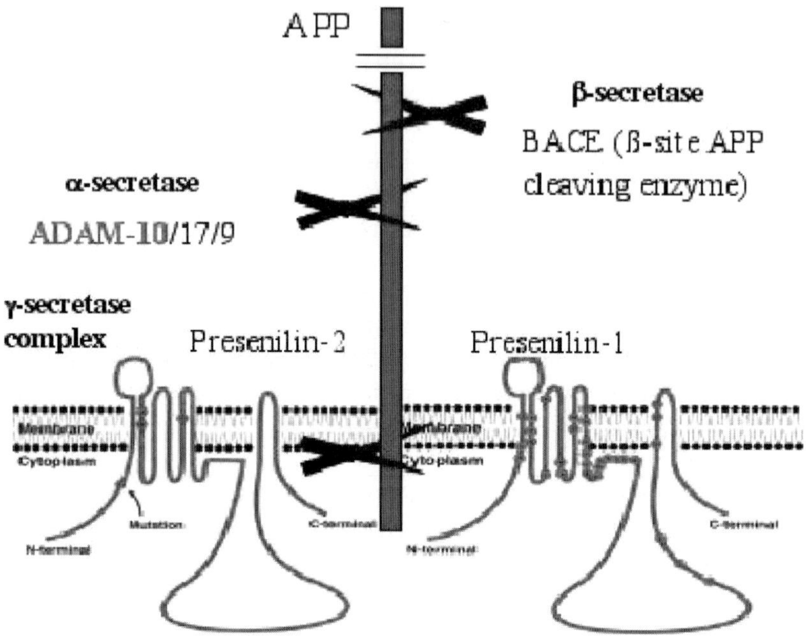


*Figure 3:* APP processing by β, α and γ secretases.

However, within the brain, an alternative, apparently unphysiologic ectodomain shedding at a further N-terminal position by β-secretase leads to the release of the 37 to 42 amino acid long Aβ peptide instead of the p3 fragment, which is now accepted as the key pathogen responsible for the development of Alzheimer's disease. Whereas β-secretase does not play a role in APP cleavage outside of the brain, its approximately 10 % contribution within the brain is sufficient to cause a slow pileup of Aβ peptides that accelerates from a critical threshold during the 6th decade of life, then causing a rapidly increasing incidence of AD (up to 40 % of the population in their 9th decade).

It is not without irony, however, that the intense research during the past three decades that has led to the unequivocal identification of β- and γ-secretase, has failed so far to clearly identify α-secretase, the longest known activity cleaving APP.

Experiments based upon inhibitor profiling were performed during the '90s and have pointed towards a metalloprotease – based activity performing this task. At that point, three different labs have turned towards ADAM proteases, not the least due to their characteristic ability to cleave transmembrane proteins close to their outer membrane surface (Buxbaum *et al.* 1998, Koike *et al.* 1999, Lammich *et al.* 1999).

ADAMs 9, 10 and 17 were investigated in *in-vitro* overexpression studies and were shown to cleave APP at the α-site. Thereby, constitutive and inducible cleavage of APP appeared to be asymmetrically spread over different enzymes, with e.g. ADAM10 contributing to both, and ADAM17 only to PMA/PKC inducible shedding (Slack *et al.* 2001). Strikingly, the next and essential step to clearly identify "THE" α-secretase by investigation of APP cleavage in knockout organisms has basically failed for all of them. Deficiency for ADAM9 (Weskamp *et al.* 2002) did not modify at all APP cleavage in the mouse brain or in derived cultures, and in embryonic fibroblasts (the only cell type that could be derived from the early lethal ADAM10 deficient mice) APP processing was normal in the majority of the cell lines (Hartmann *et al.* 2002). Only for fibroblasts derived from ADAM17 deficient mice, a loss of the PKC- inducible fraction of α-secretase activity was reported (Buxbaum *et al.* 1998). However, in contrast to ADAMs 9 and 10, both of which are highly expressed in neurons throughout the CNS, ADAM17 is mainly present in endothelia and glia, but only in selected neuronal cell types in the brain, which makes a key role as an important α-secretase less likely (Karkkainen *et al.* 2000). Regrettably, the key experiment in this regard, i.e. analysis of APP cleavage in ADAM17 - deficient neurons, was never published. This gap in our knowledge is even more deplorable as such cells could have been easily derived from these mice, which die only perinatally.

Still, the conflicting results of ADAM10 (like ADAMs 9 and –17) being able to correctly cleave APP, but apparently not being necessary in knockout mouse models are stunning enough. It has to be first kept in mind that most of the original data were obtained in non-neuronal cell types, which most probably express a different profile of metalloproteases than neurons. Then, as also neurons express both ADAM9 *and* –10, a mutual compensation may obscure the physiologic contributions of both enzymes in single knockout models.

This concept of α-secretase being a protease 'team' (Hartmann *et al.* 2002, Hooper and Turner 2002) has recently been adressed by siRNA experiments. Thus, (Asai *et al.* 2003) could reduce α-cleavage in glioblastoma cells by downregulating any of the three candidates. Comparable results were obtained by (Allinson *et al.* 2003) by siRNA experiments knocking down ADAMs 10 and 17 in SH-SY5Y cells.

In a recent study by the group of F. Fahrenholz (Postina *et al.* 2004), the approach of overexpressing ADAM10 to modulate APP cleavage published earlier by the group (Lammich *et al.* 1999) was repeated *in vivo* to investigate its effect on plaque pathology in APP V717I mice. As expected, overexpression of ADAM reduced the extent of AD-like brain pathology in these mice, indicating that ADAM10 can cleave APP also under *in-vivo* conditions. Potentially more interesting, overexpression of an inactive ADAM10 could increase the Aβ load, possibly due to a dominant negative effect on endogenous ADAM10. If confirmed, this could indeed define for the first time a physiologic role of ADAM10 as one of the α-secretases.

Such a confirmation of ADAM10 as one of the *in vivo* relevant α-secretases would also be crucial to correctly evaluate the slowly mounting data on its regulation and function during AD pathogenesis and therapy. Thus, *in vitro* data indicate that ADAM10 activity towards APP can be increased by reducing cholesterol concentration (Scalia *et al.* 2001, Kojro *et al.* 2001, Buxbaum *et al.* 2002, Parvathy *et al.* 2004), which could offer one (still debated) hypothesis for the effects of statins to reduce the incidence of AD in humans (see also Ehehalt *et al.* 2003, Ledesma *et al.* 2003 and especially Abad-Rodriguez *et al.*, in press). Likewise, increasing PKC activity that is known to stimulate ADAM proteases could serve as a partial explanation of the mild positive effect of cholinergic agents in AD therapy (Pakaski and Kasa 2003, Fisher *et al.* 2003, Narahashi *et al.* 2003, Lanni *et al.* 2004).

Several attempts have been made to investigate both α-secretase activity and the expression of especially ADAM10 in the aging normal and AD brain as well as in rodents. Whereas on the mRNA level no consistent changes were reported, both ADAM10 protein expression and α-secretase activity may be reduced both in the rodent brain during aging and more specifically in human AD brains (Colciaghi *et al.* 2002, Gatta *et al.* 2002). However, given the overall paucity of data, the considerable variation of models used and the lack of controls that include other candidate α-secretases, it appears way too early to draw major conclusions on this issue.

In summary, even if ADAM10 is still an often-documented α-secretase candidate, a definitive proof is lacking, as it is the case for the other candidates. On the other hand, the preservation of APP α-cleavage in ADAM10 knockout MEFs clearly shows that one or more alternative α-secretases are still out there, which makes it fairly unlikely that ADAM10 acts alone in the brain.

## 4.4 Ephrins

Ephrins and their receptor tyrosine kinase Eph receptors form a bi-directional signaling system of membrane proteins active in tissue compartmentalization events such as somitogenesis, as well as in angiogenesis and axon guidance. The ephrin ligands fall into two classes, 'A' ephrins being anchored by a GPI anchor, and 'B' ephrins by a classical transmembrane domain.

Based upon earlier observations indicating that kuzbanian plays a role in Drosophila axon outgrowth and patterning (Hattori *et al.* 2000) analyzed the dynamics of key axon repellants, A-class ephrins and their receptors upon transfection into Neuro2A cells, non-neuronal cell lines and primary murine neuron cultures. The apparent paradox of a high-affinity receptor/ligand system subserving a rapid contact-repellant function was first resolved by the observation that receptor binding caused a rapid release of the ephrin-A2 ectodomain from the cell membrane by cleavage within its peptide chain, not its GPI anchor. They further showed that the cleavage was inhibited by overexpression of a dominant-negative ADAM10 lacking the protease domain, but could be activated by overexpression of a full-length construct.

This mechanism to turn "attraction into repulsion" (Pasquale 2000) is not exclusive to the development of axonal connections in the nervous system. B-class ephrins, i.e. ephrins with a transmembrane protein configuration, play key roles in endothelial cell migration during vascular development, and share sequence motifs with A-class ephrins that may serve as a metalloprotease recognition site (Hattori *et al.* 2000). As these proteins may be important targets for e.g. antiangiogenic cancer therapy, further analysis of a possible regulated processing by disintegrin metalloproteases will certainly be a rewarding topic.

## 4.5 Robo/Slit

Slit proteins and their receptors of the roundabout (robo) family form a second repulsive axon guidance system originally discovered in Drosophila. Slit proteins are secreted by midline glia cells in Drosophila and inhibit axonal midline crossing. Mutations of both *leak*, the gene encoding for the robo-2 receptor, and kuzbanian cause an abnormal midline crossing. The latter thereby inhibits the clearance of the receptor from axonal membranes, pointing towards a direct function of kuzbanian as a robo sheddase (Schimmelpfeng *et al.* 2001, McFarlane 2003).

Axon guidance by slit proteins and their receptors is highly conserved during evolution, and in mammals has even taken over additional functions, including the control of neuronal and again endothelial migration.

Taken together with the data on Notch cleavage by ADAM10, it is intriguing that ADAM10 is essential for the correct function of at least two additional signaling pathways (based upon ephrins and slit proteins) that control blood vessel biology. Further research will be needed to clearly define how the severe vasculogenesis defect caused by ADAM10 deficiency (Hartmann *et al.* 2002) is created and what the potential therapeutical role of ADAM10, e.g. for antiangiogenesis, could be in the future.

## 4.6 CD44

CD44 is a major cell surface receptor for several extracellular matrix components including hyaluronan and is implicated in a wide variety of biological processes including tumor metastasis and invasion (Thomas *et al.* 1992, Gunthert *et al.* 1996). Several groups reported that CD44 is susceptible to sequential proteolytic cleavages within its ectodomain and intramembranous domain (Kajita *et al.* 2001, Okamoto *et al.* 2001, Murakami *et al.* 2003). The release of the CD44 soluble ectodomain is mediated by membrane-associated metalloproteinases. CD44 cleavage plays a crucial role in CD44-mediated tumor cell migration. The cleavage event is regulated by multiple pathways, such as the activation of protein kinase C, extracellular $Ca^{2+}$ influx, hyaluronan oligosaccharides, the Ras oncogene, and Rho family small GTPases (Kawano *et al.* 2000, Okamoto *et al.* 2001, Sugahara *et al.* 2003). Kajita and coworkers (Kajita *et al.* 2001) reported that in migratory cells including invasive tumor cells CD44 forms a complex with the membrane-type 1 MMP, which is able to cleave within the ectodomain of CD44. However, recently two groups have demonstrated conclusively that $Ca^{2+}$ influx as well as hyaluronan fragment-induced CD44 cleavage is independent of MT1-MMP processing but is mediated through ADAM10 (Murai *et al.* 2004, Nagano *et al.* 2004).

Evidence for ADAM10 as the central sheddase for CD44 and as a mediator of tumor cell migration has accumulated by RNA interference and metalloproteinase inhibitor experiments after induction of cleavage using an anti-CD44 monoclonal antibody (Murai *et al.* 2004). Depletion of ADAM10 in mouse embryonic fibroblasts and downregulation through siRNA inhibited the $Ca^{2+}$ influx-induced cell detachment from matrix in U251MG cells (Nagano *et al.* 2004). $Ca^{2+}$ influx activated ADAM10 by regulating the association between calmodulin and ADAM10, leading to CD44 ectodomain cleavage. Nagano and coworkers also showed in the same study that phorbol ester stimulation activated ADAM17 through the activation of PKC and the small GTPase Rac, inducing proteolysis of CD44. ADAM10 as well as ADAM17 contributed to CD44-dependent cancer cell migration on hyaluronic acid substrates but they are differentially regulated in response to

$Ca^{2+}$ influx and PKC activation. It is tempting to speculate that the existence of independent-differential pathways for CD44 cleavage may be of benefit for the development of specific anti-cancer drugs with little side effects on e.g. wound healing, lymphocyte homing and cell proliferation.

## 4.7 EGF

When epidermal growth factor (EGF) and its relatives (e.g. heparin-binding (HB)-EGF, TGFα, betacellulin, amphiregulin, epiregulin and epigen) bind their receptors (epidermal growth factor receptor EGFR), they trigger a rich network of signalling pathways, culminating in responses ranging from cell division to death, motility to adhesion (Yarden and Sliwkowski 2001). Ligands of the EGFR comprise a family of functionally and structurally related integral membrane proteins that can be proteolytically processed and released from cells (Harris *et al.* 2003). Under a variety of circumstances a metalloproteinase activity is needed for activation of EGFR signaling. Metalloproteinase inhibitor studies revealed the importance of these shedding events in several physiological and pathophysiological processes such as migration of a mammary epithelial cell line (Dong *et al.* 1999), activation of TGFα during mouse development (Peschon *et al.* 1998), in transactivation of EGFR by G protein-coupled receptors (Lemjabbar and Basbaum 2002), in wound healing and heart development (Yamazaki *et al.* 2003).

First hints that ADAM metalloproteinases are involved in EGFR-ligand shedding emerged from the similarity of the perinatal lethal phenotype of ADAM17 deficient mice and mice deficient for TGFα, HB-EGF and the EGF-receptor (Mann *et al.* 1993, Miettinen *et al.* 1995, Jackson *et al.* 2003). Using deficient MEF cells ADAM17 could be directly implicated in the shedding of TGFα, HB-EGF and amphiregulin (Peschon *et al.* 1998, Merlos-Suarez *et al.* 2001, Sunnarborg *et al.* 2002). Yan and colleagues (Yan *et al.* 2002) demonstrated that Kuzbanian, the ADAM10 orthologue in *Drosophila* is able to stimulate G protein coupled receptor dependent transactivation of EGFR. Transfection of a protease domain-deleted Kuzbanian, or blocking endogenous KUZ by antisense oligonucleotides, suppressed the transactivation (Yan *et al.* 2002). Convincing evidence that apart from ADAM17 also ADAM10 is critically involved in the shedding of EGFR-ligands has emerged recently by a systematic genetic approach using mouse embryonic cells lacking candidate-releasing enzymes (ADAM 9, 10, 12, 15, 17, and 19). Interestingly, ADAM10 emerged as the main protease involved in the constitutive shedding of EGF and betacellulin, and ADAM17 as the major PMA-stimulated and constitutive membrane sheddase of epiregulin, TGFα, amphiregulin, and HB-EGF in these cells (Sahin *et al.*

2004). Despite their structural similarity it will be now interesting to unravel the specific protein domains which are essential both in mediating substrate specificity and regulation of ADAM17 and ADAM10. Inhibitors for both proteases have been discussed to be attractive drug targets that may modulate the functional activation and signaling of EGFR-ligands and therefore could be functional in prevention of certain forms of cancer.

Interestingly, the cleavage of pro heparin-binding epidermal growth factor and activation of the epidermal growth factor receptor by ADAM10 was recently linked to the lung pathology in cystic fibrosis patients. ADAM10 is apparently activated under pathological conditions by bacterial lipoteichoic acid and the platelet-activating factor receptor. ADAM10-dependent EGF receptor activation may be linked to the overproduction of mucus thereby leading to morbidity and mortality by obstructing airflow and shielding bacteria from antibiotics (Lemjabbar and Basbaum 2002).

## 4.8 Interleukin 6 receptor

Interleukin-6 (IL-6) is a pleiotropic cytokine that plays a major role in a variety of human diseases. IL-6 activates cells by binding to the membrane-bound IL-6 receptor (IL-6R) and subsequent formation of a glycoprotein 130 homodimer. Homodimerization of the gp130 triggers activation of several intracellular signaling pathways including the Janus kinase/STAT, Ras/mitogen-activated protein kinase and phospatidylinositol 3-kinase pathways (Heinrich *et al.* 2003). A soluble form (sIL-6R) of the IL-6R is able to bind IL-6, and the complex of IL-6 and the sIL-6R activates target cells expressing gp130 in a process termed "trans signaling" (Rose-John and Heinrich 1994). There are increasing evidences that the soluble form of the IL-6R plays a major role in the observed pathophysiological effects of IL-6 like in chronic intestinal inflammation.

The soluble form of the IL-6 receptor is generated either by alternative splicing or by proteolytic release of the ectodomain of the membrane-bound IL-6R. Activation of protein kinase C can trigger the shedding of the IL-6R and it can be inhibited by hydroxamic acid-based inhibitors indicating the involvement of one or more metalloproteinases in this ectodomain shedding event (Mullberg *et al.* 1995).

Matthews and coworkers (Matthews *et al.* 2003) showed recently that cholesterol depletion of cell membranes with methyl-β-cyclodextrin increased IL-6R shedding. This effect was independent of protein kinase C activation. Shedding of the IL-6R caused by cholesterol depletion is highly dependent on the metalloproteinases ADAM17 and ADAM10. Compared with ADAM10 deficient MEF cells, wild type fibroblasts and ADAM10-transfected ADAM10$^{-/-}$ fibroblasts showed stronger constitutive and induced shedding of the IL-6 receptor. It is known that cholesterol depletion causes disruption of lipid rafts and an increase of the fluidity of the plasma membrane. It may be possible that increased IL-6R shedding by ADAM17 and ADAM10 in cholesterol-depleted cells is due to an increased accessibility of the protease to the receptor cleavage site. Alternatively, substrate and enzyme association may be facilitated by the different lipid phase composition. Cholesterol lowering drugs are discussed to be effective in the prevention of atherosclerosis and cardiovascular diseases (Group 2002, 2002). HMG-CoA reductase inhibitors also exert direct anti-inflammatory effects (Scalia *et al.* 2001). In light of the recent observations (Matthews *et al.* 2003) it is conceivable that the anti-inflammatory response after cholesterol depletion is partially mediated by the activities of ADAM10 and ADAM17.

## 4.9 TNFα

Tumor necrosis factor-α (TNFα), a potent pro-inflammatory cytokine, is released from cells by proteolytic cleavage of a membrane-anchored precursor. Most organs of the body appear to be affected by TNFα, and the cytokine serves a variety of functions, many of which are not yet fully understood. It possesses both growth stimulating and growth inhibitory properties, and it appears to have self-regulatory properties as well. The pathological activities of TNFα have attracted much attention. For instance, although TNFα causes necrosis of some types of tumors, it promotes the growth of other types of tumor cells. High levels of TNFα correlate with increased risk of mortality. TNFα participates in both inflammatory disorders of inflammatory and non-inflammatory origin. It has also been shown to have a critical role in autoimmune disorders such as rheumatoid arthritis and Crohn´s disease (Vassalli 1992). TNFα exhibits chronic effects as well as resulting in acute pathologies. If TNFα remains in the body for a long time, it loses its anti-tumor activity. This can occur due to shedding of TNF receptors by tumor cells. Prolonged overproduction of TNFα also results in a condition known as cachexia, which is characterized by anorexia, net catabolism, weight loss and anemia and which occurs in illnesses such as cancer and AIDS.

ADAM17 is considered to be a major sheddase for TNFα (Moss *et al.* 1997) since ADAM17 knockout mice showed a defect in TNFα shedding in T cells (Black *et al.* 1997). Biochemical studies in different laboratories revealed that in addition to ADAM17 additional candidate TNFα convertases exist. Purified ADAMs 9, 10 and 19 can also cleave a peptide corresponding to the TNFα cleavage site *in vitro*. Lunn and coworkers used purified ADAM10 and found that it has the ability to cleave TNFα (Lunn *et al.* 1997). Another group (Condon *et al.* 2001) used antisense oligonucleotides targeting both ADAM10 and ADAM17. The ADAM17 antisense oligonucleotides inhibited TNFα secretion in THP-1 cells, whereas the ADAM10 antisense oligonucleotides did not significantly inhibit release of TNFα. Recently, a study which used cells lacking ADAMs 9, 10, 17 (TACE) confirmed the central role of ADAM17 as the TNFα-convertase (Condon *et al.* 2001). The cleavage site of TNFα is necessary and sufficient for cleavage by ADAM17. Interestingly, the selective stimulated processing of TNFα by ADAM17 in cells depended on the presence of an appropriate cleavage site as well as the inhibitory role of the TNF-ectodomain towards other enzymes that can process this site (Condon *et al.* 2001).

## 4.10 CX3C L1

Leukocyte recruitment to inflammatory sites involves a sequence of adhesive events that are mediated by different classes of adhesion molecules expressed on the endothelium and the leukocytes. Within the chemokine family, which are known to modulate the cell adhesion between both cell types, the C3C chemokine ligand (CX3CL1) or fractalkine has been identified to induce cell adhesion (Bazan *et al.* 1997). CX3CL1 exists as a membrane-expressed protein and as a soluble molecule inducing chemotaxis. Transmembrane CX3CL1 is converted into its soluble form by defined proteolytic cleavage (shedding), which can be enhanced by stimulation with PMA. Whereas PMA-induced CX3CL1 shedding has been shown to involve ADAM17 (Garton *et al.* 2001), the constitutive cleavage is mediated by ADAM10 (Hundhausen *et al.* 2003). Overexpression of ADAM10 in COS-7 cells enhanced constitutive cleavage of CX3CL1 and in murine fibroblasts deficient of ADAM10 constitutive CX3CL1 cleavage was markedly reduced. Hundhausen and coworkers found that THP-1 cells firmly adhere to CX3CL1 expressing ECV-304 cells. Detachment of the cells could be achieved by vigorous washing. Interestingly, inhibition of ADAM10-mediated CX3CL1 shedding not only increased adhesive properties of CX3CL1-ECV-304 cells but also prevented de-adhesion of bound THP-1 cells. ADAM10 may therefore be an important regulator of the recruitment of monocytic cells to CX3CL1-expressing cell layers. Within inflamed vascular tissue CX3CL1 is upregulated. ADAM10 is also expressed in this tissue. It may be possible that the balance between gene expression of CX3CL1 and its removal by ADAM10 from the vascular cell surface is crucial for the CX3CL1 mediated leukocyte adhesion. Following selectin-mediated rolling CX3CL1 and integrin mediated adhesion may allow more firm contact of leucocytes to the vascular tissue. Therefore transmigration of leucocytes through the vascular wall into the underlying tissue would depend on the CX3CL1-cleaving metalloproteinases ADAM10 and ADAM17 (Fig. 4).

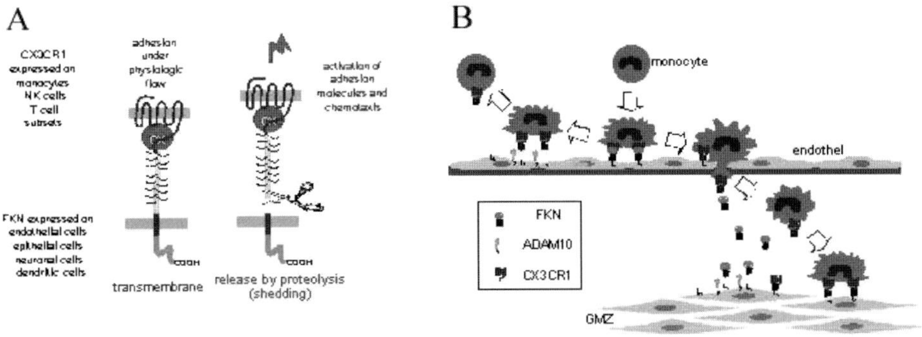


*Figure 4.* ADAM10 and its role in CX3CR1 shedding A. Transmembrane CX3CL1 (FKN, fractalkine) mediates monocyte adhesion while ADAM10 mediates cleavages of the molecule leading to the detachment of bound cells. B. Model of Adhesion and Transmigration of leucocytes according to Hudhausen et al. 2003 mediated through the vascular wall to sites of infection depends partially on the shedding activity of ADAM10.

## 4.11 CXCL16

Apart from CX3CL1 there exists a second transmembrane chemokine within the chemokine family. CXCL16 is synthesized as a transmembrane protein with its chemokine domain at the end of a mucin-rich stalk. When expressed at the cell surface, CXCL16 functions as a scavenger receptor, binding and internalizing oxidized low-density lipoprotein and bacteria (Shimaoka *et al.* 2000). As a soluble form, CXCL16 is a chemoattractant for activated $CD4^+$ and $CD8^+$ T cells through binding its receptor, CXCR6 (Matloubian *et al.* 2000). CXCL16 expressed in fibroblasts and vascular cells is converted from a transmembrane molecule into a soluble chemokine. Using inhibitors of metalloproteinases with defined potency for distinct members of the disintigrin-like metalloproteinase family and with MEF cells from ADAM10 deficient mice several lines of evidences were provided that ADAM10 is relevant for the majority of the constitutive, but not for the PMA-inducible, cleavage of the murine or human CXCL16. The constitutive shedding of CXCL16 from the endothelial cell surface is blocked by inhibitors of ADAM10 and is independent of additional inhibition of ADAM17. It appears that during inflammation in the vasculature, ADAM10 may act as a CXCL16 sheddase thereby finely controling the expression and function of CXCL16 in the inflamed tissue (Abel *et al.* 2004). Very similar findings were made in a separate study by Gough and colleagues using short interfering RNAs and a dominant negative ADAM10 mutant to demonstrate the essential role of ADAM10 in the constitutive shedding of CXCL16 in a murine macrophage cell line (Gough *et al.* 2004).

## 4.12 Prion protein

Spongiform encephalopathies are neurodegenerative diseases that are characterized by the cerebral deposition of a 33-35 kDa protein called prion protein (PRUSINER 1998). When normal prion protein ($PrP^C$) is converted into an insoluble and protease-resistant form called $PrP^{Sc}$ prion-associated pathology may occur. $PrP^C$ contains an amyloidogenic, neurotoxic sequence that is essential for conversion into $PrP^{Sc}$, the pathological isoform. During normal processing, $PrP^C$ is cleaved at a site within this sequence, and this cleavage is thought to destroy the amyloidogenic potential of the protein.

This cleavage can occur in a constitutive and phorbol ester-regulated manner. Vincent and co-workers showed that o-phenanthroline, a general zinc-metalloprotease inhibitor, as well as BB3103 and TAPI reduced the cleavage inside the 106-126 toxic domain of the protein. After overexpression of ADAM10 and ADAM17 the authors observed that ADAM10 contributed to constitutive prion cleavage whereas ADAM17 participated in regulated proteolysis. In fibroblasts deficient for ADAM10 the constitutive cleavage was severely reduced. Phorbol 12,13-dibutyrate-regulated prion proteolysis is fully abolished in TACE-deficient cells. ADAM10 and ADAM17 are therefore major protease candidates responsible for normal cleavage of $PrP^C$. These disintegrins could be putative cellular targets of a therapeutic strategy aimed at increasing normal $PrP^C$ breakdown and thereby depleting cells of the putative 106-126 "toxic" domain of $PrP^C$ (Vincent *et al.* 2001).

The role of ADAM10 in PrP proteolysis prompted an analysis of single nucleotide polymorphisms (SNPs) within ADAM10 in patients with sporadic or variant CJD. However, no relationship between genetic modification of ADAM10 and an increased risk for the disease was found indicating that ADAM10 is unlikely to confer genetic susceptibility to CJD (Budagian *et al.* 2004).

## 4.13 L1

The L1 adhesion molecule plays an important role in neuron-neuron adhesion, neurite fasciculation, synaptogenesis, neurite outgrowth and neuronal cell migration (Schachner 1997). L1 is also expressed by many human carcinomas including ovarian und uterine carcinomas suggesting a potential role of L1 in other adhesion and migration events. L1 is a 200-220 kDa type I membrane glycoprotein. L1 can undergo homophilic L1-L1 binding but it can also be released as a soluble molecule from mouse cerebellar cells in culture (Sadoul *et al.* 1988) and from mouse and human tumor cells (Gutwein *et al.* 2000). Mechtersheimer and colleges reported that the membrane-proximal cleavage of L1 involves ADAM10, since transfection with dominant-negative ADAM10 completely abolished L1 release. In L1 expressing CHO cells the authors observed a metalloproteinase activity dependent haptotactic migration on fibronectin and laminin (Mechtersheimer *et al.* 2001). Release of L1 was enhanced by phorbol ester and pervanadate and by depletion of cellular cholesterol. The constitutive cleavage and PMA incuced shedding of L1 occurs at the cell surface but treatment with cyclodextrin or pervanadate led to a release of microvesicles containing full-length L1 and the active form of ADAM10 (Gutwein *et al.* 2003). Depending on the localization of ADAM10 it appears that there are at least two pathways of L1 cleavage. In the Golgi apparatus the secretory vesicles are generated, whereas there is also direct cleavage at the cell surface (Gutwein *et al.* 2003). Interestingly, soluble L1 was detected in the majority of serum samples from patients with ovarian and uterine tumours and is therefore discussed to serve as a useful diagnostic marker (Fogel *et al.* 2003).

Despite the proposed role of ADAM10 as the major L1 sheddase which is mainly based on *in vitro* and overexpression studies the membrane shedding of L1 was not affected in ADAM10 deficient MEFs (our unpublished data) making it likely that other proteases can also contribute to L1 shedding or may be able to compensate for the loss of ADAM10 in the knockout cells.

## 4.14 Cadherins

Cadherins belong to a family of intercellular adhesion molecules that mediate calcium-dependent cell-cell adhesion through homophilic interaction. Cadherins are essential for cell recognition and tissue morphogenesis as well as for the maintenance of solid tissue (Takeichi 1991, Gumbiner 1996). Like the other classical type I cadherins, E-, P- and N-cadherin contain a large N-terminal extracellular region. The conserved cytoplasmic domains of cadherins interact with β-catenin, which in turn, binds to α-catenin and the actin cytoskeleton. The release of the extracellular domain, which contains the homophilic binding sites, is functionally of major importance for the regulation of cell adhesion, cell migration and neurite outgrowth (Paradies and Grunwald 1993). This ectodomain cleavage of N-cadherin can be inhibited with MMP inhibitors including the tissue inhibitors of metalloproteinases (TIMPs). The proteolytic ectodomain cleavage leads to a membrane bound carboxyterminal fragment which is a substrate for regulated intramembrane proteolysis (RIP) (Marambaud *et al.* 2003). We could recently demonstrate that ADAM10 is the major proteinase responsible for constitutive and regulated N-cadherin ectodomain cleavage in fibroblasts and neuronal cells. Interestingly, ADAM10 directly regulates the overall levels of N-cadherin expression at the cell surface. The ADAM10 induced N-cadherin cleavage resulted in changes in the adhesive behaviour of cells, but also in a dramatic redistribution of β-catenin from the cell surface to the cytoplasmic pool thereby influencing the expression of β-catenin downstream genes. ADAM10 is likely to play a crucial role in the regulation of cell-cell adhesion and on β-catenin signalling, thereby establishing the couple ADAM10/N-cadherin as a central switch in mediating signals from the extracellular matrix to the nucleus (P. Saftig and K. Reiss, unpublished data).

ADAM10 is also involved in the shedding of E-cadherin (our unpublished results) further increasing the importance of this protease in regulating cell adhesion in different tissues.

## 4.15 Possible further roles of ADAM10

Northern blot and RT-PCR experiments demonstrated ADAM10, ADAM12 and ADAM15 expression in resting chondrocytes derived from articular femoral head and a chondrocyte cell line. Whereas only one main transcript was found for ADAM-12 both ADAM-10 and ADAM-15 had multiple transcripts indicating possible RNA variants potentially derived from alternative splicing. ADAM-metalloproteinases including ADAM10 may be involved in joint pathology (McKie *et al.* 1997).

ADAM10 mediated ectodomain shedding may also be of relevance for gastric carcinogenesis. RT-PCR analysis of gastric biopsy specimens from patients with and without *H. pylori* showed that infection was associated with increased expression of ADAM 10 and ADAM 17. No increases in ADAM 15 and ADAM 20 were observed. High levels of ADAM 10, ADAM 17, and ADAM 20 transcripts were present in gastric carcinoma (Yoshimura *et al.* 2002).

# 5. THE ADAM10 KNOCKOUT PHENOTYPE

## 5.1 C. elegans

Studies on the effects of mutations and especially null alleles of sup-17 have focused on the interaction between sup-17 and the nematode Notch orthologue lin-12. Accordingly, two cell fate decisions in the developing gonad have received most of the attention, the first being between anchor cell (AC) and ventral uterine precursor cell (VUC) of the hermaphrodite gonad. The cells arise by lin-12 dependent lateral specification from 2 precursors termed Z1.ppp and Z4.aaa, whereby both cells have statistically the same chance to adopt either fate. Impaired lin-12 function causes both Z1.ppp and Z4.aaa cells to adopt an AC fate, and conversely, increased lin-12 function converts both into VUCs.

A second decision occurs during vulva development and controls the development of the vulval precursor cells (VPC). Normally only three of the 6 VPCs (P5.p, P6.p and P7.p) contribute to vulva formation, whereas the others will eventually join the hypodermal syncytium. Here, elevating lin-12 signalling will cause all cells to adopt the "2°" fate, which is one part of the vulval development pathway, whereas impaired lin-12 function will inhibit 2° fate selection.

For both pathways taken together, a pathologically constitutive lin-12 activity as in lin-12(d) mutations results in a distinctive phenotype characterized by an egg laying defect (Egl) caused by the absence of the AC and a multivulva (Muv) as the progeny of *each* VPC self-assembles into a pseudovulva at the ventral nematode surface.

Sup-17 was originally identified by loss-of-function mutations that suppress the development of the phenotype associated with lin-12(d) mutations (Tax *et al.* 1997, Wen *et al.* 1997). Interestingly, inactivation of sup-17 in wild-type nematodes the AC vs VUC decision occurs normally, whereas the VPC development is defective. Translated back into the biochemical function of sup-17 as a lin-12 S2 protease, this would point towards at least one additional metalloprotease able to exert S2 cleavage in specific cell types (a situation reminiscent of the putative function of ADAM17 in bone marrow – derived monocytic precursor cells (Brou *et al.* 2000)).

On the other side, sup-17 null mutants arrest development and die in late embryogenesis, which makes the overall phenotype somewhat more severe than that of a lin-12 null mutant (which results in a spectrum of phenotypes between embryonic lethality and sterility in adulthood. This points – similar to the situation e.g. in mice, see below - to additional critical functions of sup-17 during embryogenesis, which lie outside of the lin-12 pathway.

## 5.2 Drosophila melanogaster

The Notch pathway, a key target of kuzbanian function, plays an essential role by specifying neurons within neurogenic cell clusters created by the action of proneural genes (Rooke *et al.* 1996). By lateral inhibition, an emerging neural cell thereby inhibits the neural development of its neighbours, forcing them into an epidermal fate. Both the Notch- and the kuzbanian loss-of-function abolish lateral inhibition, creating an embryonic lethal neurogenic phenotype (Sotillos *et al.* 1997). Analysis of later developmental stages in mosaic flies shows that inactivation of kuzbanian again similar to a Notch loss-of-function causes the development of clustered sensory bristles instead of single bristles. However, this approach also pointed towards discrete differences between the Notch and the kuzbanian phenotype, indicating that kuzbanian as opposite to Notch does not play a role in the correct spacing of proneural clusters. Likewise in contrast to the established role of kuzbanian in Notch S2 cleavage, mutant bristle clusters in a mosaic fly develop at the interface between kuzbanian deficient and wild-type cells, but never within a clonal group of deficient cells. This would indicate that there is an additional non - cell autonomous requirement for kuzbanian in bristle development that can not be explained by its presumed role in Notch cleavage, for instance a role of kuzbanian for providing a diffusible factor from the surrounding wild-type cells available at the interface only. It has remained unclear so far, whether e.g. the Notch ligand delta could be the factor in question here, or whether other factors shed by kuzbaninan could play a role here (Rooke *et al.* 1996, Pan and Rubin 1997).

By analysis of flies bearing a 'near-null' mutation of kuzbanian and thus surviving longer than the complete knockout additional functions in nervous sytem development could be identified. Kuzbanian null neurons showed severe defects of axon extension and navigation (Fambrough *et al.* 1996, Sotillos *et al.* 1997), visible as a stalling of axon outgrowth during development. Interestingly, growth of pioneer fibers was relatively preserved, indicating either a specific function of kuzbanian for axonal navigation in a more complex environment or in axo-axonal substrate – bound migration.

Finally, by further work on adult flies bearing genetic mosaics for kuzbanian additional traits of defective Notch signalling were identified, including defective wing margin formation and wing vein patterning (Klein 2002).

## 5.3 Mus musculus

Adam10-deficient mice (Fig. 5) die at day 9.5 of embryogenesis with multiple defects of the developing central nervous system, somites, and cardiovascular system (Hartmann *et al.* 2002). In situ hybridization revealed a reduced expression of the Notch target gene, hes5, in the neural tube and an increased expression of the Notch ligand, dll1, suggesting an important role for Adam10 in Notch signalling. Since the early lethality precluded the establishment of primary neuronal cultures, amyloid precursor protein (APP) α-secretase activity was analyzed in embryonic fibroblasts and found to be preserved in 15 of 17 independently generated Adam10-deficient fibroblast cell lines, albeit at a quantitatively more variable level than in controls. These data clearly indicate that besides the currently favoured ADAM10 other proteases exist at least in MEFs that can fully compensate for ADAM10 deficiency and raise the question to what degree such a compensation phenomenon occurs in the brain as well. As no upregulation of both ADAM9 and ADAM17 was observed even on the protein level, it is even conceivable that proteases outside of the 'top three' candidates may be involved.

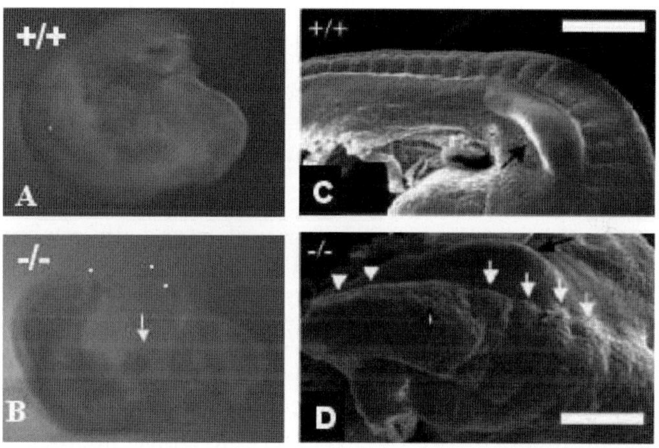


*Figure 5*: Phenotype of ADAM10 deficient mice. The embryos feature a shortened body axis and a dilated pericardium similar to PS1/2 double deficient mice. The same holds true for a defective somite segmentation, which is abrogated after the 4th to 5th pair. In contrast, heart looping is more severely affected (arrow in B), pointing towards additional ADAM10 functions.

In contrast to the still dubious situation with respect to α-secretase function, it is striking from the phenotype that ADAM10 is indispensable for mammalian development, its deficiency causing in mice a phenotype that corresponds to the complete loss of Notch signaling in invertebrates (Herreman *et al.* 1999, Hartmann *et al.* 2002). Given the recent data on the potential role of Notch ligand RIPping by ADAMs and γ-secretase, it has to be clarified in more detail on which side of this pathway ADAM10 fulfills its key functions.

For the Notch pathway, a number of human diseases have so far been identified, ranging from mainly hepatovascular malformations like in Alagille's syndrome and Fallot – type malformations of the heart to spondylocostal dysostosis, but all of these have been traced to Notch or Notch ligand mutations. It will thus be intriguing to see, whether in the future defects of Notch or Notch ligand *processing* by ADAM10 or γ-secretase contribute to human disease.

## 6. ADAM10 - A LONE PLAYER IN TISSUE DEVELOPMENT ?

To summarize the functional roles of ADAM10, it becomes apparent that the majority of its actions focus on the processing of developmentally essential signaling proteins. For at least a considerable portion of them (like Notch or the ephrins) the shedding events are absolutely essential for their function, and the loss of ADAM10 in all probability creates a kind of functional "multiple protein deficiency" phenotype.

Despite all the documented overlap in substrate specificity with other ADAMs, especially in the case of EGF-R ligands, in development ADAM10 appears to be indispensable as the "lone sheddase" in essential signaling events that are just at the verge of being understood. Invertebrate models with e.g. a complete loss of crucial axonal navigation capabilities promise likewise important roles in mammalian CNS and PNS development.

Still, the currently available knockout data only identify a small minority of them, i.e. the ones like the Notch receptors that cause the earliest lethalities in nematodes, flies and mice. Further work on conditional knockout animal models will be necessary to fully appreciate the real importance of ADAM10.

It is striking to which degree ADAM10 apparently acts in tandem with γ-secretase to initiate membrane protein RIPping, i.e. an essential signaling step beyond a mere shedding function. Thus, it is likely that many lessons for a future therapeutical exploitation of ADAM10 function may be learned from experiences with γ-secretase inhibitors and associated mouse models. Here, recent studies (Tournoy *et al*, 2004) indicate that secretase substrates known from development still exert functions in adult renewing tissues, such as skin or immune system, resulting in novel "adult" phenotypes of partial or complete secretase deficiency that may appear fully unrelated to the developmental effects. Thus, whereas impaired γ-secretase function in development causes vasculogenesis and body segmentation defects, its adult 'counterpart' results in autoimmune disease and skin tumors. From these data, it may be expected that any use of ADAM activity modulation will have to be made with great care regarding its likely side effects, and it is highly likely that this evaluation will be as difficult as for γ-secretase.

Finally, when speaking about ADAM10 function in mammals, one tends to think about emerging possible clinical applications, even if it is still a long shot from experimental work in the mouse to applications in humans. Here, it is fascinating to see the recent surge of identified ADAM10 functions in immune biology, where one could imagine that the use of selective ADAM10 inhibitors could develop into a powerful medical tool.

## ACKNOWLEDGEMENTS

The authors are indepted to Bart De Strooper, Stefan Rose-John, Andreas Ludwig and Bassem Hassan for numerous valuable discussions. Research work by P.S. and D.H. reviewed here was supported by the IAP (Interuniversity Attraction Poles program P5/19 of the Belgian Federal Science Policy Office), the APOPIS network, the Alzheimer Association and the Deutsche Forschungsgemeinschaft.

## REFERENCES

Abel, S., Hundhausen, C., Mentlein, R., Schulte, A., Berkhout, T. A., Broadway, N., Hartmann, D., Sedlacek, R., Dietrich, S., Muetze, B., Schuster, B., Kallen, K., Saftig, P., Rose-John, S. and Ludwig, A. (2004). The transmembrane CXC-chemokine ligand 16 is induced by IFN-gamma and TNF-alpha and shed by the activity of the disintegrin-like metalloproteinase ADAM10. *J Immunol* **172**: 6362-72.

Allinson, TM, Parkin, ET, Condon, TP, Schwager, SL, Sturrock, ED, Turner, AJ and Hooper, NM (2004). The role of ADAM10 and ADAM17 in the ectodomain shedding of angiotensin converting enzyme and the amyloid precursor protein. *Eur J Biochem* **271**: 2539-2547.

Anders, A., Gilbert, S., Garten, W., Postina, R. and Fahrenholz, F. (2001) Regulation of the alpha-secretase ADAM10 by its prodomain and proprotein convertases. *FASEB J* **15**: 1837-1839.

Asai, M., Hattori, C., Szabo, B., Sasagawa, N., Maruyama, K., Tanuma, S.,Ishiura, S. (2003). Putative function of ADAM9, ADAM10, and ADAM17 as APP α-secretase. *Biochem Biophys Res Commun* **301**: 231-235.

Bazan, J., Bacon, K., Hardiman, G., Wang, W., Soo, K., Rossi, D., Greaves, D., Zlotnik, A., and Schall, T. (1997) A new class of membrane-bound chemokine with a CX3C motif. *Nature* **385**: 640-644.

Bernstein, H., Bukowska, A., Krell, D., Bogerts, B., Ansorge, S. and Lendeckel, U. (2003). Comparative localization of ADAMs 10 and 15 in human cerebral cortex normal aging, Alzheimer disease and Down syndrome. *J Neurocytol* **32**: 153-160.

Black, R. A., Rauch, C. T., Kozlosky, C. J., Peschon, J. J., Slack, J. L., Wolfson, M. F., Castner, B. J., Stocking, K. L., Reddy, P., Srinivasan, S., Nelson, N., Boiani, N., Schooley, K. A., Gerhart, M., Davis, R., Fitzner, J. N., Johnson, R. and Paxton, S. (1997). A metalloproteinase disintegrin that releases tumour-necrosis factor-alpha from cells. *Nature* **385**: 729-733.

Bland, C. E., Kimberly, P. and Rand, M. D. (2003). Notch-induced proteolysis and nuclear localization of the Delta ligand. *J Biol Chem* **278**: 13607-13610.

Boissy, P., Lenhard, T. R., Kirkegaard, T., Peschon, J. J., Black, R. A., Delaisse, J., and del Carmen Ovejero, M. (2003). An assessment of ADAMs in bone cells: absence of TACE activity prevents osteoclast recruitment and the formation of the marrow cavity in developing long bones. *FEBS Lett* **553**: 257-261.

Brou, C., Logeat, F., Gupta, N., Bessia, C., LeBail, O., Doedens, J. R., Cumano, A., Roux, P., Black, R. A. and Israel, A. (2000). A novel proteolytic cleavage involved in Notch signaling: the role of the disintegrin-metalloprotease TACE. *Mol Cell* **5**: 207-216.

Budagian, V., Bulanova, E., Orinska, Z., Ludwig, A., Rose-John, S., Saftig, P., Borden, E. C., and Bulfone-Paus, S. (2004) Natural soluble IL-15Ralpha is generated by cleavage that involves the tumor necrosis factor-alpha -converting enzyme (TACE/ADAM17). *J Biol Chem.* **279**: 40368-40375.

Buxbaum, J. D., Cullen, E. I., and Friedhoff, L. T. (2002) Pharmacological concentrations of the HMG-CoA reductase inhibitor lovastatin decrease the formation of the Alzheimer beta-amyloid peptide in vitro and in patients. *Front Biosci* **7**: a50-9.

Buxbaum, J. D., Liu, K. N., Luo, Y., Slack, J. L., Stocking, K. L., Peschon, J. J., Johnson, R. S., Castner, B. J., Cerretti, D. P. and Black, R. A. (1998). Evidence that tumor necrosis factor alpha converting enzyme is involved in regulated alpha-secretase cleavage of the Alzheimer amyloid protein precursor. *J Biol Chem* **273**: 27765-27777.

Cao, Y., Zhao, Z., Gruszczynska-Biegala, J. and Zolkiewska, A. (2003). Role of metalloprotease disintegrin ADAM12 in determination of quiescent reserve cells during myogenic differentiation in vitro. *Mol Cell Biol* **23**: 6725-6738.

Chantry, A., Gregson, N. A. and Glynn, P. (1989). A novel metalloproteinase associated with brain myelin membranes. Isolation and characterization. *J Biol Chem* **264**: 21603-21607.

Colciaghi, F., Borroni, B., Pastorino, L., Marcello, E., Zimmermann, M., Cattabeni, F., Padovani, A., and Di Luca, M. (2002) alpha-Secretase ADAM10 as well as alpha APPs is reduced in platelets and CSF of Alzheimer disease patients. *Mol Med* **8**: 67-74.

Condon, T. P., Flournoy, S., Sawyer, G. J., Baker, B. F., Kishimoto, T. K. and Bennett, C. F. (2001). ADAM17 but not ADAM10 mediates tumor necrosis factor-alpha and L-selectin shedding from leukocyte membranes. *Antisense Nucleic Acid Drug Dev* **11**: 107-116.

Dong, J., Opresko, L., Dempsey, P., Lauffenburger, D., Coffey, R. and Wiley, H. (1999). Metalloprotease-mediated ligand release regulates autocrine signaling through the epidermal growth factor receptor. *Proc Natl Acad Sci U S A* **96**: 6235-6240.

Ehehalt, R., Keller, P., Haass, C., Thiele, C. and Simons, K. (2003). Amyloidogenic processing of the Alzheimer beta-amyloid precursor protein depends on lipid rafts. *J Cell Biol* **160**: 113-123.

Endres, K., Anders, A., Kojro, E., Gilbert, S., Fahrenholz, F. and Postina, R. (2003). Tumor necrosis factor-α converting enzyme is processed by proprotein-convertases to its mature form which is degraded upon phorbol ester stimulation. *Eur J Biochem* **270**: 2386-2393.

Eto, K., Puzon-McLaughlin, W., Sheppard, D., Sehara-Fujisawa, A., Zhang, X. P. and Takada, Y. (2000). RGD-independent binding of integrin alpha9beta1 to the ADAM-12 and -15 disintegrin domains mediates cell-cell interaction. *J Biol Chem* **275**: 34922-34930.

Fambrough, D., Pan, D., Rubin, G. M. and Goodman, C. S. (1996). The cell surface metalloprotease/disintegrin Kuzbanian is required for axonal extension in Drosophila. *Proc Natl Acad Sci U S A* **93**: 13233-13238.

Fisher, A., Pittel, Z., Haring, R., Bar-Ner, N., Kliger-Spatz, M., Natan, N., Egozi, I., Sonego, H., Marcovitch, I. and Brandeis, R. (2003). M1 muscarinic agonists can modulate some of the hallmarks in Alzheimer's disease: implications in future therapy. *J Mol Neurosci* **20**: 349-356.

Fogel, M., Gutwein, P., Mechtersheimer, S., Riedle, S., Stoeck, A., Smirnov, A., Edler, L., Ben-Arie, A., Huszar, M. and Altevogt, P. (2003). L1 expression as a predictor of progression and survival in patients with uterine and ovarian carcinomas. *Lancet* **362**: 869-875.

Garton, K. J., Gough, P. J., Blobel, C. P., Murphy, G., Greaves, D. R., Dempsey, P. J. and Raines, E. W. (2001). Tumor necrosis factor-α-converting enzyme (ADAM17) mediates the cleavage and shedding of fractalkine (CX3CL1). *J Biol Chem* **276**: 37993-38001.

Gatta, L. B., Albertini, A., Ravid, R. and Finazzi, D. (2002). Levels of β-secretase BACE and α-secretase ADAM10 mRNAs in Alzheimer hippocampus. *Neuroreport* **13**: 2031-2033.

Gough, P. J., Garton, K. J., Wille, P. T., Rychlewski, M., Dempsey, P. J. and Raines, E. W. (2004). A disintegrin and metalloproteinase 10-mediated cleavage and shedding regulates the cell surface expression of CXC chemokine ligand 16. *J Immunol* **172**: 3678-3685.

Group 2002 (2002). MRC/BHF Heart Protection Study of cholesterol lowering with simvastatin in 20,536 high-risk individuals: a randomised placebo-controlled trial. *Lancet* **360**: 7-22.

Gumbiner, B. (1996). Cell adhesion: the molecular basis of tissue architecture and morphogenesis. *Cell* **84**: 345-357.

Gunthert, A., Strater, J., von, R. U., Henne, C., Joos, S., Koretz, K., Moldenhauer, G., Krammer, P. and Moller, P. (1996). Early detachment of colon carcinoma cells during CD95(APO-1/Fas)-mediated apoptosis. I. De-adhesion from hyaluronate by shedding of CD44. *J Cell Biol* **134**: 1089-1096.

Gutwein, P., Mechtersheimer, S., Riedle, S., Stoeck, A., Gast, D., Joumaa, S , Zentgraf, H., Fogel, M., and Altevogt, D. 2003 Feb, ADAM10-mediated cleavage of L1 adhesion molecule at the cell surface and in released membrane vesicles. FASEB J **17**: 292-4.

Gutwein, P., Mechtersheimer, S., Riedle, S., Stoeck, A., Gast, D., Joumaa, S., Zentgraf, H., Fogel, M.,Altevogt, D. P. (2003) ADAM10-mediated cleavage of L1 adhesion molecule at the cell surface and in released membrane vesicles. *FASEB J* **17**: 292-294.

Gutwein, P., Oleszewski, M., Mechtersheimer, S., Agmon-Levin, N., Krauss, K., and Altevogt, P. (2000). Role of Src kinases in the ADAM-mediated release of L1 adhesion molecule from human tumor cells. *J Biol Chem* **275**: 15490-15497.

Harris, R., Chung, E. and Coffey, R. (2003. EGF receptor ligands. *Exp Cell Res* **284**: 2-13.

Hartmann, D., de Strooper, B., Serneels, L., Craessaerts, K., Herreman, A., Annaert, W., Umans, L., Lubke, T., Lena Illert, A., von Figura, K., and Saftig, P. (2002) The disintegrin/metalloprotease ADAM 10 is essential for Notch signalling but not for alpha-secretase activity in fibroblasts. *Hum Mol Genet* **11**: 2615-2624.

Hattori, M., Osterfield, M., and Flanagan, J. G. (2000). Regulated cleavage of a contact-mediated axon repellent. *Science* **289**: 1360-1365.

Heinrich, P., Behrmann, I., Haan, S., Hermanns, H., Muller-Newen, G., and Schaper, F. (2003). Principles of interleukin (IL)-6-type cytokine signalling and its regulation. *Biochem J* **374**: 1-20.

Herreman, A., Hartmann, D., Annaert, W., Saftig, P., Craessaerts, K., Serneels, L., Umans, L., Schrijvers, V., Checler, F., Vanderstichele, H., Baekelandt, V., Dressel, R., Cupers, P., Huylebroeck, D., Zwijsen, A., Van Leuven, F. and De Strooper, B. (1999). Presenilin 2 deficiency causes a mild pulmonary phenotype and no changes in amyloid precursor protein processing but enhances the embryonic lethal phenotype of presenilin 1 deficiency. *Proc Natl Acad Sci U S A* **96**: 11872-11877.

Herreman, A., Serneels, L., Annaert, W., Collen, D., Schoonjans, L. and De Strooper, B. (2000). Total inactivation of gamma-secretase activity in presenilin-deficient embryonic stem cells. *Nat Cell Biol* **2**: 461-462.

Hooper, N. M., and Turner, A. J. (2002). The search for alpha-secretase and its potential as a therapeutic approach to Alzheimer s disease. *Curr Med Chem* **9**: 1107-1119.

Howard, L., Lu, X., Mitchell, S., Griffiths, S. and Glynn, P. (1996). Molecular cloning of MADM: a catalytically active mammalian disintegrin-metalloprotease expressed in various cell types. *Biochem J* **317**: 45-50.

Hundhausen, C., Misztela, D., Berkhout, T. A., Broadway, N., Saftig, P., Reiss, K., Hartmann, D., Fahrenholz, F., Postina, R., Matthews, V., Kallen, K., Rose-John, S. and Ludwig, A. (2003). The disintegrin-like metalloproteinase ADAM10 is involved in constitutive cleavage of CX3CL1 (fractalkine) and regulates CX3CL1-mediated cell-cell adhesion. *Blood* **102**: 1186-1195.

Jackson, L., Qiu, T., Sunnarborg, S., Chang, A., Zhang, C., Patterson, C. and Lee, D. (2003). Defective valvulogenesis in HB-EGF and TACE-null mice is associated with aberrant BMP signaling. *EMBO J* **22**: 2704-2716.

Kajita, M., Itoh, Y., Chiba, T., Mori, H., Okada, A., Kinoh, H. and Seiki, M. (2001) Membrane-type 1 matrix metalloproteinase cleaves CD44 and promotes cell migration. *J Cell Biol* **153**: 893-904.

Kamal, A., Almenar-Queralt, A., LeBlanc, J. F., Roberts, E. A., and Goldstein, L. S. (2001). Kinesin-mediated axonal transport of a membrane compartment containing beta-secretase and presenilin-1 requires APP. *Nature* **414**: 643-648.

Karkkainen, I., Rybnikova, E., Pelto-Huikko, M. and Huovila, A. P. (2000). Metalloprotease-disintegrin (ADAM) genes are widely and differentially expressed in the adult CNS. *Mol Cell Neurosci* **15**: 547-560.

Kawano, Y., Okamoto, I., Murakami, D., Itoh, H., Yoshida, M., Ueda, S. and Saya, H. (2000). Ras oncoprotein induces CD44 cleavage through phosphoinositide 3-OH kinase and the rho family of small G proteins. *J Biol Chem* **275**: 29628-29635.

Kimberly, W. T., Zheng, J. B., Guenette, S. Y. and Selkoe, D. J. (2001). The intracellular domain of the beta-amyloid precursor protein is stabilized by Fe65 and translocates to the nucleus in a notch-like manner. *J Biol Chem* **276**: 40288-40292.

Klein, T. (2002). Kuzbanian is required cell autonomously during Notch signalling in the Drosophila wing. *Dev Genes Evol* **212**: 251-255.

Koike, H., Tomioka, S., Sorimachi, H., Saido, T. C., Maruyama, K., Okuyama, A., Fujisawa-Sehara, A., Ohno, S., Suzuki, K. and Ishiura, S. (1999) Membrane-anchored metalloprotease MDC9 has an alpha-secretase activity responsible for processing the amyloid precursor protein. *Biochem J* **343**: 371-375.

Kurisaki, T., Masuda, A., Sudo, K., Sakagami, J., Higashiyama, S., Matsuda, Y., Nagabukuro, A., Tsuji, A., Nabeshima, Y., Asano, M., Iwakura, Y. and Sehara-Fujisawa, A. (2003). Phenotypic analysis of Meltrin alpha (ADAM12)-deficient mice: involvement of Meltrin alpha in adipogenesis and myogenesis. *Mol Cell Biol* **23**: 55-61.

Lammich, S., Kojro, E., Postina, R., Gilbert, S., Pfeiffer, R., Jasionowski, M., Haass, C., and Fahrenholz, F. (1999). Constitutive and regulated α-secretase cleavage of Alzheimer's amyloid precursor protein by a disintegrin metalloprotease. *Proc Natl Acad Sci U S A* **96**: 3922-3927.

Lanni, C., Mazzucchelli, M., Porrello, E., Govoni, S. and Racchi, M. (2004). Differential involvement of protein kinase C alpha and epsilon in the regulated secretion of soluble amyloid precursor protein. *Eur J Biochem* **271**: 3068-3075.

Ledesma, M. D., Abad-Rodriguez, J., Galvan, C., Biondi, E., Navarro, P., Delacourte, A., Dingwall, C. and Dotti, C. G. (2003). Raft disorganization leads to reduced plasmin activity in Alzheimer's disease brains. *EMBO Rep* **4**: 1190-1196.

Lemjabbar, H. and Basbaum, C. (2002). Platelet-activating factor receptor and ADAM10 mediate responses to Staphylococcus aureus in epithelial cells. *Nat Med* **8**: 41-46.

Lieber, T., Kidd, S. and Young, M. W. (2002). kuzbanian-mediated cleavage of Drosophila Notch. *Genes Dev* **16**: 209-221.

Lunn, C., Fan, X., Dalie, B., Miller, K., Zavodny, P., Narula, S. and Lundell, D. (1997). Purification of ADAM 10 from bovine spleen as a TNFα convertase. *FEBS Lett* **400**: 333-335.

Mann, G., Fowler, K., Gabriel, A., Nice, E., Williams, R. and Dunn, A. (1993). Mice with a null mutation of the TGF alpha gene have abnormal skin architecture, wavy hair, and curly whiskers and often develop corneal inflammation. *Cell* **73**: 249-261.

Marambaud, P., Wen, P., Dutt, A., Shioi, J., Takashima, A., Siman, R. and Robakis, N. (2003) A CBP binding transcriptional repressor produced by the PS1/epsilon-cleavage of N-cadherin is inhibited by PS1 FAD mutations. *Cell* **114**: 635-645.

Matloubian, M., David, A., Engel, S., Ryan, J. and Cyster, J. (2000). A transmembrane CXC chemokine is a ligand for HIV-coreceptor Bonzo. *Nat Immunol* **1**: 298-304.

Matthews, V., Schuster, B., Schutze, S., Bussmeyer, I., Ludwig, A., Hundhausen, C., Sadowski, T., Saftig, P., Hartmann, D., Kallen, K. and Rose-John, S. (2003). Cellular cholesterol depletion triggers shedding of the human interleukin-6 receptor by ADAM10 and ADAM17 (TACE). *J Biol Chem* **278**: 38829-38839.

McFarlane, S. (2003). Metalloproteases: carving out a role in axon guidance. *Neuron* **37**: 559-562.

McKie, N., Edwards, T., Dallas, D. J., Houghton, A., Stringer, B., Graham, R., Russell, G. and Croucher, P. I. (1997). Expression of members of a novel membrane linked metalloproteinase family (ADAM) in human articular chondrocytes. *Biochem Biophys Res Commun* **230**: 335-339.

Mechtersheimer, S., Gutwein, P., Agmon-Levin, N., Stoeck, A., Oleszewski, M., Riedle, S., Postina, R., Fahrenholz, F., Fogel, M., Lemmon, V. and Altevogt, P. (2001) Ectodomain shedding of L1 adhesion molecule promotes cell migration by autocrine binding to integrins. *J Cell Biol* **155**: 661-673.

Merlos-Suarez, A., Ruiz-Paz, S., Baselga, J. and Arribas, J. (2001). Metalloprotease-dependent protransforming growth factor-alpha ectodomain shedding in the absence of tumor necrosis factor-alpha-converting enzyme. *J Biol Chem* **276**: 48510-48517.

Miettinen, P., Berger, J., Meneses, J., Phung, Y., Pedersen, R., Werb, Z. and Derynck, R. (1995). Epithelial immaturity and multiorgan failure in mice lacking epidermal growth factor receptor. *Nature* **376**: 337-341.

Mishra-Gorur, K., Rand, M. D., Perez-Villamil, B. and Artavanis-Tsakonas, S. (2002). Down-regulation of Delta by proteolytic processing. *J Cell Biol* **159**: 313-324.

Moss, M., Jin, S., Milla, M., Bickett, D., Burkhart, W., Carter, H., Chen, W., Clay, W., Didsbury, J., Hassler, D., Hoffman, C., Kost, T., Lambert, M., Leesnitzer, M., McCauley, P., McGeehan, G., Mitchell, J., Moyer, M., Pahel, G. and Rocque, W. (1997). Cloning of a disintegrin metalloproteinase that processes precursor tumour-necrosis factor-alpha. *Nature* **385**: 733-736.

Mullberg, J., Durie, F., Otten-Evans, C., Alderson, M., Rose-John, S., Cosman, D., Black, R. and Mohler, K. (1995) A metalloprotease inhibitor blocks shedding of the IL-6 receptor and the p60 TNF receptor. *J Immunol* **155**: 5198-5205.

Mumm, J. S., Schroeter, E. H., Saxena, M. T., Griesemer, A., Tian, X., Pan, D. J., Ray, W. J. and Kopan, R. (2000). A ligand-induced extracellular cleavage regulates gamma-secretase-like proteolytic activation of Notch1. *Mol Cell* **5**: 197-206.

Murai, T., Miyazaki, Y., Nishinakamura, H., Sugahara, K. N., Miyauchi, T., Sako, Y., Yanagida, T. and Miyasaka, M. (2004). Engagement of CD44 promotes Rac activation and CD44 cleavage during tumor cell migration. *J Biol Chem* **279**: 4541-4550.

Murakami, D., Okamoto, I., Nagano, O., Kawano, Y., Tomita, T., Iwatsubo, T., De, S. B., Yumoto, E. and Saya, H. (2003). Presenilin-dependent gamma-secretase activity mediates the intramembranous cleavage of CD44. *Oncogene* **22**: 1511-1516.

Nagano, O., Murakami, D., Hartmann, D., De Strooper, B., Saftig, P., Iwatsubo, T., Nakajima, M., Shinohara, M. and Saya, H. (2004). Cell-matrix interaction via CD44 is independently regulated by different metalloproteinases activated in response to extracellular Ca2+ influx and PKC activation. *J Cell Biol* **165**: 893-902.

Narahashi, T., Marszalec, W., Moriguchi, S., Yeh, J. Z. and Zhao, X. (2003). Unique mechanism of action of Alzheimer's drugs on brain nicotinic acetylcholine receptors and NMDA receptors. *Life Sci* **74**: 281-291.

Nye, J. S. (1997). Developmental signaling: notch signals Kuz it's cleaved. *Curr Biol* **7**: R716-720.

Okamoto, I., Kawano, Y., Murakami, D., Sasayama, T., Araki, N., Miki, T., Wong, A. and Saya, H. (2001). Proteolytic release of CD44 intracellular domain and its role in the CD44 signaling pathway. *J Cell Biol* **155**: 755-762.

Pakaski, M. and Kasa, P. (2003). Role of acetylcholinesterase inhibitors in the metabolism of amyloid precursor protein. *Curr Drug Targets CNS Neurol Disord* **2**: 163-171.

Pan, D. and Rubin, G. M. (1997). Kuzbanian controls proteolytic processing of Notch and mediates lateral inhibition during Drosophila and vertebrate neurogenesis. *Cell* **90**: 271-280.

Pandiella, A. and Massague, J. (1991). Multiple signals activate cleavage of the membrane transforming growth factor-alpha precursor. *J Biol Chem* **266**: 5769-5773.

Paradies, N., and Grunwald, G. (1993) Purification and characterization of NCAD90, a soluble endogenous form of N-cadherin, which is generated by proteolysis during retinal development and retains adhesive and neurite-promoting function. *J Neurosci Res* **36**: 33-45.

Parvathy, S., Ehrlich, M., Pedrini, S., Diaz, N., Refolo, L., Buxbaum, J. D., Bogush, A., Petanceska, S. and Gandy, S. (2004). Atorvastatin-induced activation of Alzheimer's alpha secretase is resistant to standard inhibitors of protein phosphorylation-regulated ectodomain shedding. *J Neurochem* **90**: 1005-1010.

Pasquale, E. (2000). Neurobiology. Turning attraction into repulsion. *Science* **289**: 1308-1310.

Peschon, J. J., Slack, J. L., Reddy, P., Stocking, K. L., Sunnarborg, S. W., Lee, D. C., Russell, W. E., Castner, B. J., Johnson, R. S., Fitzner, J. N., Boyce, R. W., Nelson, N., Kozlosky, C. J., Wolfson, M. F., Rauch, C. T., Cerretti, D. P.,Paxton, R., March, C. J; and Black, R. (1998), An essential role for ectodomain shedding in mammalian development. *Science* **282**: 1281-1284.

Postina, R., Schroeder, A., Dewachter, I., Bohl, J., Schmitt, U., Kojro, E., Prinzen, C., Endres, K., Hiemke, C., Blessing, M., Flamez, P., Dequenne, A., Godaux, E., van Leuven, F. and Fahrenholz, F. (2004) A disintegrin-metalloproteinase prevents amyloid plaque formation and hippocampal defects in an Alzheimer disease mouse model. *J Clin Invest* **113**: 1456-1464.

Prusiner, S. (1998). Prions. *Proc Natl Acad Sci U S A* **95**: 13363-13383.

Qi, H., Rand, M. D., Wu, X., Sestan, N., Wang, W., Rakic, P., Xu, T. and Artavanis-Tsakonas, S. (1999). Processing of the notch ligand delta by the metalloprotease Kuzbanian. *Science* **283**: 91-4.

Rooke, J., Pan, D., Xu, T. and Rubin, G. M. (1996). KUZ, a conserved metalloprotease-disintegrin protein with two roles in Drosophila neurogenesis. *Science* **273**: 1227-1231.

Rose-John, S. and Heinrich, P. (1994). Soluble receptors for cytokines and growth factors: generation and biological function. *Biochem J* **300**: 281-290.

Sabo, S. L., Ikin, A. F., Buxbaum, J. D. and Greengard, P. (2001). The Alzheimer amyloid precursor protein (APP) and FE65, an APP-binding protein, regulate cell movement. *J Cell Biol* **153**: 1403-1414.

Sabo, S. L., Ikin, A. F., Buxbaum, J. D. and Greengard, P. (2003). The amyloid precursor protein and its regulatory protein, FE65, in growth cones and synapses in vitro and in vivo. *J Neurosci* **23**: 5407-5415.

Sadoul, K., Sadoul, R., Faissner, A. and Schachner, M. (1988). Biochemical characterization of different molecular forms of the neural cell adhesion molecule L1. *J Neurochem* **50**: 510-521.

Sahin, U., Weskamp, G., Kelly, K., Zhou, H., Higashiyama, S., Peschon, J., Hartmann, D., Saftig, P. and Blobel, C. P. (2004). Distinct roles for ADAM10 and ADAM17 in ectodomain shedding of six EGFR ligands. *J Cell Biol* **164**: 769-779.

Scalia, R., Gooszen, M., Jones, S., Hoffmeyer, M., Rimmer, D. 3., Trocha, S., Huang, P., Smith, M., Lefer, A. and Lefer, D. (2001). Simvastatin exerts both anti-inflammatory and cardioprotective effects in apolipoprotein E-deficient mice. *Circulation* **103**: 2598-2603.

Schachner, M. (1997) Neural recognition molecules and synaptic plasticity. *Curr Opin Cell Biol* **9**: 627- 634.

Schimmelpfeng, K., Gogel, S. and Klambt, C. (2001). The function of leak and kuzbanian during growth cone and cell migration. *Mech Dev* **106**: 25-36.

Seals, D. F. and Courtneidge, S. A. (2003). The ADAMs family of metalloproteases: multidomain proteins with multiple functions. *Genes Dev* **17**: 7-30.

Selkoe, D. J. (2001) Alzheimer's disease: genes, proteins, and therapy. *Physiol Rev* **81**: 741-766.

Shimaoka, T., Kume, N., Minami, M., Hayashida, K., Kataoka, H., Kita, T. and Yonehara, S. (2000). Molecular cloning of a novel scavenger receptor for oxidized low density lipoprotein, SR-PSOX, on macrophages. *J Biol Chem* **275**: 40663-40666.

Sisodia, S. S. (2002). Biomedicine. A cargo receptor mystery APParently solved? *Science* **295**: 805-807.

Six, E., Ndiaye, D., Laabi, Y., Brou, C., Gupta-Rossi, N., Israel, A. and Logeat, F. (2003). The Notch ligand Delta1 is sequentially cleaved by an ADAM protease and gamma-secretase. *Proc Natl Acad Sci U S A* **100**: 7638-7643.

Slack, B. E., Ma, L. K. and Seah, C. C. (2001). Constitutive shedding of the amyloid precursor protein ectodomain is up-regulated by tumour necrosis factor-alpha converting enzyme. *Biochem J* **357**: 787-794.

Smith, K. M., Gaultier, A., Cousin, H., Alfandari, D., White, J. M. and DeSimone, D. W. (2002). The cysteine-rich domain regulates ADAM protease function in vivo. *J Cell Biol* **159**: 893-902.

Sotillos, S., Roch, F. and Campuzano, S. (1997). The metalloprotease-disintegrin Kuzbanian participates in Notch activation during growth and patterning of Drosophila imaginal discs. *Development* **124**: 4769-4779.

Sugahara, K., Murai, T., Nishinakamura, H., Kawashima, H., Saya, H. and Miyasaka, M. (2003). Hyaluronan oligosaccharides induce CD44 cleavage and promote cell migration in CD44-expressing tumor cells. *J Biol Chem* **278**: 32259-32265.

Sunnarborg, S. W., Hinkle, C. L., Stevenson, M., Russell, W. E., Raska, C. S., Peschon, J. J., Castner, B. J., Gerhart, M. J., Paxton, R. J., Black, R. A. and Lee, D. C. (2002). Tumor necrosis factor-alpha converting enzyme (TACE) regulates epidermal growth factor receptor ligand availability. *J Biol Chem* **277**: 12838-12845.

Takeichi, M. (1991). Cadherin cell adhesion receptors as a morphogenetic regulator. *Science* **251**: 1451-5.

Tax, F. E., Thomas, J. H., Ferguson, E. L. and Horvitz, H. R. (1997). Identification and characterization of genes that interact with lin-12 in Caenorhabditis elegans. *Genetics* **147**: 1675-1695.

Thomas, L., Byers, H., Vink, J. and Stamenkovic, I. (1992). CD44H regulates tumor cell migration on hyaluronate-coated substrate. *J Cell Biol* **118**: 971-977.

Vassalli, P. (1992). The pathophysiology of tumor necrosis factors. *Annu Rev Immunol* **10**: 411-452.

Vincent, B., Paitel, E., Saftig, P., Frobert, Y., Hartmann, D., De Strooper, B., Grassi, J., Lopez-Perez, E. and Checler, F. (2001). The disintegrins ADAM10 and TACE contribute to the constitutive and phorbol ester-regulated normal cleavage of the cellular prion protein. *J Biol Chem* **276**: 37743-37746.

Wen, C., Metzstein, M. M. and Greenwald, I. (1997) SUP-17, a Caenorhabditis elegans ADAM protein related to Drosophila KUZBANIAN, and its role in LIN-12/NOTCH signalling. *Development* **124**: 4759-4767.

Weskamp, G., Cai, H., Brodie, T. A., Higashyama, S., Manova, K., Ludwig, T. and Blobel, C. P. (2002). Mice lacking the metalloprotease-disintegrin MDC9 (ADAM9) have no evident major abnormalities during development or adult life. *Mol Cell Biol* **22**: 1537-1544.

Wolfsberg, T. G., Primakoff, P., Myles, D. G. and White, J. M. (1995). ADAM, a novel family of membrane proteins containing A Disintegrin And Metalloprotease domain: multipotential functions in cell-cell and cell-matrix interactions. *J Cell Biol* **131**: 275-278.

Wolfsberg, T. G., Straight, P. D., Gerena, R. L., Huovila, A. P., Primakoff, P., Myles, D. G. and White, J. M. (1995). ADAM, a widely distributed and developmentally regulated gene family encoding membrane proteins with a disintegrin and metalloprotease domain. *Dev Biol* **169**: 378-383.

Yamazaki, K., Mizui, Y., Sagane, K. and Tanaka, I. (1997). Assignment of a disintegrin and metalloproteinase domain 10 (Adam10) gene to mouse chromosome 9. *Genomics* **46**: 528.

Yamazaki, K., Mizui, Y. and Tanaka, I. (1997). Radiation hybrid mapping of human ADAM10 gene to chromosome 15. *Genomics* **45**: 457-459.

Yamazaki, S., Iwamoto, R., Saeki, K., Asakura, M., Takashima, S., Yamazaki, A., Kimura, R., Mizushima, H., Moribe, H., Higashiyama, S., Endoh, M., Kaneda, Y., Takagi, S.,

Itami, S., Takeda, N., Yamada, G. and Mekada, E. (2003). Mice with defects in HB-EGF ectodomain shedding show severe developmental abnormalities. *J Cell Biol* **163**: 469-475.

Yan, Y., Shirakabe, K. and Werb, Z. (2002). The metalloprotease Kuzbanian (ADAM10) mediates the transactivation of EGF receptor by G protein-coupled receptors. *J Cell Biol* **158**: 221-6.

Yarden, Y. and Sliwkowski, M. (2001). Untangling the ErbB signalling network. *Nat Rev Mol Cell Biol* **2**: 127-137.

Yoshimura, T., Tomita, T., Dixon, M. F., Axon, A. T. R., Robinson, P. A. and Crabtree, J. E. (2002). ADAMs (a disintegrin and metalloproteinase) messenger RNA expression in Helicobacter pylori-infected, normal, and neoplastic gastric mucosa. *J Infect Dis* **185**: 332-340.

# Chapter 6

# ADAM12

## *The long and the short of it*

Ulla M. Wewer[1], Reidar Albrechtsen [1] and Eva Engvall[2]
[1]*Institute of Molecular Pathology, University of Copenhagen, 2100 Copenhagen, Denmark*
[2]*The Burnham Institute, La Jolla, California 92037, USA*

Abstract: ADAM12 is one of only a few ADAMs that exist in two forms: the prototype transmembrane form, ADAM12-L, and an alternatively spliced secreted form, ADAM12-S. ADAM 12 is an active metalloprotease, and substrates include IGFPB-3 and –5 and membrane-anchored pro-growth factor HB-EGF. ADAM12 interacts with integrin and syndecan adhesion receptors via the disintegrin and cysteine-rich domains, and influences cell shape, cytoskeleton, and the organization of the extracellular matrix. ADAM12 is expressed mainly during development and differentiation, in remodelling tissues, and in fast growing tissues such as placenta and malignant tumors. These key features of ADAM12 may have direct implications for clinical medicine. Mice deficient in ADAM12 are born fully developed, but approximately 30% of the null mice die within 1 week of birth. Surviving adult null mice appear phenotypically normal, but have reduced amounts of adipose and muscle tissue, in accord with a role for ADAM12 in myogenesis and adipogenesis. This role is further supported by ADAM12 promoting adipogenesis and myogenesis in wild type and dystrophin-deficient *mdx* mice. We are left with many questions about the function of ADAM12 and in particular about the molecular mechanisms such as 1) why is the prodomain still attached to ADAM12 after cleavage; 2) what are the physiological substrates of ADAM12; we know a few but probably not all; 3) how is the disintegrin and cysteine-rich domain regulating syndecans and integrins; we know it binds but we need to know the cause and the effect; and 4) what are the physiological ligands of the cytoplasmic domain, and importantly, when and where in the cell do they bind. In conclusion, ADAM12 appears to exert many cellular activities which may be strictly regulated in a cell type and context–specific manner.

Key words: ADAM, metalloprotease, integrin, syndecan, tumorigenesis, adipogenesis myogenesis, regeneration, transgenic mice



*N. M. Hooper and U. Lendeckel (eds.), The ADAM Family of Proteases,* 123-146.

# 1. ADAM12: NAME AND FAMILY

ADAM12, like other ADAMs, is a multidomain protein with protease, cell adhesion, fusion, and signaling activities. Mouse ADAM12, initially called meltrin-α, was first described in 1995 as a transmembrane protein involved in muscle cell fusion (Yagami-Hiromasa *et al.* 1995). Afterwards, the human homologue was cloned (Gilpin *et al.* 1998). ADAM12 is a member of the large family of ADAM proteins, of which more than 30 have been identified; 19 ADAM genes are currently known in humans. There are several excellent reviews about ADAMs; see, for example, Moss and Bartsch 2004; White 2003; Seals and Courtneidge 2003; Blobel 2000, Schlöndorff and Blobel 1999.

ADAMTSs (a disintegrin and metalloprotease with thrombospondin motifs) belong to a closely related protein family, with 19 known members (Tang 2001, Apte *et al* 2004); these proteins are not anchored to the cell membrane, as most of the ADAMs are. The domain organization of ADAMTSs is similar to that of the extracellular portions of ADAMs but in addition contain multiple copies of thrombospondin 1-like repeats. Several ADAMTSs have important functions; ADAMTS2, for example, has been identified as a procollagen N-proteinase, and mutations in the *ADAMTS2* gene cause Ehlers-Danlos syndrome type VII (Colige *et al.* 1997). ADAMTS-1, ADAMTS-4, and ADAMTS-5 have been identified as aggrecanases, cleaving the core protein of the proteoglycan aggrecan; they play a role in the early stages of cartilage destruction in arthritis (Nagase and Kashiwagi 2003). ADAMTS-13 cleaves hyperreactive "unusually large von Willebrand factor multimers", and mutations in the ADAMTS13 gene cause familial thrombotic thrombocytopenic purpura (Moake 2004).

ADAMs share structural similarity with snake venom metalloproteases (SVMPs), which are secreted in crotalid and viperid venoms (reviewed in Bjarnason and Fox 1995; Jia *et al.* 1996, Huang *et al.* 1998). The disintegrin domain of SVMPs contains an Arg-Gly-Asp (RGD) sequence that disrupts integrin function, particularly that involved in platelet aggregation. All SVMPs are proteolytically active, and extracellular matrix proteins are among their identified substrates. The disintegrin and metalloprotease activities of SVMPs result in severe hemorrhage and tissue damage in snakebite victims.

ADAMs, ADAMTSs, and SVMPs constitute the adamalysin/reprolysin family, which is part of the larger metzincin superfamily of zinc-dependent proteases. Other metzincins include matrix metalloproteases or matrixins, involved in the remodelling of the extracellular matrix in development and many diseases; astacins, a widespread family of metalloproteases including bone morphogenic/tolloid-like enzymes, meprins, and hatching enzymes;

and serralysins, which are bacterial proteinases (for review see Hooper 1994).

## 2. ADAM12: ORIGIN OF THE GENE

Genes encoding ADAMs are dispersed on different mammalian chromosomes. Human ADAM12 is encoded by a single gene located on chromosome 10q26.3 (Gilpin *et al.* 1998), where ADAM8 also resides (Yoshiyama *et al.* 1997). The mouse *ADAM12* gene has been localized to mouse chromosome 7, and rat *ADAM12* to rat chromosome 1 (Kurisaki *et al.* 2003).

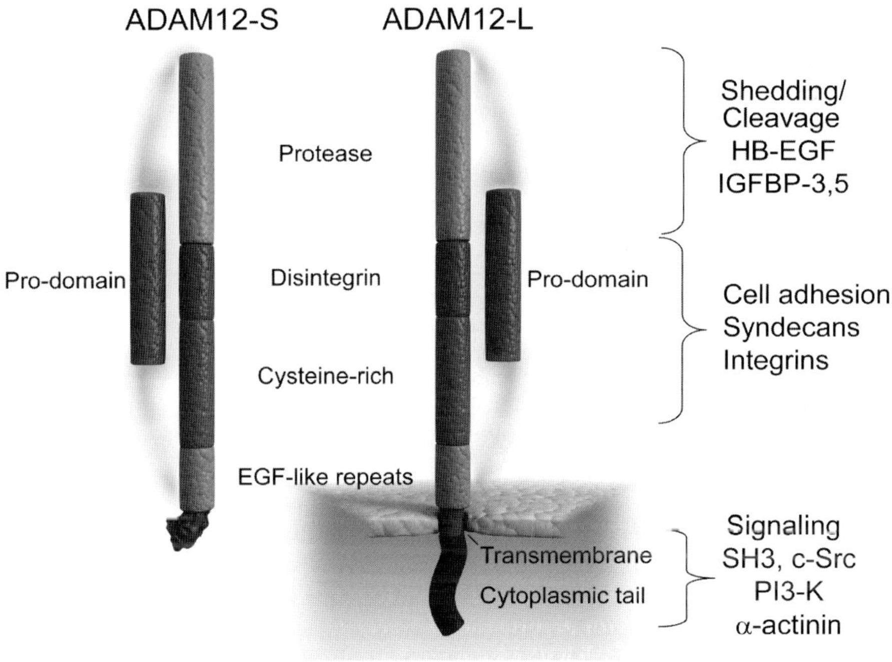


*Figure 1.* Human ADAM12 with two splice variants, ADAM12-S and ADAM12-L. Human ADAM12, unlike mouse ADAM12, may undergo alternative splicing. The long form, termed ADAM 12-L, is the prototype transmembrane form; ADAM12-S is a shorter secreted form lacking transmembrane and cytoplasmic domains (Gilpin et al. 1998). These alternative forms of human ADAM12 may allow ADAM12 a more variable function in humans than in mice.

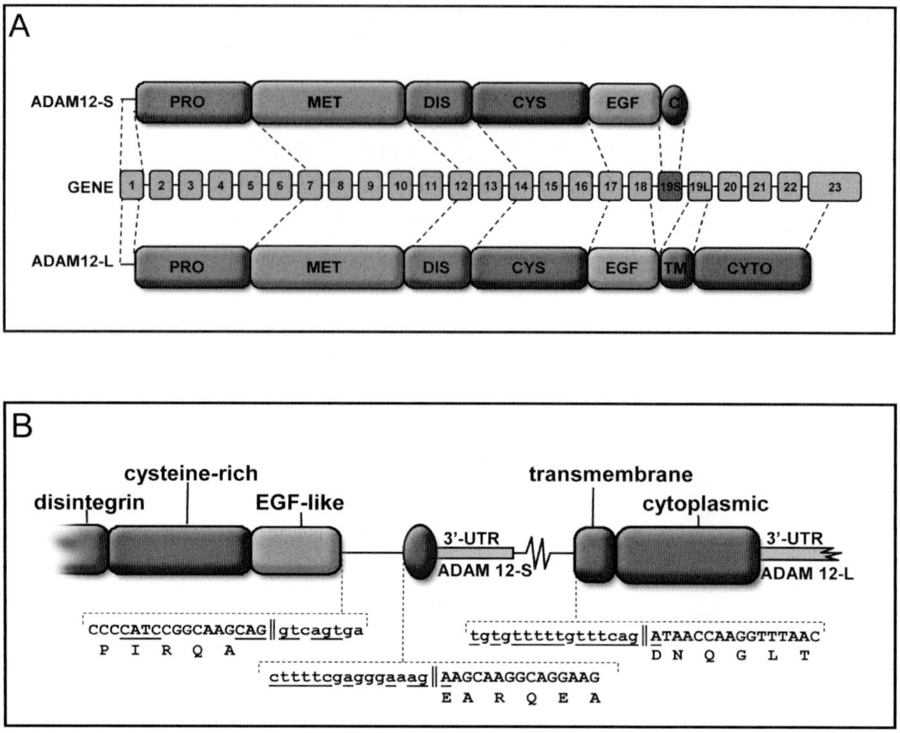


*Figure 2.* Genomic organization of the human ADAM12 gene at chromosome 10q26.3. (A) The first 18 exons (exons 1–18) are common to both splice variants. The ADAM12-L-specific COOH-terminus is encoded by 5 separate exons, exons 19-L through 23. The human ADAM12-S-specific COOH-terminus is encoded by a single exon (exon 19-S) inserted between the common exon 18 and exon 19-L. (B) A close up view of the splicing site is reproduced and modified from Gilpin *et al.* 1998.

The human *ADAM12* gene has two distinct splice variants, producing the long ADAM12-L form and the shorter ADAM12-S form (Gilpin *et al.* 1998) (Fig. 1). Sequencing of the gene revealed an intron of approximately 1 kb located at the point of divergence of the two variants. This intron is followed by a single exon encoding the ADAM12-S-specific COOH-terminus. An intron of approximately 2 kb follows before the ADAM12-L-specific exons (Gilpin *et al.* 1998). The genomic organization of the entire human *ADAM12* gene was recently revealed through sequencing of the human genome. The size of the gene from exon 1 through exon 23 is approximately 350,000 kb. The exon–intron organization of the gene and the domain organization of the protein are shown in Fig. 2. GenBank information: Full-length cDNA sequences of human ADAM12-L (AF023476), human ADAM12-S (AF023477), and mouse ADAM12 (D50411) are available. Among the many known ADAMs, ADAM12 forms a subgroup together with ADAM19 and ADAM33 – based on total sequence similarity (Fig 3).

*Figure 3.* Evolutionary tree. Protein evolution of one branch of the ADAM family. Length of branches corresponds to average number of evolutionary amino acid changes from the common ancestor per amino acid site. ADAM19 and ADAM33 are the most closely related to ADAM12.

# 3. ADAM12: PROTEIN STRUCTURE AND ACTIVITIES

The domain organization of ADAM12, like that of the other members of the ADAM family, contains a pro-domain, a metalloprotease domain, a disintegrin domain, a cysteine-rich+EGF-like domain, a transmembrane domain, and a 179-amino-acid cytoplasmic domain. ADAM12 is proteolytically competent in contrast to some other ADAMs. ADAM12 is highly conserved between species, with the metalloprotease, disintegrin, and cysteine-rich domains being the most conserved (Table 1). To generate Table 1, various genomes were searched to identify ADAM12 or its domains. The genes or gene fragments identified were then used to search the human genome to confirm that they represent ADAM12 and not any other ADAM. Results are presented as percent amino acid identity to human ADAM12.

*Table 1.* Conservation of domains in ADAM12-L from different species

| Species | pro-domain | metallo-protease | disintegrin | cysteine-rich | cyto-plasmic |
|---|---|---|---|---|---|
| Human | 100 | 100 | 100 | 100 | 100 |
| Cow | 71 | 91 | 90 | 94 | 71 |
| Rat | 66 | 92 | 91 | 94 | 72 |
| Mouse | 64 | 92 | 91 | 93 | 68 |
| Quail | 64 | 88 | 90 | 90 | 50 |
| Chicken | -- | 90 | 90 | 90 | 50 |
| Fugu | 40 | 79 | 85 | 88 | 70 |
| Tetraodonnig | 40 | 79 | 48 | 88 | 33 |

The ADAM12-S splice variant lacks the transmembrane and cytoplasmic domains; instead, its EGF-like domain is followed by a stretch of 33 amino acids that shows no similarity to any other known protein sequences in the databases (see also above).

## 3.1 Biosynthesis, processing, and maturation of ADAM12

The first 28 residues of human ADAM12 encode a typical signal peptide; the remaining ADAM12-L polypeptide consists of 881 amino acids, and that of ADAM12-S consists of 718 amino acids. ADAM12 is released from the endoplasmic reticulum as a proform (zymogen), with the prodomain keeping the metalloprotease inactive (Gilpin *et al.* 1998; Loechel *et al.* 1999). Transfections and mutational analyses have demonstrated that the prodomain is required for exit from the endoplasmic reticulum and that the cytoplasmic tail is somehow rate-limiting for this process (Hougaard *et al.* 2000; Cao *et al.* 2002). Human ADAM12 has five potential N-linked glycosylation sites N*X*(S/T); mouse ADAM12 has three additional sites. During transit in the trans-Golgi network, the prodomain is cleaved at the furin cleavage site (RHKR↓$E^{208}$) by furin or a related endopeptidase (Loechel *et al.* 1999); importantly, it is not released afterward but remains noncovalently associated with the rest of the molecule (Loechel *et al.* 2000, Kawaguchi *et al.* 2002).

The ADAM12 protein is delivered to the cell membrane as an apparently constitutively active metalloprotease, often referred to as the "mature form" (Loechel *et al.* 1999). On a typical Western blot, using an antibody recognizing the cysteine-rich domain, human ADAM12-L from total cell extract appears as 110- and 90-kDa bands; ADAM12-S appears as 90- and

70-kDa bands (Fig. 4). For each variant, the upper band represents the uncleaved proform and the lower band the mature form. The noncovalently associated prodomain of the mature form of purified ADAM12 dissociates in SDS–PAGE and can be visualized by Comasssie Blue staining (Fig. 4) or by using a domain-specific antibody as a 27-kDa band (Loechel *et al.* 2000; Kawaguchi *et al.* 2002). N-terminal sequencing of the 70-kDa (ETLKA) and 27-kDa (RGVSL) bands confirmed the identity of these bands and of the predicted furin cleavage site (Loechel *et al.* 2000). The nature and site of the association of the prodomain with the rest of the molecule have not been determined.

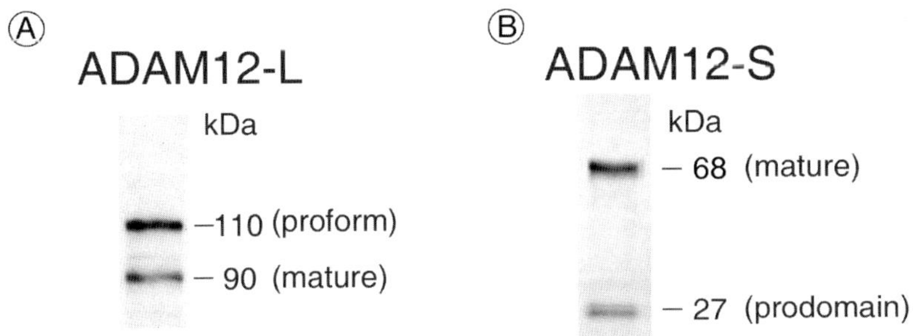


*Figure 4.* ADAM12-S and ADAM12-L protein by gel eletrophoresis. (A) Human ADAM12-L (AF23476) was produced in COS-7 cells, and the ADAM12-L protein was extracted in lysis buffer and separated on SDS-PAGE. Proteins were transferred to nitrocellulose paper and reacted with polyclonal antibodies to the cysteine rich domain (rb 122) followed by secondary antibodies. The 110 kDa proform and the 90 kDa mature form were visualized by the SuperSignal detection system essentially as described by Hougaard *et al.* (2000). The prodomain is not detected with this antibody. (B) Human ADAM12-S was expressed in 293-EBNA cells, and the ADAM12-S protein purified essentially as described by (Loechel *et al.* 2000) from the cell culture supernatant using FPLC, gelatin-Sepharose (to eliminate any contaminating fibronectin), SP-Sepharose ion exchange chromatography, and concanavalin A affinity chromatography. The purified protein was separated by SDS-PAGE under reducing conditions and the gel stained with Coomassie Blue. The 70 kDa mature form and the 27 kDa prodomain are seen.

Several studies have shown that ADAMs, including ADAM12, are mainly located inside cells, indicating that translocation of ADAM12 from intracellular storage to the cell surface might be restricted and regulated (Hougaard *et al.* 2000, Kadota *et al.* 2000, Cao *et al.* 2002). We found that PKCε activation induces ADAM12 translocation to the surface of RD rhabdomyosarcoma cells, and that the catalytic activity of PKCε is required for this translocation (Sundberg *et al.* in press ). This result agrees well with several other studies in which treatment of cells with PMA, a general PKC activator, results in increased ADAM protease activity at the cell surface (Izumi et al. 1998). ADAMs are considered "sheddases", i.e. they function in the proteolytic shedding of several ligands of growth factor receptors (Kheradmand and Werb 2002, Sahin *et al.* 2004). However, as will be described below, ADAMs in general and ADAM12 in particular, have other functions as well.

## 3.2 The prodomain

The prodomain of ADAM12 keeps the metalloprotease domain inactive inside the cell (Loechel *et al.* 1999); it also acts as a chaperone to facilitate folding and/or transport of ADAM12 (Cao *et al.* 2002). It is not yet known whether the prodomain has any function outside the cell, but we hypothesize that this relatively large domain has specific and hitherto unknown functions in the extracellular space, not necessarily confined to regulating the accessibility or proteolytic activity of ADAM12.

## 3.3 The metalloprotease domain

Members of the metzincin superfamily of proteolytic enzymes, including ADAM12, contain the distinctive zinc-binding consensus sequence motif HE*XX*H in their catalytic sites. In ADAM12, the catalytic site sequence is H<u>E</u>LGHNFGMNHD including the glutamate residue (underlined) necessary for proteolytic activity, and the "Met-turn," essential for the structural integrity of the zinc-binding site (Stöcker and Bode 1995; Gilpin *et al.* 1998). ADAM12 metalloprotease is kept in its latent state by a so-called "cysteine switch" mechanism of activation/repression: the critical cysteine residue $C^{179}$ is located in the prodomain, and it complexes to the active zinc atom at the catalytic site in the metalloprotease domain. Following cleavage of the prodomain in the trans-Golgi network, the cysteine dissociates from the zinc atom in the catalytic site, uncovering it for engagement in substrate cleavage (Loechel *et al.* 1998 and 1999). The appearance of ADAMs as constitutively active proteases outside the cell is a feature distinguishing ADAMs from matrix metalloproteases, which are secreted as proforms and become activated outside the cell.

### 3.3.1 Characteristics of ADAM12 protease

To determine and modulate ADAM12-S protease activity experimentally, we produced the short form, ADAM12–S, in transfected COS-7 cells and 293-EBNA cells. Cell supernatant or purified protein was then used in an $\alpha_2$-macroglobulin ($\alpha$2M) trapping assay (proteases are covalently bound to α2M). The principle of the α2M assay and its usefulness for identification of catalytically active forms of proteases have been described previously (Nagase *et al.* 1994; Sottrup Jensen *et al.* 1994). When this assay was used to characterize protease activity, the following results were obtained (Loechel *et al.* 1998, 1999, 2000, 2001)

- The normally inactive proform of ADAM12 became proteolytically active if the $C^{179}$ is replaced with another amino acid by site-directed mutagenesis.
- The inactive proform can be activated chemically by alkylation with *N*-ethylmaleimide.
- Mutation of the critical glutamate residue at the catalytic site ($E^{351} \rightarrow Q$ mutation) produces a protease-inactive ADAM12.
- Protease activity can be inhibited chemically by 1,10-phenanthroline and EDTA.
- Protease activity can be activated by copper.

### 3.3.2 Substrates for ADAM12 protease activity

Three substrates have been identified: (**1**) ADAM12-S can cleave IGFBP-3 and -5, and TIMP-3 (tissue inhibitor of metalloprotease) inhibits this reaction (Loechel *et al.* 2000); (**2**) ADAM12-L can shed pro-heparin-binding epidermal growth factor (proHB-EGF), resulting in transactivation of epidermal growth factor receptor (Asakura *et al.* 2002, Kurisaki *et al.* 2003); and (**3**) ADAM12-L can shed placental leucine aminopeptidase (P-LAP) (Ito *et al.* 2004)

ADAM12-S, via its cysteine-rich domain, binds to and cleaves the insulin growth factor binding protein-3 and -5 (IGFBP-3 and-5) (Shi *et al.*, 2000; Loechel *et al.* 2000). This activity of ADAM12 may regulate the bioavailability of IGFs. Proteolysis of IGFBP-3 is not increased in the presence of IGF-I or -II, suggesting that ADAM12-S is not an IGF-dependent protease. Protease activity towards IGFBP-3 and -5 was inhibited by a 10-fold molar excess of TIMP-3, but not by TIMP-1 or -2.

Recently, Asakura *et al.* (2003) showed that cleavage of the membrane-bound proHB-EGF by ADAM12 releases soluble HB-EGF, which in turn activates the epidermal growth factor receptor (EGFR) and promotes cardiac hypertrophy. An inhibitor, KB-R7785 ([4-(*N*-hydroxyamino)-2*R*-isobutyl-3*S*-methylsuccinyl]-L-phenylglycine-*N*-methylamide), was found to bind directly to ADAM12 and to block HB-EGF shedding. Interestingly, in mice with cardiac hypertrophy, inhibiting ADAM12 shedding of HB-EGF with KB-R57785 reduced heart disease, suggesting a new therapeutic strategy for cardiac hypertrophy (Asakura *et al.* 2003). The identification of ADAM12 as a distinct HB-EGF sheddase was further supported by the finding that

## 4.3 ADAM12 and cancer

A large proportion of human carcinomas express ADAM12; normal epithelial cells express much less (Iba *et al.* 1999, Le Pabic *et al.* 2003 and Albrechtsen *et al.* unpublished data). Tumor cells exhibit intense cytoplasmic immunostaining of ADAM12, and in some areas cell surface localization is also seen (Fig. 6). The presence of ADAM12 was also demonstrated by RT–PCR in breast carcinoma tissue (Iba *et al.* 1999) and in liver metastases of colon carcinoma (Le Pabic *et al.* 2003). MALDI–TOF mass spectrometry revealed ADAM12 in the urine of breast cancer patients (Roy *et al.* 2004). Lower urinary ADAM12 levels were found in patients with early stages of disease and higher levels in patients with more advanced disease. The highest ADAM12 levels were found in urine of patients with metastatic disease. Thus, urinary ADAM12 may prove to be useful as an additional noninvasive diagnostic or prognostic test for breast cancer and perhaps other cancers (Roy *et al.* 2004).

Other examples of ADAM12 expression in human conditions characterized by abnormal growth have been described. ADAM12 was found to be specifically upregulated in aggressive fibromatosis, or desmoid tumor, using Affymetrix GeneChip arrays (Skubitz *et al.* 2004).

## 4.4 ADAM12 in adipogenesis

Experiments with mice deficient in ADAM12 or overexpressing ADAM12 show that ADAM12 is not required for normal embryogenesis, but that it appears to modulate mesenchymal cell differentiation, including both adipogenesis and myogenesis.

### 4.4.1 ADAM12 null mice

Mice deficient in ADAM12 were generated by Sehara-Fujisawa´s group (Kurisaki *et al.* 2003). The numbers of homozygous null mice born were those predicted by the Mendelian ratio; however, approximately 30% of the null mice died within 1 week of birth. The reason for this perinatal death was not clear. Surviving adult null mice appeared phenotypically normal, but some mice had reduced amounts of interscapular brown adipose tissue as well as reduced neck and interscapular muscle tissue. The authors concluded that ADAM12 might be involved in adipogenesis and myogenesis through a linked developmental pathway. This agrees well with the data obtained from transgenic mice overexpressing ADAM12 (see below).

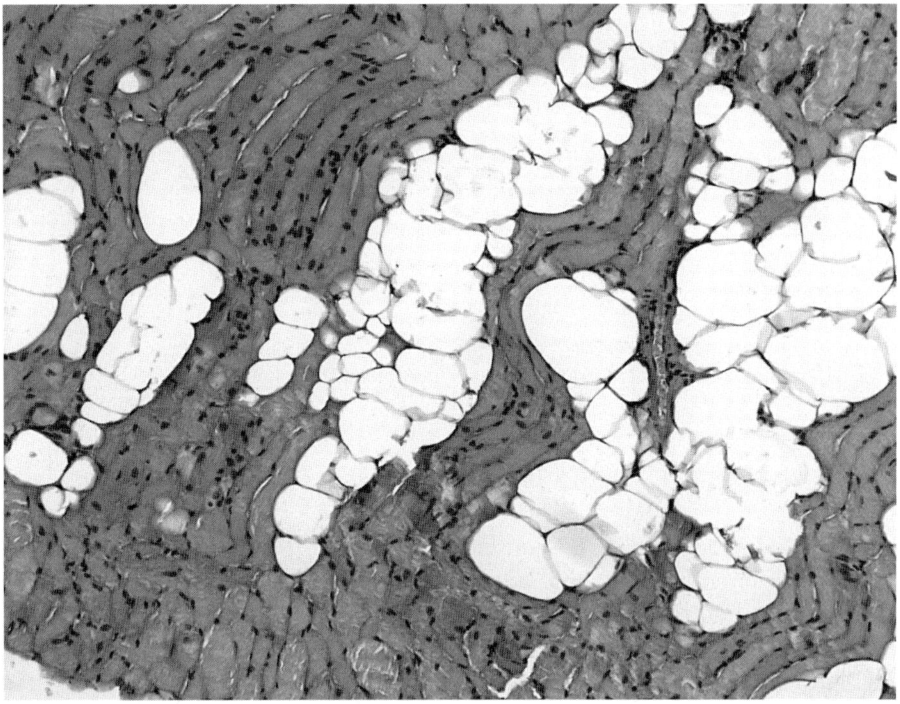

*Figure 7.* ADAM12 and adipogenesis. Skeletal muscle from adult transgenic mice expressing ADAM12-S under the MCK promoter. Clusters of adipocytes are present amongst the muscle cells. (Reproduced from Kawaguchi et al. 2002 with permission).

### 4.4.2 ADAM12 transgenic mice

We generated mice overexpressing various forms of ADAM12 under the muscle creatine kinase promoter (MCK): ADAM12-S, ADAM12-S Δpro–met (ADAM12S lacking pro- and metalloprotease domains), ADAM12-L, and ADAM12-L Δcyt (ADAM12L lacking the cytoplasmic domain) (Kawaguchi *et al.* 2002; Kronqvist *et al.* 2002, Moghadaszadeh *et al.* 2003). The MCK promoter (Lee *et al.* 1993) becomes active in the later stages of mouse embryogenesis in skeletal muscle and to a lesser degree in cardiac muscle. Mice expressing ADAM12-S or ADAM12-S Δpro–met under this promoter have ADAM12 in their circulation, and ADAM12 may thus influence other mouse tissues in addition to muscle. We found that ADAM12-S but not ADAM12-S Δpro–met promotes increased adipogenesis in skeletal muscle (Table 2 and Fig 7), and ADAM12-S but not ADAM12-S Δpro–met mice had increased abdominal white adipose tissue (Kawaguchi *et al.* 2002). Thus, the metalloprotease activity may be required for these effects. Likewise, ADAM12-L (Kawaguchi *et al.* 2002) and ADAM12-L

Δcyt (unpublished data) transgenes also increased adipogenesis locally in skeletal muscle. ADAM12 also promoted differentiation of preadipocytes in vitro (Kawaguchi *et al.* 2003).

*Table 2.* Role of ADAM12 in adipogenesis

| Transgenic Construct | Level of Adipogenesis |
|---|---|
| ADAM12 null mice | reduced |
| MCK-ADAM12-S | increased |
| MCK-ADAM12-S Δpro–met | no change |
| MCK-ADAM12-L | increased |
| MCK-ADAM12-L Δcyt | increased |
| 3T3-L1 preadipocyte + ADAM12 | promotes differentiation |
| 3T3-L1 preadipocyte + siRNA-ADAM12 | inhibits differentiation |

## 4.5 ADAM12 in skeletal muscle

ADAM12 appears in myotomes during mouse development and is expressed in embryonal and neonatal muscle. Intriguingly, ADAM12 is absent from adult muscle tissue but reappears during muscle regeneration (Yagami-Hiromasa *et al.* 1995, Kurisaki *et al.* 1998, Bornemann *et al.* 2000, Galliano *et al.* 2000, Kronqvist *et al.* 2002). In the rat, ADAM12 mRNA was detected by *in situ* hybridization in satellite cells under fusion conditions in regenerating rat muscle but not in satellite cells in normal adult rat muscle (Bornemann *et al.* 2000), and was most highly expressed in the so-called "reserve cells" *in vitro*, which may be the equivalent of satellite cells (Cao *et al.* 2003). The dynamic pattern of expression, points to a specific role for ADAM12 in muscle development and regeneration.

When ADAM12 was first identified, it was shown to be involved in myoblast fusion of C2 cells *in vitro* (Yagami-Hiromasa *et al.* 1995). When expression of ADAM12 was blocked by antisense techniques in murine C2 cells, myotube formation was inhibited; transfection with full-length ADAM12 also blocked myotube formation. In contrast, transfection with the disintegrin, cysteine-rich, and EGF-like domains promoted myotube formation. These same domains of human ADAM12-S Δ pro–met were also found to promote muscle cell differentiation in a nude mouse model system (Gilpin *et al.* 1998).

### 4.5.1 ADAM12 null mice

As ADAM12 is highly expressed in muscle during development and regeneration, it was surprising that disruption of the ADAM12 gene had little effect on these processes. Some of the ADAM12-deficient mice exhibited hypotrophy of the interscapular muscle. Following experimentally induced muscle injury, ADAM12-deficient mice showed no impairment of regeneration, a state where ADAM12 is normally highly expressed. Furthermore, ADAM12-deficient *mdx* mice (a model for dystrophin-deficient muscular dystrophy) had no changes in muscle regeneration or degeneration (Kurisaki *et al.* 2003). This result indicates either that ADAM12 has no role in regeneration or that the function of ADAM12 in the muscle can be replaced with another ADAM molecule.

### 4.5.2 ADAM12 transgenic mice

Mice with an *ADAM12* transgene expressed in skeletal muscle exhibited improved muscle regeneration following experimentally induced injury. This suggests that ADAM12 does indeed have a role in regeneration. Furthermore, dystrophin-deficient *mdx* mice with the *ADAM12* transgene showed less muscle pathology, as evidenced by less muscle cell necrosis and inflammation, lower levels of serum creatine kinase, and less uptake of Evans Blue dye into muscle fibers (Kronqvist *et al.* 2002). We found that ADAM12 may indirectly compensate for dystrophin deficiency in the *mdx* mice by increasing the expression and redistribution of several muscle adhesion proteins including $\alpha_7$ integrin and utrophin (Moghadaszadeh *et al.* 2003). This made us term ADAM12 a "booster gene" (Engvall and Wewer 2003). In laminin $\alpha_2$-deficient *dy* mice with the *ADAM12* transgene, we observed increased muscle regeneration, but it was not sufficient to improve the well-being of the mice (Guo *et al.*, submitted).

## 4.6 ADAM12 in heart muscle

In a mouse model of cardiac hypertrophy, inhibition of proHB-EGF shedding by a new inhibitor, KB-R7785, ameliorated the disease. Interestingly, this inhibitor bound to ADAM12, and since ADAM12 can shed proHB-EGF in the heart, these results suggest that HB-EGF shedding by ADAM12 protease may play a role in cardiac hypertrophy (see above, Asakura *et al.* 2002) and lend support to the strategy of developing specific metalloproteinase inhibitors for treatment of heart failure.

## 4.7 ADAM12 in bone

Northern blotting and RT–PCR demonstrated the presence of ADAM12 mRNA in both developing and adult bone (Yagami-Hiromasa *et al.* 1995). It was at first thought that since ADAM12 mediates myotube fusion in muscle, it might be mediating fusion of osteoclasts in bone, and some studies have

Hougaard, S., Loechel, F., Xu, X., Tajima, R., Albrechtsen, R., Wewer, U.M., 2000, Trafficking of human ADAM 12-L: retention in the trans-Golgi network. *Biochem.Biophys.Res.Commun.* **275:** 261-267.

Huang, T.F., 1998, What have snakes taught us about integrins? *Cell Mol.Life Sci.* **54:** 527-540.

Huovila, A.P., Almeida, E.A., White, J.M., 1996, ADAMs and cell fusion. *Curr.Opin.Cell Biol.* **8:** 692-699.

Iba, K., Albrechtsen, R., Gilpin, B.J., Loechel, F., Wewer, U.M., 1999, Cysteine-rich domain of human ADAM 12 (meltrin alpha) supports tumor cell adhesion. *Am.J.Pathol.* **154:** 1489-1501.

Iba, K., Albrechtsen, R., Gilpin, B., Frohlich, C., Loechel, F., Zolkiewska, A., Ishiguro, K., Kojima, T., Liu, W., Langford, J.K., Sanderson, R.D., Brakebusch, C., Fassler, R., Wewer, U.M., 2000, The cysteine-rich domain of human ADAM 12 supports cell adhesion through syndecans and triggers signaling events that lead to beta1 integrin-dependent cell spreading. *J.Cell Biol.* **149:** 1143-1156.

Inoue, D., Reid, M., Lum, L., Kratzschmar, J., Weskamp, G., Myung, Y.M., Baron, R., Blobel, C.P., 1998, Cloning and initial characterization of mouse meltrin beta and analysis of the expression of four metalloprotease-disintegrins in bone cells. *J.Biol.Chem.* **273:** 4180-4187.

Ito, N., Nomura, S., Iwase, A., Ito, T., Kikkawa, F., Tsujimoto, M., Ishiura, S., Mizutani, S., 2004, ADAMs, a disintegrin and metalloproteinases, mediate shedding of oxytocinase. *Biochem.Biophys.Res.Commun.* **314:** 1008-1013.

Izumi, Y., Hirata, M., Hasuwa, H., Iwamoto, R., Umata, T., Miyado, K., Tamai, Y., Kurisaki, T., Sehara-Fujisawa, A., Ohno, S., Mekada, E., 1998, A metalloprotease-disintegrin, MDC9/meltrin-gamma/ADAM9 and PKCdelta are involved in TPA-induced ectodomain shedding of membrane-anchored heparin-binding EGF-like growth factor. *EMBO J.* **17:** 7260-7272.

Jia, L.G., Shimokawa, K., Bjarnason, J.B., Fox, J.W., 1996, Snake venom metalloproteinases: structure, function and relationship to the ADAMs family of proteins. *Toxicon* **34:** 1269-1276.

Kadota, N., Suzuki, A., Nakagami, Y., Izumi, T., Endo, T., 2000, Endogenous meltrin alpha is ubiquitously expressed and associated with the plasma membrane but exogenous meltrin alpha is retained in the endoplasmic reticulum. *J.Biochem.* **128:** 941-949.

Kang, Q., Cao, Y., Zolkiewska, A., 2000, Metalloprotease-disintegrin ADAM 12 binds to the SH3 domain of Src and activates Src tyrosine kinase in C2C12 cells. *Biochem.J.* **352:** 883-892.

Kang, Q., Cao, Y., Zolkiewska, A., 2001, Direct interaction between the cytoplasmic tail of ADAM 12 and the Src homology 3 domain of p85alpha activates phosphatidylinositol 3-kinase in C2C12 cells. *J.Biol.Chem.* **276:** 24466-24472.

Kawaguchi, N., Xu, X., Tajima, R., Kronqvist, P., Sundberg, C., Loechel, F., Albrechtsen, R., Wewer, U.M., 2002, ADAM 12 protease induces adipogenesis in transgenic mice. *Am.J.Pathol..* **160:** 1895-1903.

Kawaguchi, N., Sundberg, C., Kveiborg, M., Moghadaszadeh, B., Asmar, M., Dietrich, N., Thodeti, C.K., Nielsen, F.C., Moller, P., Mercurio, A.M., Albrechtsen, R., Wewer, U.M., 2003, ADAM12 induces actin cytoskeleton and extracellular matrix reorganization during early adipocyte differentiation by regulating beta1 integrin function. *J.Cell Sci.* **116:** 3893-3904.

Kheradmand, F. and Werb, Z., 2002, Shedding light on sheddases: role in growth and development. *Bioessays* **24:** 8-12.

Kronqvist, P., Kawaguchi, N., Albrechtsen, R., Xu, X., Schroder, H.D., Moghadaszadeh, B., Nielsen, F.C., Frohlich, C., Engvall, E., Wewer, U.M., 2002, ADAM12 alleviates the skeletal muscle pathology in mdx dystrophic mice. *Am.J.Pathol.* **161:** 1535-1540.

Kurisaki, T., Masuda, A., Osumi, N., Nabeshima, Y., Fujisawa-Sehara, A., 1998, Spatially- and temporally-restricted expression of meltrin alpha (ADAM12) and beta (ADAM19) in mouse embryo. *Mech.Dev.* **73:** 211-215.

Kurisaki, T., Masuda, A., Sudo, K., Sakagami, J., Higashiyama, S., Matsuda, Y., Nagabukuro, A., Tsuji, A., Nabeshima, Y., Asano, M., Iwakura, Y., Sehara-Fujisawa, A., 2003, Phenotypic analysis of Meltrin alpha (ADAM12)-deficient mice: involvement of Meltrin alpha in adipogenesis and myogenesis. *Mol.Cell Biol.* **23:** 55-61.

Laigaard, J., Sorensen, T., Frohlich, C., Pedersen, B.N., Christiansen, M., Schiott, K., Uldbjerg, N., Albrechtsen, R., Clausen, H.V., Ottesen, B., Wewer, U.M., 2003, ADAM12: a novel first-trimester maternal serum marker for Down syndrome. *Prenat.Diagn.* **23:** 1086-1091.

Le Pabic, H., Bonnier, D., Wewer, U.M., Coutand, A., Musso, O., Baffet, G., Clement, B., Theret, N., 2003, ADAM12 in human liver cancers: TGF-beta-regulated expression in stellate cells is associated with matrix remodeling. *Hepatology* **37:** 1056-1066.

Lee, C.C., Pons, F., Jones, P.G., Bies, R.D., Schlang, A.M., Leger, J.J., Caskey, C.T., 1993, Mdx transgenic mouse: restoration of recombinant dystrophin to the dystrophic muscle. *Hum.Gene Ther.* **4:** 273-281.

Loechel, F., Gilpin, B.J., Engvall, E., Albrechtsen, R., Wewer, U.M., 1998, Human ADAM 12 (meltrin alpha) is an active metalloprotease. *J.Biol.Chem.* **273:** 16993-16997.

Loechel, F., Overgaard, M.T., Oxvig, C., Albrechtsen, R., Wewer, U.M., 1999, Regulation of human ADAM 12 protease by the prodomain. Evidence for a functional cysteine switch. *J.Biol.Chem.* **274:** 13427-13433.

Loechel, F., Fox, J.W., Murphy, G., Albrechtsen, R., Wewer, U.M., 2000, ADAM 12-S cleaves IGFBP-3 and IGFBP-5 and is inhibited by TIMP-3. *Biochem. Biophys. Res. Commun.* **278:** 511-515.

Loechel, F. and Wewer, U.M., 2001, Activation of ADAM 12 protease by copper. *FEBS Lett.* **506:** 65-68.

Moake, J.L., 2004, von Willebrand factor, ADAMTS-13, and thrombotic thrombocytopenic purpura. *Semin.Hematol.* **41:** 4-14.

Moghadaszadeh, B., Albrechtsen, R., Guo, L.T., Zaik, M., Kawaguchi, N., Borup, R.H., Kronqvist, P., Schroder, H.D., Davies, K.E., Voit, T., Nielsen, F.C., Engvall, E., Wewer, U.M., 2003, Compensation for dystrophin-deficiency: ADAM12 overexpression in skeletal muscle results in increased alpha 7 integrin, utrophin and associated glycoproteins. *Hum.Mol.Genet.* **12:** 2467-2479.

Mori, S., Tanaka, M., Nanba, D., Nishiwaki, E., Ishiguro, H., Higashiyama, S., Matsuura, N., 2003, PACSIN3 binds ADAM12/meltrin alpha and up-regulates ectodomain shedding of heparin-binding epidermal growth factor-like growth factor. *J.Biol.Chem.* **278:** 46029-46034.

Moss, M.L. and Bartsch, J.W., 2004, Therapeutic benefits from targeting of ADAM family members. *Biochemistry* **43:** 7227-7235.

Nagase, H., Itoh, Y., Binner, S., 1994, Interaction of alpha 2-macroglobulin with matrix metalloproteinases and its use for identification of their active forms. *Ann.N.Y.Acad.Sci.* **732:** 294-302.

Nagase, H. and Kashiwagi, M., 2003, Aggrecanases and cartilage matrix degradation. *Arthritis Res.Ther.* **5:** 94-103.

Roy R, Wewer UM, Zurakowski D, Pories SE, Moses MA. ADAM12 cleaves ECM proteins: Correlation with cancer status and stage. *J Biol Chem* in press.

Sahin, U., Weskamp, G., Kelly, K., Zhou, H.M., Higashiyama, S., Peschon, J., Hartmann, D., Saftig, P., Blobel, C.P., 2004, Distinct roles for ADAM10 and ADAM17 in ectodomain shedding of six EGFR ligands. *J.Cell Biol.* **164:** 769-779.

Schlondorff, J. and Blobel, C.P., 1999, Metalloprotease-disintegrins: modular proteins capable of promoting cell-cell interactions and triggering signals by protein-ectodomain shedding. *J.Cell Sci.* **112:** 3603-3617.

Seals, D.F. and Courtneidge, S.A., 2003, The ADAMs family of metalloproteases: multidomain proteins with multiple functions. *Genes Dev.* **17:** 7-30.

Shi, Z., Xu, W., Loechel, F., Wewer, U.M., Murphy, L.J., 2000, ADAM 12, a disintegrin metalloprotease, interacts with insulin-like growth factor-binding protein-3. *J.Biol.Chem.* **275:** 18574-18580.

Skubitz, K.M. and Skubitz, A.P., 2004, Gene expression in aggressive fibromatosis. *J.Lab Clin.Med.* **143:** 89-98.

Sottrup-Jensen, L., 1994, Role of internal thiol esters in the alpha-macroglobulin-proteinase binding mechanism. *Ann.N.Y.Acad.Sci.* **737:** 172-187.

Stocker, W. and Bode, W., 1995, Structural features of a superfamily of zinc-endopeptidases: the metzincins. *Curr.Opin.Struct.Biol.* **5:** 383-390.

Sundberg, C. Thodeti, C.K, Kveiborg, M., Larsson, C., Parker, P., Albrechtsen R., Wewer, U.M., Regulation of ADAM12 cell-surface expression by protein kinase C epsilon. *J Biol Chem* in press

Suzuki, A., Kadota, N., Hara, T., Nakagami, Y., Izumi, T., Takenawa, T., Sabe, H., Endo, T., 2000, Meltrin alpha cytoplasmic domain interacts with SH3 domains of Src and Grb2 and is phosphorylated by v-Src. *Oncogene 2000.* **19:** 5842-5850.

Tang, B.L., 2001, ADAMTS: a novel family of extracellular matrix proteases. *Int.J.Biochem.Cell Biol.* **33:** 33-44.

Thodeti, C.K., Albrechtsen, R., Grauslund, M., Asmar, M., Larsson, C., Takada, Y., Mercurio, A.M., Couchman, J.R., Wewer, U.M., 2003, ADAM12/syndecan-4 signaling promotes beta 1 integrin-dependent cell spreading through protein kinase Calpha and RhoA. *J.Biol.Chem.* **278:** 9576-9584.

Tian, B.L., Wen, J.M., Zhang, M., Xie, D., Xu, R.B., Luo, C.J., 2002, The expression of ADAM12 (meltrin alpha) in human giant cell tumours of bone. *Mol.Pathol.* **55:** 394-397.

Tsujimoto, M., Mizutani, S., Adachi, H., Kimura, M., Nakazato, H., Tomoda, Y., 1992, Identification of human placental leucine aminopeptidase as oxytocinase. *Arch.Biochem.Biophys.* **292:** 388-392.

White, J.M., 2003, ADAMs: modulators of cell-cell and cell-matrix interactions. *Curr.Opin.Cell Biol.* **15:** 598-606.

Yagami-Hiromasa, T., Sato, T., Kurisaki, T., Kamijo, K., Nabeshima, Y., Fujisawa-Sehara, A., 1995, A metalloprotease-disintegrin participating in myoblast fusion. *Nature* **377:** 652-656.

Yoshiyama, K., Higuchi, Y., Kataoka, M., Matsuura, K., Yamamoto, S., 1997, CD156 (human ADAM8): expression, primary amino acid sequence, and gene location. *Genomics* **41:** 56-62.

Zhao, Z., Gruszczynska-Biegala, J., Cheuvront, T., Yi, H., Von Der, M.H., Von Der, M.K., Kaufman, S.J., Zolkiewska, A., 2004, Interaction of the disintegrin and cysteine-rich domains of ADAM12 with integrin alpha7beta1. *Exp.Cell Res.* **298:** 28-37.

Zolkiewska, A., 1999, Disintegrin-like/cysteine-rich region of ADAM 12 is an active cell adhesion domain. *Exp.Cell Res.* **252:** 423-431.

# Chapter 7

# ADAM13 FUNCTION IN DEVELOPMENT

## *Prototypical or unique ADAM?*

Dominique Alfandari
*Department of Veterinary and Animal Sciences, University of Massachusetts, Amherst, Massachusetts, USA.*

Abstract: ADAM13 is a cell surface metalloprotease containing a disintegrin domain. It was cloned in an effort to identify ADAM proteins that function during early embryogenesis in the frog *Xenopus laevis*. ADAM13 is most similar in sequence to the meltrins (ADAM12, 19 and 33) and is expressed as a zygotic messenger RNA at the Mid Blastula Transition (MBT). The protein is expressed in Cranial Neural Crest (CNC) cells and somites during neurulation and subsequent tailbud formation. ADAM13 is a protease that can cleave itself and the extracellular matrix (ECM) protein fibronectin. Xenopus XTC cells expressing ADAM13 can remodel a fluorescent fibronectin substrate while cells expressing a dominant negative form of ADAM13 cannot. The adhesive domain of ADAM13 (DC) binds to the heparin-binding domain of fibronectin at the same site as the proteoglycan syndecan. Mutations in fibronectin that abolish syndecan binding also prevent ADAM13 association. The proteolytic activity of ADAM13 is essential for CNC migration in two of the three main pathways (Branchial and Hyoid but not Mandibular), while this activity is not essential for CNC migration *in vitro*. ADAM13 proteolytic specificity depends on the adhesive region of the extracellular domain (DC) rather than on the metalloprotease domain. Finally, ADAM13 activity is controlled by two specific regions of its cytoplasmic domain. The juxtamembrane domain controls protein level while the proline-rich carboxyl terminal region binds to proteins containing an SH3-domain. One of these SH3-containing proteins, PACSIN-2, binds to and down regulates ADAM13 proteolytic function in embryos.

Key words: Embryo, development, *Xenopus laevis*, extracellular matrix, fibronectin, Cranial Neural Crest.



*N. M. Hooper and U. Lendeckel (eds.), The ADAM Family of Proteases,* 147-169.

# 1. INTRODUCTION

Embryonic development can be described as a series of cell fate decisions that progressively restrict their potential. From the original totipotent first cell (the fertilized egg), the processes of cell division, cell determination, cell movement and cell differentiation collaborate to form a functional embryo. It is clear that what gives each specific organ its function is a combination of both specialized cell types and the three dimensional organization that shapes this organ. This organization is built during embryogenesis when cells migrate from one place to another in the embryo. It is also believed that during the adult life, the same type of cell movement may be responsible for the dissemination of tumor cells for metastasis to occur. The work described in this chapter focuses on advances made to understand the function of ADAM13 during early embryo development.

## 1.1 Basic embryology notions

The frog embryo was one of the first model systems used to study embryogenesis. The main reason why so many embryologists choose amphibians is because they generate large numbers (several hundreds) of sizeable eggs (≅1mm in diameter) that develop in water, outside of the mother, thus allowing both observations and experimental manipulations to be performed at any stage of development. Because the essential steps of embryogenesis are usually conserved in all vertebrates, this model is still used widely to advance the understanding of early embryogenesis. A few hours after fertilization the embryo is a hollow ball of cells composed of two cell types: the ectoderm which will give rise to the epidermis (skin), and the endoderm from which the digestive system is built. At this stage, the endoderm signals to the ectoderm so that ectodermal cells in contact with the endoderm become mesoderm. Mesodermal cells will form the intermediate tissues (located between the skin and the internal organs) including bones, kidney, blood and muscles. At gastrulation, both the mesoderm and the endoderm are internalized while the ectoderm expands to cover the entire external surface of the embryo. During this active phase of cell movements, the mesoderm signals to the ectoderm to form the neural tissue, from which, the brain and spinal cord will develop. These four original embryonic tissues (ectoderm, mesoderm, endoderm and neurectoderm) are the basis for all the organs of the adult, and subsequent tissue interactions will induce more specialized cell types as new functions are needed. One such interaction occurs at the interface between the neurectoderm and the ectoderm. Interactions between these tissues give rise to a new cell type called the neural crest. In the trunk, the neural crest cells migrate and produce pigment cells and ganglia of the peripheral nervous system (e.g. dorsal root ganglia). In the head, these cells (Cranial Neural Crest or CNC) also migrate and form most of the bones, cartilage and ganglia of the face. Neural crest cells have been the focus of many studies over the years because of their ability to migrate long distances, invade various areas of the embryo and differentiate into otherwise unrelated cell types. It is thought that these abilities are shared with the most invasive cancer cells. Neural crest cells are also studied in order to understand the signals that make them differentiate into a wide array of cell types. It is possible that similar signals will allow embryonic stem cells to differentiate and provide "replacement parts" for damaged tissues and organs in humans.

## 1.2 Cell migration

Cell migration has been extensively studied *in vitro* using immortalized cell lines in order to understand the molecules that play a role both in the actual locomotion but also in the control of this motility. Generally one cell needs to release its attachment from its neighbor, increase its binding to the ECM and reorganize its cytoskeletal components to initiate migration. During migration, the cell needs to constantly regulate the strength of its adhesion to the ECM so that the leading edge (front) can firmly attach to the substrate while the trailing edge (back) of the same cell can detach and allow the actual translocation. The number of proteins involved in this adhesion/de-adhesion process is large and includes receptors for the ECM (integrins), kinases and phosphatases, small G-proteins, and cytoskeletal proteins (Dzamba *et al.* 2002; Webb *et al.* 2002). While all this is sufficient for cell migration *in vitro* on a flat artificial substrate, it has long been thought that proteases, which modify the extracellular environment and act as a "snow plow" may be necessary for cell migration *in vivo* through a dense ECM (Moerman, 1999; Murphy and Gavrilovic, 1999; Nabeshima *et al.* 2002; Stetler-Stevenson and Yu, 2001). The discovery that a new family of proteins (containing A Disintegrin And a Metalloprotease; ADAM), with similarity to matrix metalloproteases (MMP), is expressed on the surface of naturally migrating cells adds further support to the "snow plow" hypothesis.

## 1.3 Proteins containing A Disintegrin And Metalloprotease domains (ADAM)

ADAM was originally identified as a cellular family of proteins sharing common domains and organization with snake venom metalloproteases

(Seals and Courtneidge, 2003; Weskamp and Blobel, 1994; Wolfsberg *et al.* 1995). In snake venom, a domain called disintegrin binds with high affinity to integrins present at the surface of platelets, preventing coagulation. This disintegrin domain is associated with a metalloprotease domain that can degrade components of endothelial tissue and produce lesions. The combination of these two domain functions produces internal bleeding that can result in the death of the victim (Gutierrez and Rucavado, 2000; Jia *et al.* 1996; Matsui *et al.* 2000). The first identified cellular ADAMs were found on the surface of mature spermatozoa so that a role as a counter receptor for integrins present on the egg cell surface was rapidly proposed (Evans *et al.* 1995). Over 33 ADAMs have now been identified and about 50% of these have a metalloprotease catalytic active site sequence (HExxHxxGxxH, single letter amino acid code) compatible with a proteolytic function (Black and White, 1998). The domain organization of cellular ADAMs is conserved even in the proteins with no proteolytic activity. The N-terminal extracellular domain consists of the pro-domain, the metalloprotease domain, the disintegrin and cysteine-rich domains, and an EGF-repeat. The pro-domain interacts with the metalloprotease domain and serves both as a chaperone protein during folding and to keep the metalloprotease inactive. The pro-domain is removed, either in the Trans-Golgi Network or at the cell surface by furin-like proteases. ADAM metalloproteases can be inhibited by ion chelators such as EDTA and O-phenanthroline, protein inhibitors like the TIMPs, and small chemical inhibitors (hydroxamates) that can bind directly to the active site. Substrates of ADAM metalloproteases include cytokines, growth factors and their receptors, insulin-like growth factor binding protein (IGF-BP), prion protein, amyloid precursor protein, Notch, Delta and extracellular matrix proteins (Schlondorff and Blobel, 1999; Stone *et al.* 1999). At present, it is not clear whether ADAM metalloproteases recognize a specific amino acid sequence for cleavage or if the structure and proximity of the substrate is the critical factor. As indicated by the number of potential substrates, proteolytically active ADAMs have been implicated in many biological processes during development, cellular differentiation and pathological processes.

### 1.3.1 Meltrins

ADAM13 is part of a sub-family of ADAM called Meltrins that includes ADAM12, 19 and 33. The founding member of this family meltrin α (ADAM12) was initially discovered in myoblasts. Overexpression of a truncated form lacking both the pro- and metalloprotease domain was able to promote myoblast fusion into myotubes in the C2C12 cell line (Yagami-Hiromasa *et al.* 1995). More recently, several groups have provided evidence that ADAM12 may in fact be essential during myogenesis and muscle regeneration (Cao *et al.* 2003; Gilpin *et al.* 1998; Kronqvist *et al.* 2002; Moghadaszadeh *et al.* 2003) as well as in adipogenesis and osteogenesis (Abe *et al.* 1999; Kawaguchi *et al.* 2002; Kurisaki *et al.* 2003; Tian *et al.* 2002). ADAM19 is expressed during early mouse embryogenesis in a wide array of tissue but is most abundant in neurons, dorsal root ganglia and other peripheral nervous system ganglia in the craniofacial region (Kurisaki *et al.* 1998). ADAM19 has been shown to cleave one of the ErbB ligands, Neuregulin, from the cell surface (Shirakabe *et al.* 2001). Inactivation of ADAM19 by knockout results in mice with defects in heart formation (Kurohara *et al.* 2004; Zhou *et al.* 2004). ADAM33 was first identified as a gene responsible for increased susceptibility to asthma (Van Eerdewegh *et al.* 2002). ADAM33 is a metalloprotease, which can cleave *in vitro* the Kit ligand precursor (Zou *et al.* 2004), however the mechanism by which ADAM33 plays a role in the pathophysiology of asthma remains unknown (Cakebread *et al.* 2004).

## 2. ADAM13 CLONING AND FUNCTIONAL CHARACTERIZATION

### 2.1 ADAM13 cloning and sequence comparison

*Xenopus leavis* ADAM13 was cloned by homology PCR using conserved sequences in the metalloprotease active site and the disintegrin loop of known ADAM proteins. The PCR amplification was carried out with Neurula stage cDNA and the amplified fragment was used to screen a cDNA library to obtain full-length clones (Alfandari et al., 1997). The entire ADAM13 coding sequence is over 3000 bp and encodes a 915 amino acid protein that is closely related to human ADAM19 and ADAM33 (53% identity), and to a lesser extend to ADAM12 (50% identity). Interestingly, there is no true homologue of ADAM13 in the human and mouse genome, but ADAM19 shares some striking similarity both in its sequence and biochemical characteristics (see **2.2**). Despite these similarities, none of the known ADAMs identified in other vertebrates share ADAM13 expression pattern during early embryonic development (see **2.3**).

ADAM13 possesses all the characteristic domains of ADAM proteins including the pro-, metalloprotease, disintegrin, cysteine-rich, and EGF-repeat domains attached to a single pass transmembrane and a relatively long cytoplasmic domain. The pro-domain possesses the conserved cysteine at position 170 that has been described to maintain the metalloprotease inactive by interacting with the Zinc present in the catalytic site (Van Wart and Birkedal-Hansen, 1990). While the pro-form of ADAM13 is clearly inactive, the involvement of the pro-domain cysteine has not been investigated. The metalloprotease domain contains the consensus active-site residues for a zinc-dependent metalloprotease, HEXGHXGXXHD, where the three histidines coordinate the zinc ion and the glutamic acid, in position two, catalyzes the proteolytic cleavage. The disintegrin loop of ADAM13 shares all amino acids previously identified to promote integrin $\alpha9\beta1$ binding (Eto et al., 2000). The sequence at the tip of the disintegrin loop A-G-S does not correspond to any known integrin ligand. The cysteine-rich domain contains one hydrophobic moment and a dibasic sequence resembling a proteolytic cleavage site (KDRR). The cytoplasmic domain of ADAM13 is rich in prolines and basic residues with an expected isoelectric point of 12.5. The ADAM13 cytoplasmic domain contains 6 proline-rich repeats containing the minimal sequence PXXP that correspond to SH3 binding motifs. There are three type I repeats (K/RXXPXXP) and one type II (PXXPXK/R). The type I repeat allows binding of SH3 domains in the same orientation (N to C-term) while the type II promotes binding in the reverse orientation (Feng S, 1994). The ADAM13 cytoplasmic domain contains putative serine (4) and threonine (4) phosphorylation sites (Blom *et al.* 1999).

## 2.2 ADAM13 biochemical characterization

ADAM13 is first synthesized as a 120 kDa precursor form, which is then cleaved before expression at the cell surface in the protease active 100 kDa mature form. The mature polypeptide is highly unstable in Triton X-100 protein extract and disappears in ten minutes at 20°C in the absence of metalloprotease inhibitors. While the mature form disappears in these conditions, a 90 kDa form, recognized by antibodies to the cysteine-rich domain, accumulates (Gaultier and Alfandari, unpublished). This ADAM13 fragment contains the extracellular domain but no cytoplasmic domain. This proteolysis is not detected when a mutated form of ADAM13 (E/A) lacking catalytic activity, is used in the extract (Alfandari *et al.* 2001). Upon transfection in Cos-7 cells, ADAM13 maturation occurs and a second proteolytic cleavage generates a small polypeptide that is released from the cell surface. Its size (55 kDa) and immuno-reactivity show that this fragment contains the metalloprotease, disintegrin and cysteine-rich (MDC) domains of ADAM13. This fragment is not observed in the supernatant of Cos-7 cells transfected with the E/A mutant suggesting that it is likely generated by an auto-proteolytic process (Gaultier *et al.* 2002). Interestingly, a similar autoproteolytic cleavage in the cysteine-rich domain of ADAM19 is also observed in cells transfected with the wild-type but not the E/A ADAM19 cDNA (Kang et al., 2002). ADAM13 shed MDC polypeptide remains proteolytically active as demonstrated by its ability to bind to the suicide substrate α2-macroglobulin (α2M). The shedding of ADAM13's extracellular domain is increased by phorbol ester (PMA) and inhibited by some metalloprotease inhibitors (Marimastat, BB3103; Gaultier and Alfandari, unpublished). The exact cleavage site remains unknown but is located either near the C-terminus of the cysteine-rich domain or in the EGF repeat. Together these results show that ADAM13 is an active metalloprotease that can cleave itself both *in vivo* and *in vitro*.

The ADAM13 adhesive region is composed of both disintegrin and cysteine-rich domains (DC). This adhesive region is capable of binding to extracellular matrix proteins including fibronectin and laminin but not collagen IV. Using GST fusion proteins we have identified the second heparin-binding domain of fibronectin as the target for ADAM13 DC binding. Mutation of two amino acids, identified as critical for syndecan 4 binding to this domain (PPRR to PPTM), also abolishes ADAM13 binding to fibronectin (Gaultier *et al.* 2002). Interestingly, the shed MDC region of ADAM13 also binds to fibronectin suggesting that this active metalloprotease may associate with fibronectin and laminin-rich extracellular matrix *in vivo*. Two lines of evidence suggest that fibronectin may in fact be a substrate for ADAM13 proteolytic activity. First, when ADAM13 is immunoprecipitated in the presence of a reversible hydroxamate inhibitor (BB3103) and incubated with Xenopus plasma fibronectin, with or without the inhibitor, a shorter fragment (90 kDa), recognized by a monoclonal antibody to fibronectin appears only when the inhibitor is washed away. Second, when Xenopus fibroblast (XTC), expressing a low endogenous level of ADAM13, are placed on a fluorescent fibronectin substrate, they remodel this substrate after 24 hours in culture. In the same conditions, XTC cells transfected with the E/A mutant, which has been described as a dominant negative protein, do not remodel the substrate. In contrast, overexpression of ADAM13 in XTC cells promotes a more extensive and faster (18 hours) remodeling of the substrate (Alfandari *et al.* 2001). These results suggest that fibronectin is a substrate for ADAM13 proteolytic activity and that binding of the adhesive region of ADAM13 to fibronectin could help select and/or present the substrate to the metalloprotease domain. Other results presented below (**2.5**) further suggest that the adhesive region of ADAM13 is responsible for the specificity of ADAM13 proteolysis.

## 2.3 ADAM13 expression

In Xenopus, large amounts of mRNA and protein are stored in the oocyte and there is no de novo transcription during the first hours of development. At the thirteenth cell division, about 8 hours after fertilization (Mid Blastula Transition or MBT), the zygotic genome is activated, and gene transcription

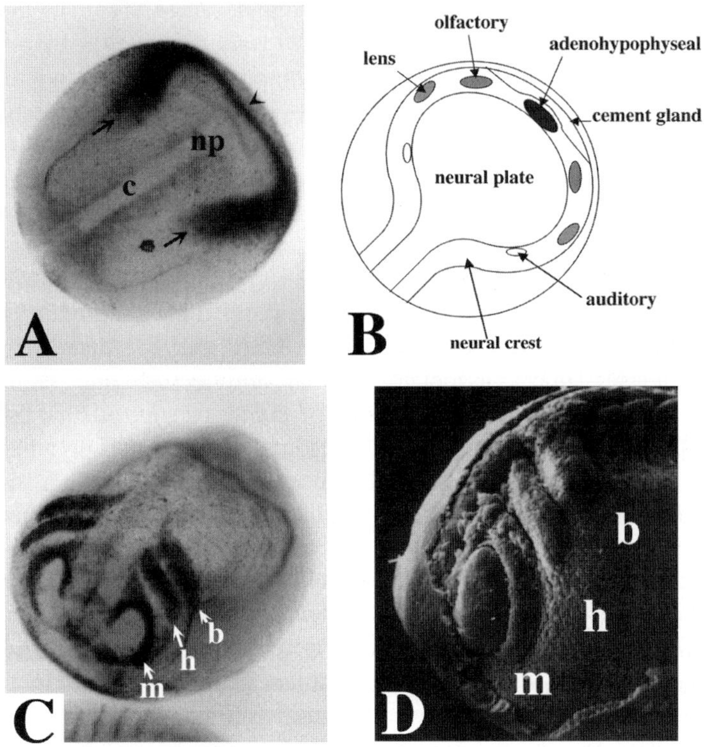


*Figure 1*. Expression of ADAM13. ADAM13 mRNA was localized by whole mount *in situ* hybridization at the neurula stage (**A**) and early tailbud stage (**C**) and is compared to a neurula fate map (**B**) and a scanning electron microphotograph (SEM) of similar stage embryos (**D**). **A**) On this dorsal view of the neurula, the ADAM13 staining (arrows) is mainly localized to the lateral and anterior sides of the neural plate (**np**) and on either side of the notochord (**c**). This staining corresponds to regions of the neural folds including the sensory placodes and the CNC (**B**) as well as the pre-somitic mesoderm. **C**) At tailbud stage, the staining is localized to the 3 streams of CNC: Mandibular (**m**), Hyoid (**h**) and Branchial (**b**). **D**) The streams of migrating CNC are clearly visible at this stage by SEM (Sadaghiani and Thiebaud, 1987).

starts (Etkin, 1988; Krieg and Melton, 1985). ADAM13 mRNA is expressed at the MBT (stage 8), but there is also a very low level of maternal expression that can be detected using RT-PCR (35 cycles). It is not clear whether this maternal mRNA is being translated at this stage as ADAM13 protein can only be detected at the beginning of gastrulation (Stage 10). The first polypeptide detected by ADAM13 specific antibodies is 120 kDa, which corresponds to the calculated molecular weight for the fully glycosylated precursor (containing the pro-domain).

During gastrulation, a second polypeptide appears at 100 kDa, a size that corresponds to the calculated molecular weight of the mature ADAM13 (lacking the pro-domain). During later embryonic development, additional smaller forms are also recognized by ADAM13-specific antibodies. The nature of these additional forms is discussed in the previous section.

The expression patterns of ADAM13 mRNA and protein are one of the most exciting features of this gene. By *in situ* hybridization, the ADAM13 mRNA is first detected at the end of gastrulation (Stage 13). At this stage, the mRNA localizes to the anterior and lateral border of the neural plate as well as two narrow bands on either side of the notochord (Fig.1). The position of these cells coincide with the presumptive territory of the cranial neural crest (CNC), the sensory placodes and the somitic mesoderm, as defined by the various fate maps of Xenopus embryos (Bauer *et al.* 1994; Eagleson and Harris, 1990; Heasman *et al.* 1985; Kumano and Smith, 2002). During later development, the expression of ADAM13 remains associated with migrating CNC cells and with some cells in the central nervous system and in the somites. By immunofluorescence, the localization of the protein can be refined in all three tissues. In the central nervous system, the protein is localized to the outer layer of the neural tube in contact with laminin. In the somites, the protein is found at the intersomitic boundary where muscle cell precursors from one somite (functional unit) are in contact with the next group. In the CNC, the ADAM13 protein is found mainly at the area of cell-cell contact rather than in direct contact with the extracellular matrix (Alfandari, unpublished observations).

## 2.4 ADAM13 function in cranial neural crest cell migration

As presented above, ADAM13 is expressed in two main regions of the embryo: the cranial neural crest (CNC) and the somites (Fig.1). To date, evidence of ADAM13 function has only been collected in the CNC. The CNC are a transient population of cells that emerge from the lateral edges of the neural plate and migrate ventrally to form muscle, bones, cartilage and ganglia of the face. In Xenopus, the initial population of CNC condenses into three main streams that migrate in separate pathways (Fig.1). The most anterior stream, called mandibular, originates from the mesencephalon and migrates around the growing optic vesicle and later to the prosencephalon. The second segment, called hyoid, arises from the anterior part of the rhombencephalon and migrates into the first branchial arch and toward the pharynx. The most posterior stream, branchial, originates from the posterior part of the rhombencephalon and migrates into the second and third branchial arches (Sadaghiani and Thiebaud, 1987). ADAM13 is expressed in CNC cells from all three segments, but it is not clear whether all the cells in each segment express the protein.

Functional analysis of ADAM13, during CNC migration, was performed using the dominant negative E/A mutant protein. While the mechanism by which this mutant acts as a dominant negative remains unclear, similar constructs have been used to analyze the function of ADAM10 during Drosophila embryogenesis (Pan and Rubin, 1997; Rooke *et al.* 1996). In these reports, the authors showed that a construct with a point mutation in the essential glutamic acid (E) within the active site was able to prevent cleavage by ADAM10 of Notch thus interfering with the Notch-Delta signaling during neurogenesis. They also showed that a mutant from which the entire metalloprotease domain was removed could act in the same way.

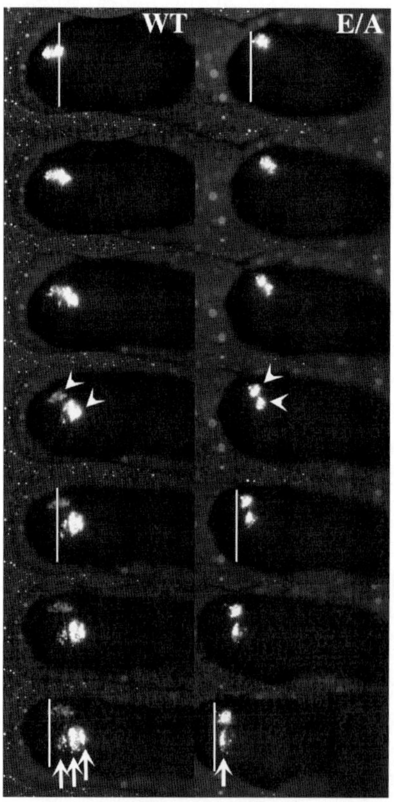


*Figure 2.* Time-lapse video microscopy of CNC migrating *in vivo*. CNC explants expressing GFP with either the wild type (left) or the E/A-ADAM13 (right) were grafted into unlabeled host embryos. The upper panels represent the grafts at t=0 and the following lower panels represent 90 minutes intervals. The embryos are oriented with the anterior to the left and dorsal up. The white arrowheads point to the CNC. At the end, the fluorescent CNC cells have migrated into 3 segments in the control embryos (left, white arrows) but only into one segment for the CNC expressing the E/A mutant (right, white arrows).

We have shown that the E/A-ADAM13 mutant can prevent fibronectin remodeling in XTC cells. *In vivo*, injection of the E/A mutant together with the wild type ADAM13 rescues phenotypes associated with ADAM13 ectopic expression (cement gland; (Alfandari et al., 2001). This assay will be described in more detail in the next paragraph (**2.5**). Together these assays show that the E/A mutant can interfere with ADAM13 function *in vivo*.

When E/A-ADAM13 mRNA is injected in embryos, migration of the CNC is perturbed as shown by *in situ* hybridization using several markers specific for neural crest cells (Xtwist, Xslug). In these experiments, CNC cells are present in the most anterior pathway (mandibular), but are not detected in both the hyoid and branchial pathways. While this study shows that the cells are missing, it does not prove that they failed to migrate into the pathways, but that they may have differentiated into another cell type that no longer expresses the CNC markers. Another problem with these experiments is that the expression of the dominant negative is not restricted to just the CNC cells, but affects a wider cell population. Because the exact mechanism of the dominant negative function is not known, it may be causing the defect in CNC position by affecting the surrounding tissues rather than the CNC cells themselves. To remedy both shortcomings, homotypic grafts between a donor embryo, expressing both GFP and either the wild type or the E/A-ADAM13 protein, and a host untreated embryo were analyzed. At neurula stage, explants corresponding to the pre-migratory CNC were transplanted from the donor embryos to the exact same site in the host embryos. Grafted cells were then followed during their migration using time-lapse fluorescent microscopy to visualize GFP expressing cells (Fig.2). These experiments showed that CNC cells expressing the E/A dominant negative form of ADAM13 failed to migrate into the hyoid and branchial pathways, but either remained in their original position or migrated into the mandibular pathway (Alfandari *et al.* 2001).

While the E/A-ADAM13 dominant negative clearly perturbs CNC cell migration, it is still unclear how it may do so. There are several hypotheses that can be proposed based on the vast neural crest literature and the growing number of substrates identified for various ADAM proteins (Fig.3). First, based exclusively on the result from ADAM13 studies, one model would be that ADAM13 on the CNC cell surface binds to and cleaves fibronectin present in the migration pathways. This cleavage could serve several functions including the opening of spaces between the tightly joined mesoderm and the epidermis allowing passage of migrating CNC cells, or the unmasking of cryptic adhesion sites on fibronectin that could promote cell migration. While ADAM13 clearly binds and cleaves fibronectin, it also binds laminin (Gaultier *et al.* 2002). Thus, laminin may also be a proteolytic substrate for ADAM13, and cryptic pro-migratory sites activated upon

that there is no positive or negative effect of this interaction *in vivo*. Similar results were obtained in co-transfection experiments in Cos-7 cells where ADAM13 proteolytic activity was directly tested in the presence or absence of wild type Src-1 (Cousin and Alfandari, unpublished). It is, nevertheless, possible that dominant negative or active forms of Src-1 may acquire the ability to bind to and control the ADAM13 function.

The second protein identified in this screen, PACSIN2, belongs to an emerging family of adaptor proteins called PCH (Pombe CDC15 Homology). The proteins of this family are characterized by an FCH domain (Fer CIP4 Homology) at the N-terminus followed by multiple coiled coil domains (also called CDC15 N-terminal domain) and an SH3 domain at their C-terminal. Between the coiled coil region and the SH3 domain, PCH proteins display other domains involved in protein-protein interaction. There are either two (PACSIN-1, Syndapin-I, Syndapin-IIs) or three (PACSIN-2, Syndapin-III) NPF motifs that are thought to mediate binding to EH (Epsin Homology), followed by one PxxP motifs (Proline rich repeats that promote SH3 binding, PACSIN-3) or PEST sequences (PST-PIP).

PCH proteins have been implicated in a wide array of biological processes including control of actin remodeling (Carnahan and Gould, 2003; Da Costa et al., 2003; Miller et al., 2004; Modregger et al., 2000; Nikki et al., 2002; Qualmann and Kelly, 2000; Spencer et al., 1997), inhibition of cell surface receptors (Li *et al.* 1998), and endocytosis (Simpson et al., 1999). In the case of PSTPIP, the actin-remodeling and receptor inhibition function both seem to depend on its capacity to recruit the tyrosine phosphatases PTP-PEST or PTP-HSCF. These phosphatases would then dephosphorylate WASP, Abl, or CD2 and inhibit their functions (Cong *et al.* 2000; Cote *et al.* 2002; Li *et al.* 1998).

Xenopus PACSIN2, like many other PCH family members, is localized to the ruffle membrane and to some vesicles, where it co-localizes with ADAM13. During embryogenesis, both proteins are expressed in CNC cells and in the pre-somitic mesoderm. PACSIN2 has a wider expression pattern than ADAM13 suggesting that ADAM13 is most likely not its only target. Using the cement gland assay described above (**2.5.1**), we have shown that overexpression of PACSIN2 rescues embryos expressing ADAM13 ectopically. In contrast, PACSIN2 lacking the SH3 domain (responsible for binding to the ADAM13 cytoplasmic domain) was incapable of rescuing the cement gland expansion, suggesting that *in vivo* binding of the two proteins was essential for the rescue (Cousin *et al.* 2000). The data accumulated on the function of PCH family proteins suggests that PACSIN2 could inhibit ADAM13 activity by at least two different mechanisms: 1) by controlling the level of ADAM13 present at the cell surface (endocytose/exocytose), or 2) by assembling a complex of regulatory proteins at the ADAM13 cytoplasmic domain. Our initial work showing that PACSIN2 does not affect the level of ADAM13 at the cell surface (Cousin *et al.* 2000) does not support the first hypothesis. One of the most likely candidates as part of a regulatory complex involving PACSIN2 is a protein tyrosine phosphatase (PTP-PEST). In mouse, this phosphatase binds to CD2BP1 (a close relative to PACSIN2), and inhibits CD2 binding to CD58 (Li et al., 1998). The domain responsible for this interaction is the CTH domain of PTP-PEST. However, the Xenopus PTP-PEST lacks this domain and does not bind to PACSIN2 (Cousin and Alfandari, 2004).

## 3. CONCLUSIONS

Is ADAM13 a curiosity of frog embryos with a unique function in cranial neural crest cell migration or is it the homologue of ADAM19? While this question cannot be answered with any certainty until the Xenopus genome is sequenced, available structural and functional data suggest that the two proteins may be functional equivalents. First, ADAM13 is clearly a meltrin with most similarity to ADAM12, 19 and 33. Partial sequencing suggests that there is a Xenopus orthologue of ADAM12 expressed at the larvae stage (Alfandari *et al.* 1997), but until the full length sequence is known, it is impossible to be certain of the identity of this other frog ADAM. In birds, orthologues for both ADAM12 and 19 have been identified while ADAM13 remains elusive (Lewis and Newgreen, personal communication). In the absence of Xenopus genome data, we will need to identify Xenopus ADAM19 or prove its absence to conclude about ADAM13 orthologues in higher vertebrates.

While ADAM13 remains unique to frogs, several other ADAM proteins have been implicated in cell migration during embryogenesis and in tissue culture assays. In the nematode *C. elegans*, one transmembrane ADAM, unc-71, and one secreted ADAM-like protein, mig-17, are involved in cell migration (Huang *et al.* 2003; Nishiwaki *et al.* 2000). The *C. elegans* unc-71 encodes an ADAM with no proteolytic activity. Mutations of unc-71 that affect the disintegrin domain perturb both axon guidance and sex myoblast migration through a yet unknown mechanism (Huang *et al.* 2003). In the same species, mig-17, which is secreted by the migrating distal tip cells, binds to the basement membrane via its disintegrin-like domain and promotes cell migration via its metalloprotease activity (Nishiwaki *et al.* 2000). While there are obvious differences between mig-17 and ADAM13 proteins (e.g. one is secreted while the other is membrane-bound), the mechanism of action for both proteins may in fact be very similar. For example, ADAM13 may shed itself from the surface of CNC cells (as shown in Cos-7 cells), bind to the extracellular matrix in the pathways and promote cell migration. Thus, these two distantly related ADAMs may be functionally equivalent between worms and frogs.

In cell culture experiments, ADAM proteins have been linked to the control of cell migration through two distinct mechanisms. In the first model, ADAM adhesive domains are used as substrate for cell migration. These experiments show that most ADAM proteins can, if placed in the right conditions (salt and concentration), promote cell adhesion and eventually cell migration through integrin binding (Nath *et al.* 2000). In a second model, ADAM proteolytic activity is responsible for the shedding of a cell surface molecule that binds to a receptor responsible for signals controlling cell migration (Gutwein *et al.* 2000; Hundhausen *et al.* 2003; Mechtersheimer *et al.* 2001). While ADAM13's adhesive domain can promote cell adhesion via a β1 integrin (Gaultier *et al.* 2002), this domain does not promote CNC cell adhesion or migration when used as a substrate. Similarly, preliminary results show that the adhesive region of ADAM13 does not affect CNC cell adhesion and migration when plated over fibronectin (Gaultier and Alfandari, unpublished). Thus, we favor the second model whereby ADAM13 would shed a protein from the cell surface that would in turn activate a signaling pathway controlling cell migration. Finally, ADAM15 appears to be involved in glomerular mesenglial cell migration possibly through remodeling of a collagen IV extracellular matrix (Martin *et al.* 2002). If this hypothesis is confirmed, it will support our model (**2.4**) by which ADAM proteolysis of extracellular matrix components is required for cell migration (Fig.3A).

## ACKNOWLEDGEMENTS

This work was supported in part by grants from the University of Massachusetts (Startup fund), the National Institute of Health in the laboratory of Dr. Douglas DeSimone at the University of Virginia (HD26402 and DE14365), and by French Granting agencies at the University Pierre and Marie Currie (UPMC, CNRS, ARC). Thanks to Drs. Alban Gaultier and Helene Cousin for their participation in the work and discussion of this manuscript. Thanks also to Anna Maria Salicioni, Joe Jerry, and Amy Burnside for their critical reading of the manuscript.

# REFERENCES

Abe, E., Mocharla, H., Yamate, T., Taguchi, Y., and Manolagas, S. C. (1999). Meltrin-alpha, a fusion protein involved in multinucleated giant cell and osteoclast formation. *Calcif Tissue Int* **64,** 508-515.

Alfandari, D., Cousin, H., Gaultier, A., Smith, K., White, J. M., Darribere, T., and DeSimone, D. W. (2001). Xenopus ADAM 13 is a metalloprotease required for cranial neural crest-cell migration. *Curr Biol* **11,** 918-30.

Alfandari, D., Wolfsberg, T. G., White, J. M., and DeSimone, D. W. (1997). ADAM 13: a novel ADAM expressed in somitic mesoderm and neural crest cells during Xenopus laevis development. *Dev Biol* **182,** 314-330.

Bauer, D. V., Huang, S., and Moody, S. A. (1994). The cleavage stage origin of Spemann's Organizer: analysis of the movements of blastomere clones before and during gastrulation in Xenopus. *Development* **120,** 1179-89.

Black, R. A., and White, J. M. (1998). ADAMs: Focus on the protease domain. *Curr. Op. in Cell Biol.* **10,** 654-659.

Blom, N., Gammeltoft, S., and Brunak, S. (1999). Sequence and structure-based prediction of eukaryotic protein phosphorylation sites. *J Mol Biol* **294,** 1351-62.

Cakebread, J. A., Haitchi, H. M., Holloway, J. W., Powell, R. M., Keith, T., Davies, D. E., and Holgate, S. T. (2004). The role of ADAM33 in the pathogenesis of asthma. *Springer Semin Immunopathol* **25,** 361-75.

Cao, Y., Zhao, Z., Gruszczynska-Biegala, J., and Zolkiewska, A. (2003). Role of metalloprotease disintegrin ADAM12 in determination of quiescent reserve cells during myogenic differentiation in vitro. *Mol Cell Biol* **23,** 6725-38.

Carnahan, R. H., and Gould, K. L. (2003). The PCH family protein, Cdc15p, recruits two F-actin nucleation pathways to coordinate cytokinetic actin ring formation in Schizosaccharomyces pombe. *J Cell Biol* **162,** 851-62.

Cong, F., Spencer, S., Cote, J. F., Wu, Y., Tremblay, M. L., Lasky, L. A., and Goff, S. P. (2000). Cytoskeletal protein PSTPIP1 directs the PEST-type protein tyrosine phosphatase to the c-Abl kinase to mediate Abl dephosphorylation. *Mol Cell* **6,** 1413-23.

Cote, J. F., Chung, P. L., Theberge, J. F., Halle, M., Spencer, S., Lasky, L. A., and Tremblay, M. L. (2002). PSTPIP Is a Substrate of PTP-PEST and Serves as a Scaffold Guiding PTP-PEST Toward a Specific Dephosphorylation of WASP. *J Biol Chem* **277,** 2973-2986.

Cousin, H., and Alfandari, D. (2004). A PTP-PEST like protein affects (alpha)5(Beta)1 integrin-dependent matrix assembly cell adhesion and migration in xenopus gastrula. *Dev Biol* **265,** 416-432.

Cousin, H., Gaultier, A., Bleux, C., Darribere, T., and Alfandari, D. (2000). PACSIN2 is a regulator of the metalloprotease/disintegrin ADAM13. *Dev. Biol.* **227,** 197-210.

Da Costa, S. R., Sou, E., Xie, J., Yarber, F. A., Okamoto, C. T., Pidgeon, M., Kessels, M. M., Mircheff, A. K., Schechter, J. E., Qualmann, B., and Hamm-Alvarez, S. F. (2003). Impairing actin filament or syndapin functions promotes accumulation of clathrin-coated vesicles at the apical plasma membrane of acinar epithelial cells. *Mol Biol Cell* **14,** 4397-413.

Dzamba, B. J., Bolton, M. A., and DeSimone, D. W. (2002). "Cell Adhesion: The integrin family of cell adhesion molecules." Oxford University Press, Oxford.

Eagleson, G. W., and Harris, W. A. (1990). Mapping of the presumptive brain regions in the neural plate of Xenopus laevis. *J Neurobiol* **21,** 427-40.

Etkin, L. D. (1988). Regulation of the mid-blastula transition in amphibians. *Dev Biol (N Y 1985)* **5,** 209-25.

Nabeshima, K., Inoue, T., Shimao, Y., and Sameshima, T. (2002). Matrix metalloproteinases in tumor invasion: role for cell migration. *Pathol Int* **52,** 255-64.

Nath, D., Slocombe, P. M., Webster, A., Stephens, P. E., Docherty, A. J., and Murphy, G. (2000). Meltrin gamma(ADAM-9) mediates cellular adhesion through alpha(6)beta(1 )integrin, leading to a marked induction of fibroblast cell motility. *J Cell Sci* **113,** 2319-28.

Nieuwkoop, P. D., and Faber, J. (1994). "Normal table of Xenopus Laevis (Daudin)." Garland Publishing, Inc., New York.

Nikki, M., Merilainen, J., and Lehto, V. P. (2002). FAP52 regulates actin organization via binding to filamin. *J Biol Chem* **277,** 11432-40.

Nishiwaki, K., Hisamoto, N., and Matsumoto, K. (2000). A metalloprotease disintegrin that controls cell migration in Caenorhabditis elegans. *Science* **288,** 2205-8.

Pan, D., and Rubin, G. M. (1997). Kuzbanian controls proteolytic processing of Notch and mediates lateral inhibition during Drosophila and vertebrate neurogenesis. *Cell* **90,** 271-280.

Pirila, E., Sharabi, A., Salo, T., Quaranta, V., Tu, H., Heljasvaara, R., Koshikawa, N., Sorsa, T., and Maisi, P. (2003). Matrix metalloproteinases process the laminin-5 gamma 2-chain and regulate epithelial cell migration. *Biochem Biophys Res Commun* **303,** 1012-7.

Qualmann, B., and Kelly, R. B. (2000). Syndapin isoforms participate in receptor-mediated endocytosis and actin organization. *J Cell Biol* **148,** 1047-62.

Rooke, J., Pan, D., Xu, T., and Rubin, G. M. (1996). KUZ, a conserved metalloprotease-disintegrin protein with two roles in Drosophila neurogenesis. *Science* **273,** 1227-1231.

Sadaghiani, B., and Thiebaud, C. H. (1987). Neural crest development in the Xenopus laevis embryo, studied by interspecific transplantation and scanning electron microscopy. *Dev Biol* **124,** 91-110.

Schlondorff, J., and Blobel, C. P. (1999). Metalloprotease-disintegrins: modular proteins capable of promoting cell-cell interactions and triggering signals by protein-ectodomain shedding. *J Cell Sci* **112,** 3603-3617.

Seals, D. F., and Courtneidge, S. A. (2003). The ADAMs family of metalloproteases: multidomain proteins with multiple functions. *Genes Dev* **17,** 7-30.

Shirakabe, K., Wakatsuki, S., Kurisaki, T., and Fujisawa-Sehara, A. (2001). Roles of Meltrin beta /ADAM19 in the processing of neuregulin. *J Biol Chem* **276,** 9352-8.

Simpson, F., Hussain, N. K., Qualmann, B., Kelly, R. B., Kay, B. K., McPherson, P. S., and Schmid, S. L. (1999). SH3-domain-containing proteins function at distinct steps in clathrin- coated vesicle formation. *Nat Cell Biol* **1,** 119-124.

Sive, H., and Bradley, L. (1996). A sticky problem: the Xenopus cement gland as a paradigm for anteroposterior patterning. *Dev Dyn* **205,** 265-280.

Sive, H. L., Hattori, K., and Weintraub, H. (1989). Progressive determination during formation of the anteroposterior axis in Xenopus laevis. *Cell* **58,** 171-180.

Smith, K. M., Gaultier, A., Cousin, H., Alfandari, D., White, J. M., and DeSimone, D. W. (2002). The cysteine-rich domain regulates ADAM protease function in vivo. *J Cell Biol* **159,** 893-902.

Spencer, S., Dowbenko, D., Cheng, J., Li, W., Brush, J., Utzig, S., Simanis, V., and Lasky, L. A. (1997). PSTPIP: a tyrosine phosphorylated cleavage furrow-associated protein that is a substrate for a PEST tyrosine phosphatase. *Journal Of Cell Biology* **138,** 845-60.

Stetler-Stevenson, W. G., and Yu, A. E. (2001). Proteases in invasion: matrix metalloproteinases. *Semin Cancer Biol* **11,** 143-52.

Stone, A. L., Kroeger, M., and Sang, Q. X. (1999). Structure-function analysis of the ADAM family of disintegrin-like and metalloproteinase-containing proteins (review). *J Protein Chem* **18,** 447-465.

Tian, B. L., Wen, J. M., Zhang, M., Xie, D., Xu, R. B., and Luo, C. J. (2002). The expression of ADAM12 (meltrin alpha) in human giant cell tumours of bone. *Mol Pathol* **55,** 394-7.

Van Eerdewegh, P., Little, R. D., Dupuis, J., Del Mastro, R. G., Falls, K., Simon, J., Torrey, D., Pandit, S., McKenny, J., Braunschweiger, K., Walsh, A., Liu, Z., Hayward, B., Folz, C., Manning, S. P., Bawa, A., Saracino, L., Thackston, M., Benchekroun, Y., Capparell, N., Wang, M., Adair, R., Feng, Y., Dubois, J., FitzGerald, M. G., Huang, H., Gibson, R., Allen, K. M., Pedan, A., Danzig, M. R., Umland, S. P., Egan, R. W., Cuss, F. M., Rorke, S., Clough, J. B., Holloway, J. W., Holgate, S. T., and Keith, T. P. (2002). Association of the ADAM33 gene with asthma and bronchial hyperresponsiveness. *Nature* **418,** 426-30.

Van Wart, H. E., and Birkedal-Hansen, H. (1990). The cysteine switch: a principle of regulation of metalloproteinase activity with potential applicability to the entire matrix metalloproteinase gene family. *Proc Natl Acad Sci U S A* **87,** 5578-82.

Webb, D. J., Parsons, J. T., and Horwitz, A. F. (2002). Adhesion assembly, disassembly and turnover in migrating cells -- over and over and over again. *Nat Cell Biol* **4,** E97-100.

Weskamp, G., and Blobel, C. P. (1994). A family of cellular proteins related to snake venom disintegrins. *Proc Natl Acad Sci U S A* **91,** 2748-2751.

Weskamp, G., Kratzschmar, J., Reid, M. S., and Blobel, C. P. (1996). MDC9, a widely expressed cellular disintegrin containing cytoplasmic SH3 ligand domains. *J Cell Biol* **132,** 717-726.

Wolfsberg, T. G., Straight, P. D., Gerena, R. L., Huovila, A. P., Primakoff, P., Myles, D. G., and White, J. M. (1995). ADAM, a widely distributed and developmentally regulated gene family encoding membrane proteins with a disintegrin and metalloprotease domain. *Dev Biol* **169,** 378-383.

Yagami-Hiromasa, T., Sato, T., Kurisaki, T., Kamijo, K., Nabeshima, Y., and Fujisawa-Sehara, A. (1995). A metalloprotease-disintegrin participating in myoblast fusion. *Nature* **377,** 652-656.

Zhou, H. M., Weskamp, G., Chesneau, V., Sahin, U., Vortkamp, A., Horiuchi, K., Chiusaroli, R., Hahn, R., Wilkes, D., Fisher, P., Baron, R., Manova, K., Basson, C. T., Hempstead, B., and Blobel, C. P. (2004). Essential role for ADAM19 in cardiovascular morphogenesis. *Mol Cell Biol* **24,** 96-104.

Zou, J., Zhu, F., Liu, J., Wang, W., Zhang, R., Garlisi, C. G., Liu, Y. H., Wang, S., Shah, H., Wan, Y., and Umland, S. P. (2004). Catalytic activity of human ADAM33. *J Biol Chem* **279,** 9818-30.

Many metalloproteases have been suggested to mediate ectodomain shedding. However, the role of ADAM 17 in shedding has been distinctively confirmed through the analysis of knockout mice and cell lines established from them. The use of these cell lines has allowed a clear advance in our knowledge of the repertoire of substrates, also known as the degradome (Lopez-Otin and Overall 2002), of ADAM 17. In this chapter, we will update data on this subject. On the other hand, several relevant questions remain obscure. How does ADAM 17 select its substrates? What are the mechanisms and factors that regulate its biosynthesis, intracellular trafficking and proteolytic activity? We will review the limited information available on these topics and comment on current efforts to elucidate them.

## 2. REGULATION OF GENE EXPRESSION

ADAM 17 mRNA was originally found in many tissues, suggesting an ubiquitous and constitutive expression typical of housekeeping proteins. However, substantial differences in ADAM 17 mRNA levels are found in different tissues and developmental stages. The highest levels of ADAM17 mRNA in adult are found in heart, placenta, skeletal muscle, pancreas, spleen, thymus, prostate, testes, ovary and small intestine. In fetal tissues the highest levels are detected in brain, lung, liver and kidney (Black *et al.* 1997). On the other hand, several studies show differences in the levels of expression of ADAM 17 in pathological conditions such as osteoarthritis (OA), rheumatoid arthritis (RA), acute myocardial infarction (AMI) and breast cancer. Finally, certain compounds can modulate the expression of ADAM 17 in cell lines, further supporting the notion that the expression of ADAM 17 can be regulated under certain conditions.

Up-regulation of human ADAM 17 mRNA has been demonstrated in human OA (Amin 1999): while ADAM 17 mRNA is apparent in human osteoarthritis-affected cartilage, no expression of ADAM 17 is found in normal cartilage. The presence of ADAM 17 mRNA in OA leads to the expression of the active protein that, in turn, increases TNF-$\alpha$ production which has a causal role in the pathogenesis of the disease (Amin 1999). An independent study also demonstrated the up-regulation of ADAM 17 and TNF-$\alpha$ mRNA in OA and in RA, compared with normal cartilage (Patel *et al.* 1998). Again, TNF-$\alpha$ produced by OA-affected cartilage is functionally active and acts in an autocrine or paracrine fashion to modulate other inflammatory/chemotactic mediators, such as Interleukin-8 (IL-8) (Patel *et al.* 1998). During acute myocardial infarction (AMI), the necrotic tissue produces pro-inflammatory cytokines locally, priming the systemic inflammatory response and resulting in further exacerbation of myocardial damage (Akatsu *et al.* 2003). The levels of ADAM 17 and TNF-α mRNA in leucocytes from patients with AMI are elevated with respect to leukocytes from healthy individuals (Akatsu *et al.* 2003). The authors of this study hypothesized that the "ADAM 17/TNF-α system" in circulating leucocytes is activated during the early stages of AMI and may have a negative impact on the clinical outcome.

A dramatic up-regulation of ADAM 17 expression has also been described in mammary tumors; however, in this case the authors analyzed protein levels and it is not known whether the up-regulation observed corresponded to differences in mRNA levels or to some post-transcriptional regulation (Borrell-Pages *et al.* 2003). ADAM 17 is required for the shedding of several ligands of the Epidermal Growth Factor Receptor (EGFR) (see Table 1), a tyrosine kinase coupled receptor that regulates multiple biological processes, such as gene expression, cellular proliferation, angiogenesis and apoptosis (Yarden and Sliwkowski 2001). Up-regulation of the EGFR and/or similar receptors has been linked to the development of several tumors, including breast cancer, making it possible that overexpression of ADAM 17 contributes to the malignant phenotype. Supporting this hypothesis, the shedding of proTGF-α, an EGFR ligand is required for the activation of the receptor, and for the development of tumors in nude mice. Thus, ADAM 17 participates in tumorigenesis, at least in some models, and an enhanced expression of the metalloprotease may contribute to the development of the malignancy (Borrell-Pages *et al.* 2003).

Classical experiments using cells in culture also indicate that the expression of ADAM 17 is regulated. Treatment with phorbol esters, well-characterized activators of the protein kinase C (PKC), or cytokines such as TNF-α enhances the levels of ADAM17 mRNA. This up-regulation is more pronounced by the addition of lipopolysaccharide (LPS) or different cytokines and growth factors, indicating that a PKC- and TNF-α-mediated pathway can regulate the expression of the metalloprotease (Wang *et al.* 2000; Bzowska *et al.* 2004). On the other hand, nitrogen-containing biphosphonates (BPs), such as zoledronic acid (ZOL), are known to inhibit osteoclast (OC)-mediated bone resorption and are used to treat malignancy-associated bone disease (Russell and Rogers 1999). Treatment of primary human Osteoblast-like (OB-like) cells with ZOL results in an increase in ADAM 17 mRNA and protein levels (Pan *et al.* 2004). This up-regulation precedes a down-modulation of the membrane-associated RANKL (also known as TRANCE/OPGL/ODF), a cytokine which plays a pivotal role in osteoclastogenesis (Pan *et al.* 2004). Since ADAM 17 is required for the shedding of RANKL (Schlondorff *et al.* 2001), the effect of ZOL on RANKL shedding could be mediated by the regulation of ADAM 17 gene expression.

Despite evidence indicating that the expression of ADAM 17 can be regulated, at least in some models and pathological conditions, very little is known about the structure and functionality of the ADAM 17 gene promoter. Analysis of a DNA fragment that comprises the promoter and the 5' upstream region of the gene encoding murine ADAM 17, revealed the existence of several binding sites for the transcription factors Activating Protein 2 (AP2) and Stimulating protein 1 (Sp1) (Mizui *et al.* 1999). This fragment lacks a consensus TATA box but includes a GC-box, responsible for transcription of widely expressed genes having TATA-less promoters (Duffy *et al.* 2003), and a CCAAT box (Mizui *et al.* 1999). A deletion analysis showed that the region essential for basal transcription of ADAM 17 is the small fragment from nucleotide –290 to –1, which includes the GC and CCAATT boxes (Mizui *et al.* 1999). The lack of TATA box and Activator Protein 1 (AP1) elements in the promoter regions is common to different ADAMs: human ADAM 10, 11, 15, 17 and mouse ADAM 17, 19 promoters. Another common feature is that all promoters contain GC-rich regions and AP2 sites (Duffy *et al.* 2003). In summary, the current limited knowledge about ADAM17 promoter, has allowed the region that supports basal transcription to be determined. Although the mechanisms of regulation of transcription for ADAM 17 are presently unknown, phorbol esters have been shown to up-regulate transcription of other genes through AP2 and Sp1 transcription factors (Biggs *et al.* 1996; Zeng *et al.* 1997). It is therefore tempting to speculate that these transcription factors are involved in the regulation of ADAM 17 expression by phorbol esters.

In summary, current evidence indicates that despite the fact that ADAM 17 is broadly distributed in a variety of human tissues, its expression can be regulated; however, very little is still known on the mechanisms mediating this regulation. In several pathologies, the levels of expression of the metalloprotease are augmented, but it remains unclear whether this up-regulation is a cause or a consequence of the pathological condition. The development of transgenic mice will help to solve this question.

## 3. BIOSYNTHESIS AND INTRACELLULAR TRAFFICKING

As is the case for other members of the family, ADAM 17 is synthesized as a type I transmembrane protein containing the following identifiable domains: prodomain, catalytic, disintegrin, epidermal growth factor like (EGF-like) domain, transmembrane and cytoplasmic. This primary translation product enters the Endoplasmic Reticulum (ER) as a stably folded zymogen, in which the prodomain inhibits the catalytic domain.

Although, by analogy with other metzincins, it has been proposed that the mechanism of inhibition involves the cystein-switch model, recent reports indicate that the odd cysteine of the prodomain is not required for inhibition of ADAM 17 (Gonzales *et al.* 2004).

The maturation of ADAM 17 is tightly linked to its correct trafficking from the ER to the late Golgi, where the proteolytic removal of the prodomain takes place (reviewed in Zhou *et al.* 1999). Different evidence suggests that these initial steps in the trafficking of proADAM 17 are highly controlled. Independent mutations of ADAM 17 lead to the accumulation of the metalloprotease in the early secretory pathway (Borroto *et al.* 2003; Villanueva de la Torre *et al.* 2004). On the other hand, analysis of the processing of ADAM 17 in a variety of breast cancer cell lines showed that the ratio between proADAM 17 and ADAM 17 is variable and that these differences correlate with the rate of transport of proADAM17 from the early secretory pathway to late Golgi (Ruiz-Paz, S, unpublished results). Thus, these results point to the existence of a mechanism that regulates the rate of transport of ADAM 17 from the ER to the Golgi system. At least one of the factors mediating this transport seems to be saturable: transient overexpression of ADAM 17 invariably leads to the accumulation of the proform in the early secretory pathway, while little or no processed form can be detected (Vincent *et al.* 2001; Fan *et al.* 2003; Li and Fan 2004; Villanueva de la Torre *et al.* 2004). In agreement with this notion, accumulation of proADAM 17 increases in a linear fashion by transfection with increasing amounts of ADAM 17 cDNA (Slack *et al.* 2001; Peiretti *et al.* 2003b). In contrast with the full-length molecule, some deletion constructs of ADAM 17 seem to traffic efficiently when overexpressed. ADAM 17 lacking the cytoplasmic and transmembrane domains is efficiently secreted in insect cells (Milla *et al.* 1999). Similarly, a construct encoding the pro and the catalytic domain of human ADAM 17, was readily secreted by cell lines (Maskos *et al.* 1998). Thus, it is possible that the cytoplasmic domain plays an inhibitory role in the correct progression through the secretory pathway of overexpressed ADAM 17. Future characterization of the trafficking of additional ADAM 17 deletion constructs will help to clarify this issue.

The prodomain and metalloprotease domain of all ADAMs are separated by a typical furin-like recognition site (Black *et al.* 1997; Moss *et al.* 1997). Several pieces of evidence show that, indeed, furin is, at least in part, responsible for the constitutive removal of the prodomain of proADAM 17. *In vitro*, furin cleaves proADAM 17 at the predicted site (Peiretti *et al.* 2003a). Furthermore, the processing of ADAM 17 is reduced in cells devoid of furin activity and can be augmented by transfection with the proprotein convertase (Borroto *et al.* 2003). However, the fact that in the absence of furin ADAM 17 is still processed, albeit to a lesser extent, clearly shows that other furin-like convertases can remove the prodomain of the metalloprotease (Zhou *et al.* 1999; Endres *et al.* 2003; Srour *et al.* 2003). In summary, although furin clearly participates in the processing of ADAM 17, in its absence other convertases can remove the prodomain of the metalloprotease.

After the proteolytic removal of the prodomain, mature ADAM 17 exits the late Golgi and rapidly reaches the cell surface, where, in contrast with other ADAMs, only ADAM 17 devoid of prodomain can be detected (Schlondorff *et al.* 2000). Since the shedding of several substrates can be increased by different compounds, additional regulatory mechanisms probably control the activity of mature ADAM 17 at the cell surface.

## 4. DEGRADOME OF ADAM 17

Given the structural and functional diversity of cell surface proteins that are shed and the high number of Zn-dependent metalloproteases potentially involved, an obvious early hypothesis implicated various Zn-dependent metalloproteases in ectodomain shedding. Each one would be endowed with the ability to cleave a group of substrates: for example, ADAM 17 would be involved in shedding of proTNF-α and similar proteins (Black *et al.* 1997; Moss *et al.* 1997). The generation of mouse cells genetically deficient for active ADAM 17 (ADAM $17^{\Delta Zn/\Delta Zn}$) (Black *et al.* 1997), allowed this theory to be tested. Unexpectedly, the shedding of a diversity of proteins unrelated to proTNF-α is affected in ADAM 17 knockout cells (see Table 1), invalidating, at least partially, the initial hypothesis, and pointing to a central role of this particular member of the ADAM family in ectodomain shedding.

Independent genetic evidence confirmed this notion. M1 and M2 are CHO mutant cell lines initially isolated because of their defective proTGF-α shedding (Arribas and Massague 1995; Arribas *et al.* 1996). Characterization of the metalloprotease in ectodomain shedding of these mutant cells showed a generalized defect in ectodomain shedding similar to that of ADAM 17ΔZn/ΔZn fibroblasts (Arribas *et al.* 1996; Merlos-Suarez *et al.* 2001; Tsou *et al.* 2001; Borrell-Pages *et al.* 2003; Borroto *et al.* 2003; Le Gall *et al.* 2003). Despite initial evidence indicating otherwise (Merlos-Suarez *et al.* 1998; Borroto *et al.* 2003), it has been recently shown that M1 and M2 cells have mutations affecting ADAM 17 (Li and Fan 2004; Villanueva de la Torre *et al.* 2004). Since transfection of wild type ADAM 17 rescues the

*Table 1.* Current ADAM 17 degradome

| Substrates of ADAM 17 | Functional category | Cleavage site | Reference |
|---|---|---|---|
| ProTNF-α | Cytokine | SPLAQA ↓ VRSSSR | (Mezyk *et al.* 2003) |
| Fractalkine | | Juxtamembrane sequence | (Garton *et al.* 2001) |
| TRANCE | | IVGPQR ↓ FSGAPA | (Mohan *et al.* 2002) |
| ProTGF-α | EGF family ligand | HADLLA ↓ VVAASQ | (Althoff *et al.* 2001; Sahin *et al.* 2004) |
| ProNRGα-2C | | Not established | (Montero *et al.* 2000) |
| ProHB-EGF | | CHGLTP ↓ VENPLYTYDH | (Hinkle *et al.* 2004; Sahin *et al.* 2004) |
| Proamphiregulin | | CGEK ↓ SMSTHEDDKDLSK | (Hinkle *et al.* 2004; Sahin *et al.* 2004) |
| Betacellulin | | CERVDLFY ↓ LQQDRGQ | (Hinkle *et al.* 2004; Sahin *et al.* 2004) |
| Epiregulin | | CEHFFL ↓ TVHQPLSKEY | (Hinkle *et al.* 2004; Sahin *et al.* 2004) |
| p 75 TNF-α RII | Receptors | APGAV ↓ HLPQP | (Mezyk *et al.* 2003) |
| p 55 TNF-α R | | PQIEN ↓ VKGTE | (Mezyk *et al.* 2003) |
| CD30 | | Cysteine-rich domains 2 and 5 | (Hansen *et al.* 2000; Hansen *et al.* 2004) |
| IL-6R α | | TSLPVQ ↓ DSSSVP | (Mullberg *et al.* 1994) |
| IL-1R II | | Not established | (Reddy *et al.* 2000) |
| GHR | | MSP ↓ FTCEEDFR | (Wang *et al.* 2002) |
| HER4 | | HGLSLPVENRLYTYDH[a] | (Mohan *et al.* 2002) |
| Notch | | PYKIEA ↓ VKSEPV | (Mohan *et al.* 2002) |
| CD40 | | Not established | (Contin *et al.* 2003) |
| P75 $^{NTR}$ | | MGSSQP ↓ VVTRGT | (Weskamp *et al.* 2004) |
| CD44 | | Not established | (Nagano *et al.* 2004) |
| IL-15Ralpha | | Not established | (Budagian *et al.* 2004) |
| Hybrid receptor SorLA | | Not established | (Guo *et al.* 2002) |
| LDL receptor | | Not established | (Guo *et al.* 2002) |
| AXL receptor Tyrosine kinase | | Not established | (Guo *et al.* 2002) |
| L-selectin | Adhesion molecule | QKLDK ↓ SFSMI | (Mezyk *et al.* 2003) |
| Collagen XVII | | Not established | (Franzke *et al.* 2002) |
| VCAM-1 | | Not established | (Garton *et al.* 2003) |
| MUC-1 | Others | QYKTEA ↓ ASRYNL | (Thathiah *et al.* 2003; Thathiah and Carson 2004), |
| β-APP | | VHHQK ↓ LVFFA | (Mezyk *et al.* 2003) |
| Cellular Prion protein | | Not established | (Vincent *et al.* 2001) |
| GP of Ebola virus | | VDKTLPD ↓ QGDNDNWW | (Dolnik *et al.* 2004) |

[a] The physiological cleavage site has not been established conclusively

mutant phenotype (Li and Fan 2004; Villanueva de la Torre *et al.* 2004), it can be concluded that both cell lines are also functional knockout for ADAM 17, further supporting a crucial role of the metalloprotease in ectodomain shedding.

In addition to the genetic approaches, a variety of *in vitro* analysis has been carried out to characterize the degradome of ADAM 17 using the soluble recombinant protein and model peptides. However, in general, these approaches have lead to less consistent results because of several limitations: on the one hand, the activity of ADAM17 appears to require a cellular environment; on the other hand, the conformation of the substrates also seems to determine the effectiveness of the shedding reaction.

Apparently ADAM 17 must be anchored to the plasma membrane to display full activity in cells (Reddy *et al.* 2000) and it appears to act only on substrates expressed within the same cell (Reddy *et al.* 2000). Co-expression of full-length ADAM 17 with proTNF-α in mammalian cells efficiently produced soluble TNF-α, while the truncated soluble form has no effect on the shedding of the cytokine (Itai *et al.* 2001). Furthermore, when ADAM 17-expressing cells are co-cultured with cells expressing proTNF-α, no significant enhancement of proTNF-α shedding is observed (Itai *et al.* 2001). These results open the possibility that in the context of an intact cell, membrane-bound ADAM 17 interacts with other proteins and/or acquires a different conformation that increases its affinity for certain substrates (Reddy *et al.* 2000).

Studies on the cleavage of model synthetic peptides by recombinant ADAM 17 are well exemplified by a recent report on the *in vitro* specificity of several ADAMs, including ADAM 17 (Fourie *et al.* 2003). ADAM 17 does not show preference for peptides containing the cleavage site of L-Selectin or P75 TNF receptor, two substrates of the metalloprotease, as judged by the phenotype of ADAM 17 knock out cells (Arribas *et al.* 1996; Peschon *et al.* 1998). Similarly, recombinant soluble ADAM 17 displays little or no activity against peptides containing the cleavage site of HER4, APP or IL-6R (Mohan *et al.* 2002), well characterized substrates of the metalloprotease, according to genetic evidence. Therefore, it appears that assays with model peptides do not always reflect the *in vivo* situation, probably because the secondary, tertiary and/or quaternary structures of some substrates are required for proper recognition by ADAM 17.

In conclusion, although several *in vitro* approaches aimed to identify members of the degradome of ADAM 17 have been undertaken, the results obtained in cell-free systems have to be cautiously interpreted. In contrast, the results obtained with ADAM $17^{\Delta Zn/\Delta Zn}$ cells and the mutant CHO cell lines are consistent and point to a central role of ADAM 17 in ectodomain shedding. To date, genetic evidence indicates that nearly 30 diverse transmembrane proteins are substrates of the metalloprotease (see Table 1) and presumably, more substrates are to be described. Thus, as discussed below, a crucial question is how ADAM 17 recognizes its substrates.

## 4.1 How does ADAM 17 select its substrates?

The characterization of the degradome of ADAM 17 clearly points to an unusually wide substrate specificity. However, since most cell surface proteins are not shed by the metalloprotease, there must be a common feature recognized in all substrates. Despite the effort of many labs, we are far from understanding the properties of this "common feature".

No common recognizable pattern around the cleavage site of the substrates of ADAM 17 described to date can be identified (see Table 1). Furthermore, mutational analysis of the cleavage site of several substrates, such as IL-6R or p55R revealed relaxed sequence specificities surrounding the cleavage site (Brakebusch *et al.* 1994; Mullberg *et al.* 2000).

Some authors have proposed that the extracellular juxtamembrane region is essential to confer susceptibility to shedding: introduction of such sequence from substrates of ADAM 17 allow constitutive shedding to chimeric reporter molecules (see, for example Arribas *et al.* 1997; Althoff *et al.* 2000). However, mutational analysis of these juxtamembrane stalk regions suggest that an unidentified structural feature, rather than the presence or absence of a specific amino acid sequence, determines the susceptibility and the efficiency of the shedding of ADAM 17 substrates (Hinkle *et al.* 2004). Several mutations do not affect shedding, indicating that such modifications allow the preservation of the structural integrity of the juxtamembrane domain (see, for example, Zhao *et al.* 2001). In contrast, in certain chimeric molecules the overall structure of the juxtamembrane region may be disrupted. That would explain why the exchange of the TNF-α cleavage sequence for that of IL-6R results in a chimeric protein that is resistant to shedding (Althoff *et al.* 2000). Similarly, some deletions of the Fractalkine (FKN) juxtamembrane domain abrogate phorbol ester-induced shedding in some contexts but not others (Garton *et al.* 2001). In summary, current evidence indicates that short extracellular juxtamembrane regions are recognized by ADAM 17 but the relevant feature is still unknown.

The formation of stable complexes between ADAM 17 and some of its substrates has also been suggested as a mechanism to mediate recognition. Sequences outside the juxtamembrane region would participate in the generation of these complexes, which would mediate specificity and, in addition, could enhance the efficiency of the cleavage. For example, the transient formation of a complex between ADAM 17 and the growth hormone receptor (GHR), precedes the shedding of the latter (Schantl *et al.*

These authors suggest that the phosphorylation of ADAM 17 in T735 could participate in the regulation of the maturation of ADAM 17, enhancing the production of the active form of the enzyme (Diaz-Rodriguez *et al.* 2002). In contrast with these results, other authors have shown that the cytoplasmic domain of ADAM 17 is phosphorylated only in serine (Fan *et al.* 2003). Mutagenesis analysis identified Ser819 as the major target for phosphorylation in response to fibroblast growth factor and serum and Ser791 as the major site for growth factor-induced dephosphorylation (Fan *et al.* 2003). The authors suggest that the growth factor-induced changes in the phosphorylation state of the ADAM 17 cytoplasmic domain did not regulate ADAM 17 activity because several mutations of the phosphorylation sites do not decrease the shedding of proTGF-α (Fan *et al.* 2003). In agreement with these results, it has been shown that ADAM 17 devoid of the cytoplasmic domain is able to shed TNF-α in cells stimulated with phorbol esters, suggesting that the cytoplasmic domain is not required for its phorbol ester-induced ectodomain shedding activity (Reddy *et al.* 2000). Thus, although several stimuli regulate the phosphorylation of the cytoplasmic domain of ADAM 17, the significance of this phosphorylation is still unclear.

Recently, a functional interaction between the disintegrin/cysteine-rich region of ADAM 17 and α5β1 integrin in membrane ruffles and focal adhesion has been found (Bax *et al.* 2004). It has been hypothesized that this interaction may regulate or target the metalloprotease activity of ADAM 17 (Bax *et al.* 2004). Since ADAM 17 interacts with the integrin through its disintegrin/cysteine rich region, analysis of the activity of ADAM 17 deletion mutants will likely help to test this hypothesis.

In conclusion, different physiological and non-physiological compounds activate ectodomain shedding. It has recently been established that MAPK are involved; however, the relevant targets of these kinases remain elusive, as is the mechanism that mediates the regulation of the activity of ADAM 17. Clearly, much more work should be done on this important area to identify the factors involved in the regulation of ADAM 17 activity and to elucidate the relevant mechanisms.

## 5.3 Constitutive shedding

Several substrates of ADAM 17 are also constitutively shed from untreated cells, and, in some case, this type of shedding seems to be independent of ADAM 17. For example, the basal shedding of APP is unaffected in cells derived from knockout ADAM 17 mice (Buxbaum *et al.* 1998). The candidate to act as an enzyme with constitutive α-secretase activity is ADAM 10 (Lammich *et al.* 1999), although experiments with cells derived from ADAM 10 knockout mice did not produce clear results (Hartmann *et al.* 2002). The reason for the relative inactivity of ADAM 17 in unstimulated cells is unknown, some authors have suggested that the majority of ADAM 17 molecules may be complexed with inhibitors (Doedens *et al.* 2003). In contrast with APP, the constitutive shedding of proTGF-α seems to depend also on the activity of ADAM 17. It has been shown that p38 MAPK signaling regulates this basal shedding (Fan and Derynck 1999). However, other metalloproteases are endowed with the ability to shed proTGF-α, although their physiological meaning is unknown (Merlos-Suarez *et al.* 2001).

## 5.4 Turning-off

Approximately 3 hours after treatment with phorbol esters, cells recover normal levels of proTGF-α at the cell surface, indicating that activated ADAM 17 is turned-off ≈ 2 hours after activation (Arribas *et al.* 1997). This result is in agreement with those showing that treatment of cells with phorbol esters during 1.5 h, negatively regulates the amount of the mature form of ADAM 17 (Doedens and Black 2000; Endres *et al.* 2003). Since ADAM 17 possesses an internalization motive (YESL) in its cytoplasmic domain and the effect of phorbol esters on the amount of mature ADAM 17 was inhibited by blocking endocytosis, it is possible that the effect of phorbol esters on ADAM 17 maturation depends on vesicle formation and endocytosis (Doedens and Black 2000). However, it is not clear if the mechanism of down-modulation observed with PMA is general because in cells activated with acetylcholine the levels of ADAM 17 remain unchanged (Endres *et al.* 2003).

# 6. PHYSIOLOGICAL ROLES OF ADAM 17

In addition to its well characterized role in proTNF-α shedding (reviewed in Moss and Lambert 2002), characterization of knockout mice show that the shedding of EGFR ligands by ADAM 17 is physiologically relevant. The phenotype of mice homozygous for an inactivating ADAM 17 mutation (ADAM 17 $^{Zn/Zn}$) is similar to that of TGF-α $^{-/-}$ mice (Luetteke *et al.* 1993; Mann *et al.* 1993). Both knockout mice display an identical wavy hair and open-eye-at-birth phenotype. Furthermore, similarities between ADAM 17 and EGFR knockout mice are also apparent (Miettinen *et al.* 1995; Sibilia and Wagner 1995). Thus, the shedding of proTGF-α and, likely, other EGFR ligands seems to be crucial for the activation of the EGFR. The phenotypic similarities between ADAM 17 $^{Zn/Zn}$, Egfr $^{-/-}$, and EGF family knockout mice accentuate the importance of soluble EGFR ligands, and thus proteolytic processing, for normal development.

## 6.1 EGFR transactivation

In addition to its ligands, a variety of external stimuli such as ultraviolet (UV)-and γ- radiation, various oxidants such as hydrogen peroxide, permanganate and heavy metal ions can activate the EGFR (Knebel *et al.* 1996; Weiss *et al.* 1997). Furthermore, activation of receptors apparently unrelated to the EGFR can also lead to the activation of the latter: the EGFR becomes rapidly activated upon stimulation of cells with the GPCR agonists endothelin-1, Lysophosphatidic acid (LPA) and thrombin, suggesting that there is a mechanism of cross-talk between GPCR and the EGFR pathways (Daub *et al.* 1996). The signals generated by either transactivation of the EGFR or direct EGF binding are indistinguishable (Daub *et al.* 1997). In 1999, the mechanism of transactivation become apparent when Prenzel *et al.* (1999) showed that the activation of GPCR induced the metalloprotease-dependent cleavage of proHB-EGF and subsequently, the EGFR activation. The identity of the metalloproteinase(s) involved and the mechanism of activation upon GPCR stimulation remain poorly characterized. Although some reports indicate that ADAM 17 is not involved in transactivation (Yan *et al.* 2002), others indicate otherwise (Gschwind *et al.* 2003); furthermore , since several reports show that ADAM 17 is the major sheddase of EGFR ligands in a variety of cells types, the involvement of ADAM 17 in EGFR transactivation seems a reasonable hypothesis (Peschon *et al.* 1998; Merlos-Suarez *et al.* 2001; Hart *et al.* 2004; Sahin *et al.* 2004).

## 6.2 Other transactivating pathways

Additional transactivating pathways have been described: osmotic and oxidative stress can activate ADAM 17 and induce the shedding of proHB-EGF causing the EGFR activation (Fischer *et al.* 2004). In fact, ADAM 17 appears to be generally involved in stress-stimulated shedding events (Zhang *et al.* 2001) but the mechanism involved is unknown. Stimulation of certain cells with TNF-α stimulates DNA replication by causing release of TGF-α that, in turn, activates the EGFR and multiple downstream intracellular signaling cascades that are required for DNA replication (Argast *et al.* 2004). Although the metalloprotease involved has not been characterize, ADAM 17 is an obvious candidate.

## 7. CONCLUSION

Since its discovery, ADAM 17 has been the subject of intense study. As a result, there have been considerable advances in the determination of its degradome and it has become clear that this ADAM constitutes a major sheddase that acts on a variety of transmembrane proteins. However, other important aspects of the biology of this metalloprotease remain poorly characterized. Particularly, we do not know how ADAM 17 selects its substrates or how its activity is regulated. Since the characterization of knockout mice show the important role of ADAM 17 during development and several evidences indicate the participation of the metalloprotease in the progression of different diseases, the elucidation of pendent questions seems of critical importance.

## ACKNOWLEDGEMENTS

We wish to thank Sonia Martínez Arca, Belén Santiago Josefat, Teresa Villanueva de la Torre and Joan Josep Bech Serrra for critical reading of the manuscript. This work was supported by grants from La Marató de TV3, Fondo de Investigación Sanitaria (PI021003) and the EMBO Young Investigator Programme (YIP) to J. A.

## REFERENCES

Akatsu, T.,Nakamura, M.,Satoh, M.Hiramori, K., 2003, Increased mRNA expression of tumour necrosis factor-alpha and its converting enzyme in circulating leucocytes of patients with acute myocardial infarction. *Clin Sci (Lond)*. **105**: (1): 39-44.

Althoff, K.,Reddy, P.,Voltz, N.,Rose-John, S.Mullberg, J., 2000, Shedding of interleukin-6 receptor and tumor necrosis factor alpha. Contribution of the stalk sequence to the cleavage pattern of transmembrane proteins. *Eur J Biochem*. **267**: (9): 2624-31.

Althoff, K.,Mullberg, J.,Aasland, D.,Voltz, N.,Kallen, K.,Grotzinger, J.Rose-John, S., 2001, Recognition sequences and structural elements contribute to shedding susceptibility of membrane proteins. *Biochem J*. **353**: (Pt 3): 663-72.

Amin, A. R., 1999, Regulation of tumor necrosis factor-alpha and tumor necrosis factor converting enzyme in human osteoarthritis. *Osteoarthritis Cartilage*. **7**: (4): 392-4.

Argast, G. M.,Campbell, J. S.,Brooling, J. T.Fausto, N., 2004, Epidermal growth factor receptor transactivation mediates tumor necrosis factor-induced hepatocyte replication. *J Biol Chem*.**279**: (33): 34530-6

Arribas, J. and Massague, J., 1995, Transforming growth factor-alpha and beta-amyloid precursor protein share a secretory mechanism. *J Cell Biol*. **128**: (3): 433-41.

Arribas, J.,Coodly, L.,Vollmer, P.,Kishimoto, T. K.,Rose-John, S.Massague, J., 1996, Diverse cell surface protein ectodomains are shed by a system sensitive to metalloprotease inhibitors. *J Biol Chem.* **271**: (19): 11376-82.

Arribas, J.,Lopez-Casillas, F.Massague, J., 1997, Role of the juxtamembrane domains of the transforming growth factor-alpha precursor and the beta-amyloid precursor protein in regulated ectodomain shedding. *J Biol Chem.* **272**: (27): 17160-5.

Arribas, J. and Borroto, A., 2002, Protein ectodomain shedding. *Chem Rev.* **102**: (12): 4627-38.

Arribas, J. and Merlos-Suarez, A., 2003, Shedding of plasma membrane proteins. *Curr Top Dev Biol.* **54**: 125-44.

Bax, D. V.,Messent, A. J.,Tart, J.,Van Hoang, M.,Kott, J.,Maciewicz, R. A.Humphries, M. J., 2004, Integrin alpha 5beta 1 and TNFalpha converting enzyme (TACE/ADAM-17) interact in vitro and colocalise in migrating HeLa cells. *J Biol Chem.* **279**: (21): 22377-86

Biggs, J. R.,Kudlow, J. E.Kraft, A. S., 1996, The role of the transcription factor Sp1 in regulating the expression of the WAF1/CIP1 gene in U937 leukemic cells. *J Biol Chem.* **271**: (2): 901-6.

Black, R. A.,Rauch, C. T.,Kozlosky, C. J.,Peschon, J. J.,Slack, J. L.,Wolfson, M. F.,Castner, B. J.,Stocking, K. L.,Reddy, P.,Srinivasan, S.,Nelson, N.,Boiani, N.,Schooley, K. A.,Gerhart, M.,Davis, R.,Fitzner, J. N.,Johnson, R. S.,Paxton, R. J.,March, C. J.Cerretti, D. P., 1997, A metalloproteinase disintegrin that releases tumour-necrosis factor-alpha from cells. *Nature.* **385**: (6618): 729-33.

Blobel, C. P., 2000, Remarkable roles of proteolysis on and beyond the cell surface. *Curr Opin Cell Biol.* **12**: (5): 606-12.

Borrell-Pages, M.,Fernandez-Larrea, J.,Borroto, A.,Rojo, F.,Baselga, J.Arribas, J., 2000, The carboxy-terminal cysteine of the tetraspanin L6 antigen is required for its interaction with SITAC, a novel PDZ protein. *Mol Biol Cell.* **11**: (12): 4217-25.

Borrell-Pages, M.,Rojo, F.,Albanell, J.,Baselga, J.Arribas, J., 2003, TACE is required for the activation of the EGFR by TGF-alpha in tumors. *Embo J.* **22**: (5): 1114-24.

Borroto, A.,Ruiz-Paz, S.,de la Torre, T. V.,Borrell-Pages, M.,Merlos-Suarez, A.,Pandiella, A.,Blobel, C. P.,Baselga, J.Arribas, J., 2003, Impaired trafficking and activation of tumor necrosis factor-alpha-converting enzyme in cell mutants defective in protein ectodomain shedding. *J Biol Chem.* **278**: (28): 25933-9.

Bosenberg, M. W.,Pandiella, A.Massague, J., 1992, The cytoplasmic carboxy-terminal amino acid specifies cleavage of membrane TGF alpha into soluble growth factor. *Cell.* **71**: (7): 1157-65.

Brakebusch, C.,Varfolomeev, E. E.,Batkin, M.Wallach, D., 1994, Structural requirements for inducible shedding of the p55 tumor necrosis factor receptor. *J Biol Chem.* **269**: (51): 32488-96.

Budagian, V.,Bulanova, E.,Orinska, Z.,Ludwig, A.,Rose-John, S.,Saftig, P.,Borden, E. C.Bulfone-Paus, S., 2004, Natural soluble IL-15Ralpha is generated by cleavage that involves the tumor necrosis factor-alpha -converting enzyme (TACE/ADAM17). *J Biol Chem.* **279**: (39): 40368-75

Buxbaum, J. D.,Gandy, S. E.,Cicchetti, P.,Ehrlich, M. E.,Czernik, A. J.,Fracasso, R. P.,Ramabhadran, T. V.,Unterbeck, A. J.Greengard, P., 1990, Processing of Alzheimer beta/A4 amyloid precursor protein: modulation by agents that regulate protein phosphorylation. *Proc Natl Acad Sci U S A.* **87**: (15): 6003-6.

Buxbaum, J. D.,Oishi, M.,Chen, H. I.,Pinkas-Kramarski, R.,Jaffe, E. A.,Gandy, S. E.Greengard, P., 1992, Cholinergic agonists and interleukin 1 regulate processing and secretion of the Alzheimer beta/A4 amyloid protein precursor. *Proc Natl Acad Sci U S A.* **89**: (21): 10075-8.

Buxbaum, J. D.,Liu, K. N.,Luo, Y.,Slack, J. L.,Stocking, K. L.,Peschon, J. J.,Johnson, R. S.,Castner, B. J.,Cerretti, D. P.Black, R. A., 1998, Evidence that tumor necrosis factor alpha converting enzyme is involved in regulated alpha-secretase cleavage of the Alzheimer amyloid protein precursor. *J Biol Chem.* **273**: (43): 27765-7.

Bzowska, M.,Jura, N.,Lassak, A.,Black, R. A.Bereta, J., 2004, Tumour necrosis factor-alpha stimulates expression of TNF-alpha converting enzyme in endothelial cells. *Eur J Biochem.* **271**: (13): 2808-20.

Contin, C.,Pitard, V.,Itai, T.,Nagata, S.,Moreau, J. F.Dechanet-Merville, J., 2003, Membrane-anchored CD40 is processed by the tumor necrosis factor-alpha-converting enzyme. Implications for CD40 signaling. *J Biol Chem.* **278**: (35): 32801-9.

Daub, H.,Weiss, F. U.,Wallasch, C.Ullrich, A., 1996, Role of transactivation of the EGF receptor in signalling by G-protein-coupled receptors. *Nature.* **379**: (6565): 557-60.

Daub, H.,Wallasch, C.,Lankenau, A.,Herrlich, A.Ullrich, A., 1997, Signal characteristics of G protein-transactivated EGF receptor. *Embo J.* **16**: (23): 7032-44.

Diaz-Rodriguez, E.,Montero, J. C.,Esparis-Ogando, A.,Yuste, L.Pandiella, A., 2002, Extracellular signal-regulated kinase phosphorylates tumor necrosis factor alpha-converting enzyme at threonine 735: a potential role in regulated shedding. *Mol Biol Cell.* **13**: (6): 2031-44.

Doedens, J. R. and Black, R. A., 2000, Stimulation-induced down-regulation of tumor necrosis factor-alpha converting enzyme. *J Biol Chem.* **275**: (19): 14598-607.

Doedens, J. R.,Mahimkar, R. M.Black, R. A., 2003, TACE/ADAM-17 enzymatic activity is increased in response to cellular stimulation. *Biochem Biophys Res Commun.* **308**: (2): 331-8.

Dolnik, O.,Volchkova, V.,Garten, W.,Carbonnelle, C.,Becker, S.,Kahnt, J.,Stroher, U.,Klenk, H. D.Volchkov, V., 2004, Ectodomain shedding of the glycoprotein GP of Ebola virus. *Embo J.* **23**: (10): 2175-84

Duffy, M. J.,Lynn, D. J.,Lloyd, A. T.O'Shea, C. M., 2003, The ADAMs family of proteins: from basic studies to potential clinical applications. *Thromb Haemost.* **89**: (4): 622-31.

Endres, K.,Anders, A.,Kojro, E.,Gilbert, S.,Fahrenholz, F.Postina, R., 2003, Tumor necrosis factor-alpha converting enzyme is processed by proprotein-convertases to its mature form which is degraded upon phorbol ester stimulation. *Eur J Biochem.* **270**: (11): 2386-93.

Fan, H. and Derynck, R., 1999, Ectodomain shedding of TGF-alpha and other transmembrane proteins is induced by receptor tyrosine kinase activation and MAP kinase signaling cascades. *Embo J.* **18**: (24): 6962-72.

Fan, H.,Turck, C. W.Derynck, R., 2003, Characterization of growth factor-induced serine phosphorylation of tumor necrosis factor-alpha converting enzyme and of an alternatively translated polypeptide. *J Biol Chem.* **278**: (20): 18617-27.

Fischer, O. M.,Hart, S.,Gschwind, A.,Prenzel, N.Ullrich, A., 2004, Oxidative and osmotic stress signaling in tumor cells is mediated by ADAM proteases and heparin-binding epidermal growth factor. *Mol Cell Biol.* **24**: (12): 5172-83.

Fourie, A. M.,Coles, F.,Moreno, V.Karlsson, L., 2003, Catalytic activity of ADAM8, ADAM15, and MDC-L (ADAM28) on synthetic peptide substrates and in ectodomain cleavage of CD23. *J Biol Chem.* **278**: (33): 30469-77.

Franzke, C. W.,Tasanen, K.,Schacke, H.,Zhou, Z.,Tryggvason, K.,Mauch, C.,Zigrino, P.,Sunnarborg, S.,Lee, D. C.,Fahrenholz, F.Bruckner-Tuderman, L., 2002, Transmembrane collagen XVII, an epithelial adhesion protein, is shed from the cell surface by ADAMs. *Embo J.* **21**: (19): 5026-35.

Garton, K. J.,Gough, P. J.,Blobel, C. P.,Murphy, G.,Greaves, D. R.,Dempsey, P. J.Raines, E. W., 2001, Tumor necrosis factor-alpha-converting enzyme (ADAM17) mediates the cleavage and shedding of fractalkine (CX3CL1). *J Biol Chem.* **276**: (41): 37993-8001.

Garton, K. J.,Gough, P. J.,Philalay, J.,Wille, P. T.,Blobel, C. P.,Whitehead, R. H.,Dempsey, P. J.Raines, E. W., 2003, Stimulated shedding of vascular cell adhesion molecule 1 (VCAM-1) is mediated by tumor necrosis factor-alpha-converting enzyme (ADAM 17). *J Biol Chem.* **278**: (39): 37459-64.

Gechtman, Z.,Alonso, J. L.,Raab, G.,Ingber, D. E.Klagsbrun, M., 1999, The shedding of membrane-anchored heparin-binding epidermal-like growth factor is regulated by the Raf/mitogen-activated protein kinase cascade and by cell adhesion and spreading. *J Biol Chem.* **274**: (40): 28828-35.

Gonzales, P. E.,Solomon, A.,Miller, A. B.,Leesnitzer, M. A.,Sagi, I.Milla, M. E., 2004, Inhibition of the TNFalpha converting enzyme (TACE) by its Pro domain. *J Biol Chem.*

Gschwind, A.,Hart, S.,Fischer, O. M.Ullrich, A., 2003, TACE cleavage of proamphiregulin regulates GPCR-induced proliferation and motility of cancer cells. *Embo J.* **22**: (10): 2411-21.

Guo, L.,Eisenman, J. R.,Mahimkar, R. M.,Peschon, J. J.,Paxton, R. J.,Black, R. A.Johnson, R. S., 2002, A proteomic approach for the identification of cell-surface proteins shed by metalloproteases. *Mol Cell Proteomics.* **1**: (1): 30-6.

Hansen, H. P.,Dietrich, S.,Kisseleva, T.,Mokros, T.,Mentlein, R.,Lange, H. H.,Murphy, G.Lemke, H., 2000, CD30 shedding from Karpas 299 lymphoma cells is mediated by TNF-alpha-converting enzyme. *J Immunol.* **165**: (12): 6703-9.

Hansen, H. P.,Recke, A.,Reineke, U.,Von Tresckow, B.,Borchmann, P.,Von Strandmann, E. P.,Lange, H.,Lemke, H.Engert, A., 2004, The ectodomain shedding of CD30 is specifically regulated by peptide motifs in its cysteine-rich domains 2 and 5. *Faseb J.* **18**: (7): 893-5.

Hart, S.,Fischer, O. M.Ullrich, A., 2004, Cannabinoids induce cancer cell proliferation via tumor necrosis factor alpha-converting enzyme (TACE/ADAM17)-mediated transactivation of the epidermal growth factor receptor. *Cancer Res.* **64**: (6): 1943-50.

Hartmann, D.,de Strooper, B.,Serneels, L.,Craessaerts, K.,Herreman, A.,Annaert, W.,Umans, L.,Lubke, T.,Lena Illert, A.,von Figura, K.Saftig, P., 2002, The disintegrin/metalloprotease ADAM 10 is essential for Notch signalling but not for alpha-secretase activity in fibroblasts. *Hum Mol Genet.* **11**: (21): 2615-24.

Hinkle, C. L.,Sunnarborg, S. W.,Loiselle, D.,Parker, C. E.,Stevenson, M.,Russell, W. E.Lee, D. C., 2004, Selective roles for TACE/ADAM17 in the shedding of the epidermal growth factor receptor ligand family. The juxtamembrane stalk determines cleavage efficiency. *J Biol Chem.* **279**: (23): 24179-88

Hooper, N. M.,Karran, E. H.Turner, A. J., 1997, Membrane protein secretases. *Biochem J.* **321 ( Pt 2)**: 265-79.

Hung, A. Y. and Sheng, M., 2002, PDZ domains: structural modules for protein complex assembly. *J Biol Chem.* **277**: (8): 5699-702.

Itai, T.,Tanaka, M.Nagata, S., 2001, Processing of tumor necrosis factor by the membrane-bound TNF-alpha-converting enzyme, but not its truncated soluble form. *Eur J Biochem.* **268**: (7): 2074-82.

Jolly-Tornetta, C. and Wolf, B. A., 2000, Protein kinase C regulation of intracellular and cell surface amyloid precursor protein (APP) cleavage in CHO695 cells. *Biochemistry.* **39**: (49): 15282-90.

Kahn, J.,Walcheck, B.,Migaki, G. I.,Jutila, M. A.Kishimoto, T. K., 1998, Calmodulin regulates L-selectin adhesion molecule expression and function through a protease-dependent mechanism. *Cell.* **92**: (6): 809-18.

Knebel, A.,Rahmsdorf, H. J.,Ullrich, A.Herrlich, P., 1996, Dephosphorylation of receptor tyrosine kinases as target of regulation by radiation, oxidants or alkylating agents. *Embo J.* **15**: (19): 5314-25.

Lammich, S.,Kojro, E.,Postina, R.,Gilbert, S.,Pfeiffer, R.,Jasionowski, M.,Haass, C.Fahrenholz, F., 1999, Constitutive and regulated alpha-secretase cleavage of Alzheimer's amyloid precursor protein by a disintegrin metalloprotease. *Proc Natl Acad Sci U S A.* **96**: (7): 3922-7.

Lanni, C.,Mazzucchelli, M.,Porrello, E.,Govoni, S.Racchi, M., 2004, Differential involvement of protein kinase C alpha and epsilon in the regulated secretion of soluble amyloid precursor protein. *Eur J Biochem.* **271**: (14): 3068-75.

Le Gall, S. M.,Auger, R.,Dreux, C.Mauduit, P., 2003, Regulated cell surface pro-EGF ectodomain shedding is a zinc metalloprotease-dependent process. *J Biol Chem.* **278**: (46): 45255-68.

Li, X. and Fan, H., 2004, Loss of ectodomain shedding due to mutations in the metalloprotease and cysteine-rich/disintegrin domains of the tumor necrosis factor-alpha converting enzyme (TACE). *J Biol Chem.* **279**: (26): 27365-75

Lopez-Otin, C. and Overall, C. M., 2002, Protease degradomics: a new challenge for proteomics. *Nat Rev Mol Cell Biol.* **3**: (7): 509-19.

Luetteke, N. C.,Qiu, T. H.,Peiffer, R. L.,Oliver, P.,Smithies, O.Lee, D. C., 1993, TGF alpha deficiency results in hair follicle and eye abnormalities in targeted and waved-1 mice. *Cell.* **73**: (2): 263-78.

Mann, G. B.,Fowler, K. J.,Gabriel, A.,Nice, E. C.,Williams, R. L.Dunn, A. R., 1993, Mice with a null mutation of the TGF alpha gene have abnormal skin architecture, wavy hair, and curly whiskers and often develop corneal inflammation. *Cell.* **73**: (2): 249-61.

Maskos, K.,Fernandez-Catalan, C.,Huber, R.,Bourenkov, G. P.,Bartunik, H.,Ellestad, G. A.,Reddy, P.,Wolfson, M. F.,Rauch, C. T.,Castner, B. J.,Davis, R.,Clarke, H. R.,Petersen, M.,Fitzner, J. N.,Cerretti, D. P.,March, C. J.,Paxton, R. J.,Black, R. A.Bode, W., 1998, Crystal structure of the catalytic domain of human tumor necrosis factor-alpha-converting enzyme. *Proc Natl Acad Sci U S A.* **95**: (7): 3408-12.

Massague, J. and Pandiella, A., 1993, Membrane-anchored growth factors. *Annu Rev Biochem.* **62**: 515-41.

Matthews, V.,Schuster, B.,Schutze, S.,Bussmeyer, I.,Ludwig, A.,Hundhausen, C.,Sadowski, T.,Saftig, P.,Hartmann, D.,Kallen, K. J.Rose-John, S., 2003, Cellular cholesterol depletion triggers shedding of the human interleukin-6 receptor by ADAM10 and ADAM17 (TACE). *J Biol Chem.* **278**: (40): 38829-39.

McGeehan, G. M.,Becherer, J. D.,Bast, R. C., Jr.,Boyer, C. M.,Champion, B.,Connolly, K. M.,Conway, J. G.,Furdon, P.,Karp, S.,Kidao, S.et al., 1994, Regulation of tumour necrosis factor-alpha processing by a metalloproteinase inhibitor. *Nature.* **370**: (6490): 558-61.

Merlos-Suarez, A.,Fernandez-Larrea, J.,Reddy, P.,Baselga, J.Arribas, J., 1998, Pro-tumor necrosis factor-alpha processing activity is tightly controlled by a component that does not affect notch processing. *J Biol Chem.* **273**: (38): 24955-62.

Merlos-Suarez, A.,Ruiz-Paz, S.,Baselga, J.Arribas, J., 2001, Metalloprotease-dependent protransforming growth factor-alpha ectodomain shedding in the absence of tumor necrosis factor-alpha-converting enzyme. *J Biol Chem.* **276**: (51): 48510-7.

Mezyk, R.,Bzowska, M.Bereta, J., 2003, Structure and functions of tumor necrosis factor-alpha converting enzyme. *Acta Biochim Pol.* **50**: (3): 625-45.

Miettinen, P. J.,Berger, J. E.,Meneses, J.,Phung, Y.,Pedersen, R. A.,Werb, Z.Derynck, R., 1995, Epithelial immaturity and multiorgan failure in mice lacking epidermal growth factor receptor. *Nature*. **376**: (6538): 337-41.

Milla, M. E.,Leesnitzer, M. A.,Moss, M. L.,Clay, W. C.,Carter, H. L.,Miller, A. B.,Su, J. L.,Lambert, M. H.,Willard, D. H.,Sheeley, D. M.,Kost, T. A.,Burkhart, W.,Moyer, M.,Blackburn, R. K.,Pahel, G. L.,Mitchell, J. L.,Hoffman, C. R.Becherer, J. D., 1999, Specific sequence elements are required for the expression of functional tumor necrosis factor-alpha-converting enzyme (TACE). *J Biol Chem*. **274**: (43): 30563-70.

Mizui, Y.,Yamazaki, K.,Sagane, K.Tanaka, I., 1999, cDNA cloning of mouse tumor necrosis factor-alpha converting enzyme (TACE) and partial analysis of its promoter. *Gene*. **233**: (1-2): 67-74.

Mohan, M. J.,Seaton, T.,Mitchell, J.,Howe, A.,Blackburn, K.,Burkhart, W.,Moyer, M.,Patel, I.,Waitt, G. M.,Becherer, J. D.,Moss, M. L.Milla, M. E., 2002, The tumor necrosis factor-alpha converting enzyme (TACE): a unique metalloproteinase with highly defined substrate selectivity. *Biochemistry*. **41**: (30): 9462-9.

Montero, J. C.,Yuste, L.,Diaz-Rodriguez, E.,Esparis-Ogando, A.Pandiella, A., 2000, Differential shedding of transmembrane neuregulin isoforms by the tumor necrosis factor-alpha-converting enzyme. *Mol Cell Neurosci*. **16**: (5): 631-48.

Montero, J. C.,Yuste, L.,Diaz-Rodriguez, E.,Esparis-Ogando, A.Pandiella, A., 2002, Mitogen-activated protein kinase-dependent and -independent routes control shedding of transmembrane growth factors through multiple secretases. *Biochem J*. **363**: (Pt 2): 211-21.

Moss, M. L.,Jin, S. L.,Milla, M. E.,Bickett, D. M.,Burkhart, W.,Carter, H. L.,Chen, W. J.,Clay, W. C.,Didsbury, J. R.,Hassler, D.,Hoffman, C. R.,Kost, T. A.,Lambert, M. H.,Leesnitzer, M. A.,McCauley, P.,McGeehan, G.,Mitchell, J.,Moyer, M.,Pahel, G.,Rocque, W.,Overton, L. K.,Schoenen, F.,Seaton, T.,Su, J. L.,Becherer, J. D.et al., 1997, Cloning of a disintegrin metalloproteinase that processes precursor tumour-necrosis factor-alpha. *Nature*. **385**: (6618): 733-6.

Moss, M. L. and Lambert, M. H., 2002, Shedding of membrane proteins by ADAM family proteases. *Essays Biochem*. **38**: 141-53.

Mullberg, J.,Schooltink, H.,Stoyan, T.,Heinrich, P. C.Rose-John, S., 1992, Protein kinase C activity is rate limiting for shedding of the interleukin-6 receptor. *Biochem Biophys Res Commun*. **189**: (2): 794-800.

Mullberg, J.,Oberthur, W.,Lottspeich, F.,Mehl, E.,Dittrich, E.,Graeve, L.,Heinrich, P. C.Rose-John, S., 1994, The soluble human IL-6 receptor. Mutational characterization of the proteolytic cleavage site. *J Immunol*. **152**: (10): 4958-68.

Mullberg, J.,Althoff, K.,Jostock, T.Rose-John, S., 2000, The importance of shedding of membrane proteins for cytokine biology. *Eur Cytokine Netw*. **11**: (1): 27-38.

Nagano, O.,Murakami, D.,Hartmann, D.,De Strooper, B.,Saftig, P.,Iwatsubo, T.,Nakajima, M.,Shinohara, M.Saya, H., 2004, Cell-matrix interaction via CD44 is independently regulated by different metalloproteinases activated in response to extracellular Ca2+ influx and PKC activation. *J Cell Biol*. **165**: (6): 893-902.

Nelson, K. K.,Schlondorff, J.Blobel, C. P., 1999, Evidence for an interaction of the metalloprotease-disintegrin tumour necrosis factor alpha convertase (TACE) with mitotic arrest deficient 2 (MAD2), and of the metalloprotease-disintegrin MDC9 with a novel MAD2-related protein, MAD2beta. *Biochem J*. **343 Pt 3**: 673-80.

Pan, B.,Farrugia, A. N.,To, L. B.,Findlay, D. M.,Green, J.,Lynch, K.Zannettino, A. C., 2004, The nitrogen-containing bisphosphonate, zoledronic acid, influences RANKL expression in human osteoblast-like cells by activating TNF-alpha converting enzyme (TACE). *J Bone Miner Res.* **19**: (1): 147-54.

Pandiella, A. and Massague, J., 1991, Cleavage of the membrane precursor for transforming growth factor alpha is a regulated process. *Proc Natl Acad Sci U S A.* **88**: (5): 1726-30.

Patel, I. R.,Attur, M. G.,Patel, R. N.,Stuchin, S. A.,Abagyan, R. A.,Abramson, S. B.Amin, A. R., 1998, TNF-alpha convertase enzyme from human arthritis-affected cartilage: isolation of cDNA by differential display, expression of the active enzyme, and regulation of TNF-alpha. *J Immunol.* **160**: (9): 4570-9.

Peiretti, F.,Canault, M.,Deprez-Beauclair, P.,Berthet, V.,Bonardo, B.,Juhan-Vague, I.Nalbone, G., 2003a, Intracellular maturation and transport of tumor necrosis factor alpha converting enzyme. *Exp Cell Res.* **285**: (2): 278-85.

Peiretti, F.,Deprez-Beauclair, P.,Bonardo, B.,Aubert, H.,Juhan-Vague, I.Nalbone, G., 2003b, Identification of SAP97 as an intracellular binding partner of TACE. *J Cell Sci.* **116**: (Pt 10): 1949-57.

Peschon, J. J.,Slack, J. L.,Reddy, P.,Stocking, K. L.,Sunnarborg, S. W.,Lee, D. C.,Russell, W. E.,Castner, B. J.,Johnson, R. S.,Fitzner, J. N.,Boyce, R. W.,Nelson, N.,Kozlosky, C. J.,Wolfson, M. F.,Rauch, C. T.,Cerretti, D. P.,Paxton, R. J.,March, C. J.Black, R. A., 1998, An essential role for ectodomain shedding in mammalian development. *Science.* **282**: (5392): 1281-4.

Prenzel, N.,Zwick, E.,Daub, H.,Leserer, M.,Abraham, R.,Wallasch, C.Ullrich, A., 1999, EGF receptor transactivation by G-protein-coupled receptors requires metalloproteinase cleavage of proHB-EGF. *Nature.* **402**: (6764): 884-8.

Reddy, P.,Slack, J. L.,Davis, R.,Cerretti, D. P.,Kozlosky, C. J.,Blanton, R. A.,Shows, D.,Peschon, J. J.Black, R. A., 2000, Functional analysis of the domain structure of tumor necrosis factor-alpha converting enzyme. *J Biol Chem.* **275**: (19): 14608-14.

Russell, R. G. and Rogers, M. J., 1999, Bisphosphonates: from the laboratory to the clinic and back again. *Bone.* **25**: (1): 97-106.

Sahin, U.,Weskamp, G.,Kelly, K.,Zhou, H. M.,Higashiyama, S.,Peschon, J.,Hartmann, D.,Saftig, P.Blobel, C. P., 2004, Distinct roles for ADAM10 and ADAM17 in ectodomain shedding of six EGFR ligands. *J Cell Biol.* **164**: (5): 769-79.

Schantl, J. A.,Roza, M.,Van Kerkhof, P.Strous, G. J., 2004, The growth hormone receptor interacts with its sheddase, the tumour necrosis factor-alpha-converting enzyme (TACE). *Biochem J.* **377**: (Pt 2): 379-84.

Schlondorff, J. and Blobel, C. P., 1999, Metalloprotease-disintegrins: modular proteins capable of promoting cell-cell interactions and triggering signals by protein-ectodomain shedding. *J Cell Sci.* **112 ( Pt 21)**: 3603-17.

Schlondorff, J.,Becherer, J. D.Blobel, C. P., 2000, Intracellular maturation and localization of the tumour necrosis factor alpha convertase (TACE). *Biochem J.* **347 Pt 1**: 131-8.

Schlondorff, J.,Lum, L.Blobel, C. P., 2001, Biochemical and pharmacological criteria define two shedding activities for TRANCE/OPGL that are distinct from the tumor necrosis factor alpha convertase. *J Biol Chem.* **276**: (18): 14665-74.

Sibilia, M. and Wagner, E. F., 1995, Strain-dependent epithelial defects in mice lacking the EGF receptor. *Science.* **269**: (5221): 234-8.

Slack, B. E.,Ma, L. K.Seah, C. C., 2001, Constitutive shedding of the amyloid precursor protein ectodomain is up-regulated by tumour necrosis factor-alpha converting enzyme. *Biochem J.* **357**: (Pt 3): 787-94.

Srour, N.,Lebel, A.,McMahon, S.,Fournier, I.,Fugere, M.,Day, R.Dubois, C. M., 2003, TACE/ADAM-17 maturation and activation of sheddase activity require proprotein convertase activity. *FEBS Lett.* **554**: (3): 275-83.

suggesting that prodomain removal is required for proteolytic activity in the ADAMs. There are three conserved cysteine residues in the prodomain of ADAM19, and based upon sequence alignment, it is highly possible that $C^{133}$ in mADAM19, or $C^{132}$ in hADAM19, as shown in Figure 1, serves as the "cysteine-switch" that maintains ADAM19 in its latent, zymogen form.

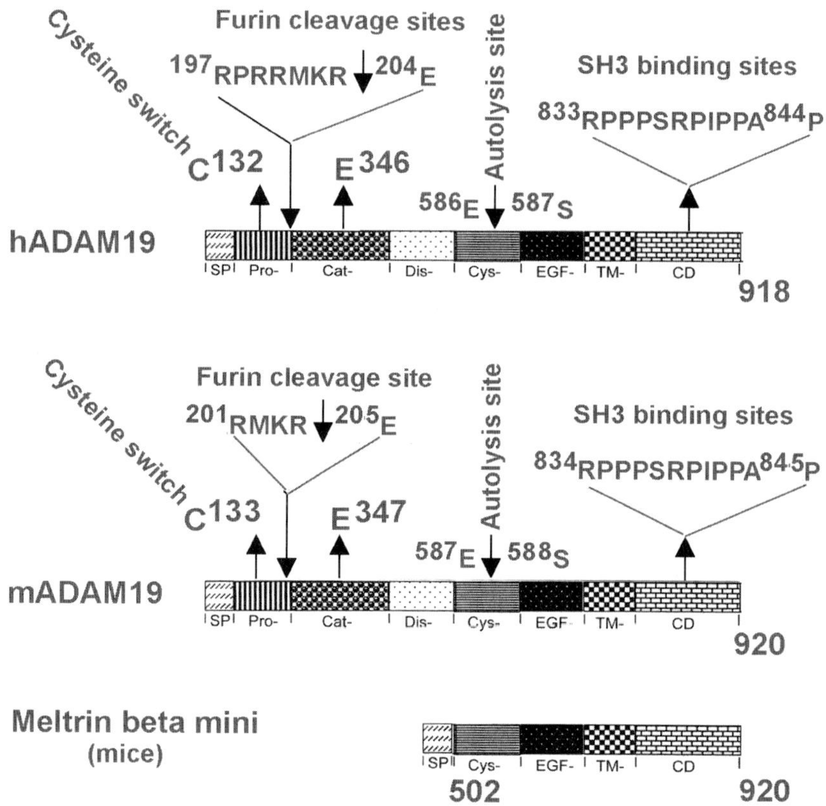


*Figure 1.* A schematic illustration of hADAM19, mADAM19, and meltrin beta mini. SP: Signal peptide; Pro-: Prodomain; Cat-: Catalytic domain; Dis-: Disintegrin domain; Cys-: Cysteine-rich domain; EGF-: EGF-like domain; TM-: Transmembrane domain; CD: Cytoplasmic domain. The cysteine residue (C) that functions as the "cysteine switch" within Pro-, the furin cleavage site(s) between Pro- and Cat-, the glutamate residue (E) within the HEXXHXXGXXHD consensus sequence in Cat-, the autolysis site within Cys-, and the SH3 binding sites within CD are labelled accordingly.

## 3.2 Metalloprotease Domain

All members of the metzincin family, including MMPs, astacins, serralysins, and the catalytic ADAMs (including ADAM19) share the common zinc-binding catalytic site consensus sequence, HEXXHXXGXXH, that is the hallmark of the metzincin superfamily of zinc peptidases. Furthermore, active ADAM family members also share a highly conserved aspartate residue following the terminal histidine of this sequence (Bergers and Coussens 2000; Fritsche *et al.* 2000; Inoue *et al.* 1998; Stocker *et al.* 1995; Wei *et al.* 2001). It may be assumed that both mADAM19 and hADAM19 are active enzymes *in vivo* and *in vitro* based upon the following evidence: 1) null mice of ADAM19 showed abnormal heart development (Kurohara *et al.* 2004; Zhou *et al.* 2004); 2) shedding of beta type NRG (Shirakabe *et al.* 2001) or TRANCE (Chesneau *et al.* 2003; Zhou *et al.* 2004) was enhanced by mADAM19 overexpression in cell culture systems; 3) hADAM19 is capable of effecting intracellular autolysis in its overexpressed cells (Kang *et al.* 2002c, 2004); 4) myelin basic protein, the insulin B chain, and α2M were cleaved by soluble ADAM19 *in vitro* (Chesneau *et al.* 2003; Kang *et al.* 2002b, 2002c); 5) soluble mADAM19 is able to cleave peptides corresponding to the known cleavage sites of TNF-α, TRANCE, and KL-1 (Chesneau *et al.* 2003); 6) The sequence Ac-RPLESNAV, corresponding to known processing site sequences, is cleaved at the E-S bond by soluble hADAM19 (Kang *et al.* 2002c) . As expected, the glutamate residue (E) within the HEXXHXXGXXHD consensus sequence is essential for the proteolytic activity of ADAM19, and E to A mutations result in a loss of catalytic activity (Kang *et al.* 2002c; Shirakabe *et al.* 2001). A methionine turn (met-turn) motif is also present in ADAM19, and while there is no direct evidence regarding the function of this motif in ADAM19, it may be assumed to be necessary for proteolytic activity based on the fact that this highly conserved turn serves as the basement of the active site cleft in many other members of the metzincin family.

## 3.3 Disintegrin Domain

The disintegrin domains of the ADAMs are believed to be involved in the adhesion functions mediated by the ADAMs, most likely through interactions with the integrins. However, there is no direct evidence as yet to support such a function for the disintegrin domain of ADAM19. Instead, the disintegrin domain of hADAM19 has been shown to play a key role in the proteolytic activity of hADAM19, as a specific antibody targeting the disintegrin domain inhibited substrate cleavage *in vitro* (Zhao *et al.* 2001), and a deletion mutant containing only the metalloproteinase domain of hADAM19 lacked proteolytic activity during α2M and peptide substrate assays *in vitro* (Kang and Sang, unpublished data). The notion that the disintegrin domain is involved in the proteolytic activity of ADAMs is supported by studies of TACE/ADAM17, in which the disintegrin domain is required for the shedding of IL-1R-II, while also affecting the inhibitor sensitivity of TNF-α shedding (Reddy *et al.* 2000).

## 3.4 Cysteine-rich Domain

The cysteine-rich domains of the ADAMs are believed to be similarly engaged in the adhesion activities of the ADAMs, but again, there is no direct evidence to support such a function for the cysteine-rich domain in ADAM19. However, it was recently shown (Kang *et al.* 2004) that disulfide bonds that include the cysteine residues at $C^{605}$, $C^{633}$, $C^{639}$, and $C^{643}$ within the C-fragment portion of the cysteine-rich domain following autolysis at $E^{586}$-$S^{587}$ are necessary for the association of the processed N-fragment with its C-fragment. A new processing site at $K^{543}$-$V^{544}$ is exposed when these cysteine residues are mutated individually, suggesting that intrafragment disulfide bonds contribute to the normal conformation of hADAM19, concealing this alternative processing site. In particular, all mutants with these specific cysteine residues exchanged for serines seem to exhibit increased processing, as the resulting 26 kDa fragment is more detectable in these mutants. We may therefore speculate that following autolysis at $E^{586}$-$S^{587}$, the C-fragment, containing a fraction of the cysteine-rich domain, might exert an inhibitory effect on hADAM19 processing through the involvement of interfragment disulfide bonds with the N-fragment, containing the metalloproteinase and disintegrin domains and the remainder of the cysteine-rich domain. This proposed model may provide *in vitro* evidence to support the results of *in vivo* ADAM13 studies showing that the cysteine-rich domain of ADAM13 cooperates intramolecularly with the metalloproteinase domain of ADAM13 to regulate its function, providing the first evidence that a downstream extracellular adhesive domain plays an active role in the regulation of ADAM protease function *in vivo* (Smith *et al.* 2002). This is also consistent with the report by Reddy *et al.* that the cysteine-rich domain of TACE/ADAM17 is required for the shedding of IL-1R-II, while also affecting the inhibitor sensitivity of TNF-α shedding (Reddy *et al.* 2000).

## 3.5 EGF-like Domain

There is very little information available pertaining to the function of the EGF-like domain in any of the ADAMs, and what is known about this domain does not appear to conform to the functional themes exhibited by the metalloproteinase and disintegrin domains. No work is currently found in the literature as to the possible functions of the EGF-like domain of ADAM19.

## 3.6 Transmembrane Domain

While the transmembrane domain of any protein serves as an anchor, localizing the protein in the membrane lumen or on the cell surface, the transmembrane domain of ADAM19 appears to have additional functions. For example, NRG beta1 ectodomain shedding is not enhanced if mADAM19 lacks its transmembrane domain, which is required for its localization in lipid rafts, where mADAM19-mediated shedding of NRG beta1 takes place in neurons (Wakatsuki *et al.* 2004). We have found that PMA, a common inducer of shedding, enhances the processing/shedding of hADAM19 by a mechanism that likely involves the cytoplasmic domain, transmembrane domain, or both, as PMA did not alter the shedding exhibited by soluble hADAM19 (Kang *et al.* 2002c, 2004). This is consistent with reports that the cytoplasmic domain of ADAM9 is required for PMA-induced shedding (Izumi *et al.* 1998), and that the membrane anchor of TACE is necessary for its processing of TNF-$\alpha$ (Itai *et al.* 2001; Reddy *et al.* 2000).

## 3.7 Cytoplasmic Domain

The cytoplasmic tails of the ADAMs contain at least one proline-rich SH3 binding site, suggesting that ADAMs might be linked to multiple signaling pathways. For example, ADAM15, 9, or 12 can associate with a number of different proteins including adaptors, such as Grb2, and Src family tyrosine kinases, through their cytoplasmic domains (Kang *et al.* 2001; Kang *et al.* 2000; Howard *et al.* 1999; Poghosyan *et al.* 2002; Suzuki *et al.* 2000). It has been reported that PKC delta binds *in vivo* and *in vitro* to the cytoplasmic domain of MDC9/meltrin-gamma/ADAM9, regulating the TPA-induced, ADAM9-mediated shedding of HB-EGF (Izumi *et al.* 1998), and that ADAM9 is phosphorylated in cells upon PMA treatment (Roghani *et al.* 1999). The cytoplasmic tail of ADAM19 contains two SH3 binding sites which have been shown to interact with the oncoprotein Arg-binding protein 1 (ArgBP1), the beta-subunit of coatomer protein (beta-cop), ubiquitin, and another unknown protein by a yeast two-hybrid screening (Huang *et al.* 2002). Abram *et al.* utilized a phage display screen with the fifth SH3 domain of Fish, which is a scaffolding protein and Src substrate, and isolated ADAM19 as a binding partner (Abram *et al.* 2003). In addition, the cytoplasmic domain of hADAM19 may also be involved in the PMA-mediated enhancement of processing/shedding by hADAM19, as mentioned earlier (Kang *et al.* 2002c, 2004).

# 4. ADAM19 REGULATION AND PROCESSING

The activities of the ADAMs, including those of ADAM19, are regulated at multiple levels, including transcription, translation, zymogen activation, and inhibition by tissue inhibitors of matrix metalloproteinases (TIMPs) (Brew *et al.* 2000; Kheradmand and Werb 2002; Schlondorff and Blobel 1999; Seals and Courtneidge 2003). There are multiple steps leading to full enzyme maturity.

## 4.1 Regulation of ADAM19 Transcription

The promoter of mADAM19 has been cloned and analyzed by Dr. Fujisawa-Sehara's group (Kurohara *et al.* 2000). They reported the existence of two regulatory regions in the 5' upstream region capable of functioning in the promotion of gene expression. The distal region is a negative regulatory element between –1.9 kb and –642 bp, and the proximal region is a positive regulatory element between –168 bp and –59 bp. There is a GC-box sequence (Sp-1) from –74 bp to –65 bp in the proximal region, and this Sp-1 site plays a critical role in promoter activity, as both mutation and deletion of this site resulted in considerable reduction of the reporter gene. It is worth noting the existence of other transcription sites, such as AP-2, PEA-3, p53, AP-1, and NF-kB, between –642 bp and –59 bp, suggesting that ADAM19 is regulated by multiple stimuli including PMA, TNF-α, LPS, and 1, 25-dihydroxyvitamin D3, (Abe *et al.* 1999; Fritsche *et al.* 2000, 2003; Kang *et al.* 2002c, 2004; Kurohara *et al.* 2004; Shirakabe *et al.* 2001).

## 4.2 Regulation of ADAM19 Translation

Meltrin β mini was cloned recently from mice (Kurisaki *et al.* 2002), encoding amino acid residues 438-920 of mADAM19 (AB096093). The first methionine (502) of this translated sequence lies just upstream of the cysteine-rich domain coding region, and is preceded by a termination codon downstream of the beta mini-specific exon, which is located between exons 12 and 13 of meltrin beta. The resultant amino acid sequence lacks the prodomain, metalloprotease and disintegrin domains, and this cDNA is able to express its own protein, which is distinct from the proteins expressed by meltrin beta; however, mouse meltrin beta mini should be considered as a translation product of mouse meltrin beta. It would be most intriguing to investigate the possibility of a human meltrin beta mini analogue.

## 4.3 Intracellular Processing of ADAM19

### 4.3.1 Activation of ADAM19 by Prodomain Cleavage

Two mechanisms have been reported for the removal of prodomains in ADAM zymogens. The first is an autolytic mechanism, exhibited by ADAM8 (Schlomann *et al.* 2002) and ADAM28 (Howard *et al.* 2000), as no prodomain processing is observed in their inactive forms, in which the critical glutamate residue in their metalloproteinase domains is mutated to A or Q. The second and more predominant mechanism for the removal of ADAM prodomains, however, is mediated by furin or furin-like proprotein convertases in the secretory pathway. This mechanism has been clearly demonstrated for many ADAMs, including ADAM1, 9, 12, 15, 17, 19, and ADAM-TS1, 4, 9, and 12, by way of N-terminal sequencing, specific inhibitors of furin, blockers of protein trafficking from the ER to the golgi, the use of exogenous soluble furin *in vitro,* the use of furin-deficient cell lines, and mutagenesis at the furin recognition site(s), RXK/RR, between the pro- and catalytic-domains (Cal *et al.* 2001; Cao *et al.* 2002; Gao *et al.* 2002; Kang *et al.* 2002b; Kuno *et al.* 1999; Lum and Blobel 1997; Lum *et al.* 1998; Loechel *et al.* 1999; Roghani *et al.* 1999; Schlondorff *et al.* 2000; Somerville *et al.* 2003). Typically, there is only one furin recognition site between the pro- and catalytic- domain of furin's substrates, and this includes most members in the ADAMs family (Cal *et al.* 2001; Cao *et al.* 2002; Kuno *et al.* 1999; Lum and Blobel 1997; Lum *et al.* 1998; Loechel *et al.* 1999; Roghani *et al.* 1999; Schlondorff *et al.* 2000; Somerville *et al.* 2003). However, our lab has reported (Kang *et al.* 2002b) the presence of two functional and consecutive furin recognition sites, $^{197}\underline{\text{R}}\text{P}\underline{\text{R}}^{200}\underline{\text{R}}$ and $^{200}\underline{\text{R}}\text{M}\underline{\text{K}}^{203}\underline{\text{R}}$, between the pro- and catalytic-domains of hADAM19, both of which are predicted to be efficiently cleaved by furin (Nakayama 1997). Only pro-forms of ADAM19 were detectable in a $^{199}$RA mutant lacking a furin recognition site between its pro- and catalytic-domain, whereas the mutants of both $^{196}$RA and $^{202}$RA, with intact recognition sites, were converted into active forms corresponding to the wild type. This suggests that the Arg residue at the $P_4$ site is required for intracellular hADAM19 maturation mediated by furin. Interestingly, N-terminal sequencing of wild type mature forms confirmed that the preferred intracellular cleavage site for hADAM19 activation is the one nearer to the catalytic domain, $^{200}\underline{\text{R}}\text{M}\underline{\text{K}}^{203}\underline{\text{R}}$, as was predicted (Fritsche *et al.* 2000; Wei *et al.* 2000). This motif is conserved in mice as $^{201}\underline{R}M\underline{K}^{204}\underline{R}$ (Inonu *et al.* 1998; Kurisaki *et al.* 1998). The motif farther removed from the catalytic domain in humans, $^{197}\underline{R}P\underline{R}^{200}\underline{R}$, is replaced with $^{198}QPR^{201}R$ in mice, which is not efficiently cleaved by furin, suggesting that the two consecutive, alternative furin recognition sites might be the result of evolution. Although the significance of the two alternative recognition sites in these precursors remains poorly understood, we may speculate that the processing of these zymogens is crucial for certain biological events, and if one site is mutated, the zymogens might still be activated by furin cleavage at the alternate site. Multiple furin recognition sites are also present in ADAM-TS4 (Wang *et al.* 2004). By analyzing the intracellular activation of hADAM19 we have demonstrated that the activation is dependent upon calcium and proper secretory pathway trafficking, and provided direct evidence that furin is co-localized with hADAM19 in the ER/golgi complex and/or the trans-golgi network (TGN). This colocalization between furin and hADAM19 is independent of the furin recognition site and is resistant to a variety of treatments including CMK, BFA, and A23187, that inhibit furin activity, vesicular trafficking, and calcium signaling, respectively (Kang *et al.* 2002b). These findings are consistent with the report that furin was co-localized with MMP16 independent of their apparent enzyme-substrate relationship (Kang *et al.* 2002a), and were further extended by the demonstration of co-localization between furin and ADAM-TS4 (Wang *et al.* 2004).

### 4.3.2 Intramolecular Autolytic Processing of ADAM19

Increasing evidence suggests that shedding is of vital importance for the regulation of metalloproteinase activity. Among ADAMs, ADAM8, following the autolytic removal of its prodomain, has been shown to undergo further autolytic processing to separate its metalloproteinase domain from its disintegrin and cysteine-rich (DC) domains. Subsequently, the DC domains mediate cell adhesion by dimerizing through homophilic interactions (Schlomann *et al.* 2002). ADAM13 is also processed through an autolytic mechanism, producing an active enzyme able to bind with α2M and integrins (Gaultier *et al.* 2002). In addition, truncation of mature ADAM-TS4 at its C-terminus is required for its aggrecanase activity (Gao *et al.* 2002). In the case of ADAM19, our laboratory has demonstrated that intramolecular autolytic processing at $E^{586}$-$S^{587}$ within the cysteine-rich domain is necessary for its proteolytic activity, and that the size of the residues, particularly of the side-chains at both the $E^{586}$ and $S^{587}$ sites, are extremely important for normal processing. Even delicate changes, such as $E^{586}$ to $D^{586}$ and $S^{587}$ to $T^{587}$, resulted in dramatic decreases in the processing (Kang *et al.* 2002c). More interestingly, processing or shedding occurs at an alternate position, $K^{543}$-$V^{544}$, when normal disulfide bonding patterns are disrupted either by mutation or reduction (Kang *et al.* 2004). Disulfide bonds that include the cysteine residues at $C^{605}$, $C^{633}$, $C^{639}$, and $C^{643}$ within the fraction of the cysteine-rich domain retained by the C-fragment after processing at $E^{586}$-$S^{587}$ play a critical role in concealing the alternative processing site at $K^{543}$-$V^{544}$, as alternative processing at $K^{543}$-$V^{544}$ can be induced by individually mutating these cysteine residues to serine residues. This alternate processing also produces an active enzyme, which may be indicative of ADAM19's important roles in both physiological and pathological conditions. Regarding the mechanism for these two cleavage events, both of which produce active enzymes with almost equivalent activity, we speculate that they may occur in a similar fashion, as shown by treatment with GM6001, PMA, W7, or A23187 (Kang *et al.* 2002c, 2004).

Signal pathways involved in shedding processes are poorly understood for the ADAMs, however, there are characteristics unique to the regulation of hADAM19 processing or shedding within its cysteine-rich domain which are distinguishable from those of other ADAMs, MT-MMPs, or other membrane-bound proteins. We have shown that PMA, a common inducer of shedding, also enhances the processing or shedding of hADAM19, and this enhancement likely involves the cytoplasmic domain, transmembrane domain, or both (Kang *et al.* 2002c, 2004). However, in most cases, endogenous and/or inducer-mediated shedding is independent of the cytoplasmic domain (Crowe *et al.* 1995; Diaz-Rodriguez *et al.* 2000; Kahn *et al.* 1998; Vecchi *et al.* 1998). Calcium ionophore A23187 and inhibitors of calmodulin (CaM) are potent inducers of many protein-shedding processes (Annabi *et al.* 2001; Diaz-Rodriguez *et al.* 2000; Fors *et al.* 2001; Kahn *et al.* 1998; Pandiella and Massague 1991; Vecchi *et al.* 1996; Yee *et al.* 1993), but we have found that A23187 and inhibitors of CaM block or impair the processing or shedding of hADAM19 (Kang *et al.* 2002c, 2004). On the other hand, neither tyrosine kinase, MAP kinase, phosphatase, nor PI-3K seemed to play any role in the processing of hADAM19 (Kang *et al.* 2002c), although they have been shown to participate in some shedding processes in the past (Desdouits-Magnen *et al.* 1998; Fan and Derynck 1999; Gechtman *et al.* 1999; Gutwein *et al.* 2000; Nath *et al.* 2001; Schlondorff *et al.* 2001; Vecchi *et al.* 1998).

Intriguingly, many studies have shown that potent synthetic inhibitors of metalloproteinases including TAPI, BB94, and GM6001, can block most, if not all, shedding processes, and that many shedding processes are sensitive to TIMP3 inhibition, as TIMP3 is a matrix-associated TIMP that preferably inhibits ADAMs (Amour *et al.* 1998, 2000; Arribas *et al.* 1996, 1997; Bailly *et al.* 2002; Borland *et al.* 1999; Crowe *et al.* 1995; Diaz-Rodriguez *et al.* 2000; Fitzgerald *et al.* 2000; Hargreaves *et al.* 1998; Hooper *et al.* 1997; Ilan *et al.* 2001; Kashiwagi *et al.* 2001; Loechel *et al.* 2000; Müllberg *et al.* 1997; Nath *et al.* 2001; Peschon *et al.* 1998). However, neither GM6001 nor TIMP3 inhibits the processing or shedding of hADAM19 (Kang *et al.* 2002c, 2004), although both GM6001 and TIMP3 can dramatically inhibit the activity of soluble hADAM19 against a synthetic peptide substrate *in vitro* (Kang *et al.* 2004). One possible solution for this paradox is that shedding, in most cases, occurs at membrane-proximal regions on the cell surface, which are easily accessible to hydroxamate-based inhibitors and TIMP3 (Alfalah *et al.* 2001; Arribas *et al.* 1997; Hooper *et al.* 1997; Migaki *et al.* 1995), while hADAM19 processing occurs at a region distal from the transmembrane domain in the secretory pathway, which is less accessible to GM6001 and TIMP3 molecules (Kang *et al.* 2002c, 2004).

## 4.4 Inhibition of ADAM19

As a metalloprotease, ADAM19 would presumably be inhibited by TIMPs and/or synthetic inhibitors. Indeed, TIMP3, but not TIMP1 or TIMP2, is able to dramatically inhibit the activity of soluble hADAM19 against a synthetic peptide substrate *in vitro* (Kang *et al.* 2004; Wei *et al.* 2001). However, none of the four known TIMPs is capable of blocking the cleavage of MBP by soluble mADAM19 *in vitro* (Chesneau *et al.* 2003). Synthetic hydroxamic acid-type metalloprotease inhibitors, such as BB94 and GM6001, can block ADAM19 using any protein or peptide substrate tested *in vitro* (Chesneau *et al.* 2003; Kang *et al.* 2004; Wei *et al.* 2001). In addition, our laboratory has generated separate anti-metalloprotease domain antibodies and anti-disintegrin domain antibodies using rational peptide antigens, and both of these have been proven to be functional-blocking antibodies against hADAM19 activity both *in vitro* and in cell culture systems (Zhao *et al.* 2001, unpublished data in Dr. Sang's group). As was mentioned earlier, neither GM6001 nor TIMP3 inhibits the processing or shedding of hADAM19 (Kang *et al.* 2002c, 2004).

# 5. FUNCTIONS OF ADAM19

ADAM19 has been implicated in the development of the heart, nervous system and skeletal muscles, and studies with mice null for ADAM19 are particularly compelling in their demonstration of an essential function for this enzyme in cardiovascular morphogenesis (Zhou *et al.* 2004, Kurohara *et al.* 2004). The majority of knockout mice died soon after their birth, and presumably, this was the result of multiple abnormalities that included ventricular septal defects, aberrant formation of the aortic and pulmonic valves resulting in stenosis, and defects in cardiac vasculature. While a fraction of these mice survived despite their heart defects, their tolerance for stress appeared to be reduced, as pregnancy and its associated demands on the cardiac system resulted in the death of these mice prior to term (Zhou *et al.* 2004). While ADAM19 was also highly expressed in osteoblast-like cells associated with bone production, it did not appear to be an essential component for bone growth and skeletal development in these mice, although subtle defects in development may not have been apparent given that the vast majority of these mice failed to attain truly significant levels of maturity. Similar negative findings were reported following the evaluation of the lungs in these knockout animals (Zhou *et al.* 2004).

## 5.1 Expression of ADAM19

While the proteolytic activities of ADAM19, in conjunction with the activities exerted by the disintegrin and cysteine-rich domains, may account for the essential role of ADAM19 in heart development, as shown in null-mice studies (Kurohara *et al.* 2004; Zhou *et al.* 2004), and for ADAM19's suitability as a marker for human monocyte-derived dendritic cells (Fritsche *et al.* 2000, 2003), Northern blot analysis (Wei *et al.* 2001) has revealed widespread expression of hADAM19 mRNA as summarized in Table 1.

*Table 1.* Human tissues and cell lines expressing ADAM19 mRNA

| | HIGH | INTERMEDIATE |
|---|---|---|
| HEART | Atrium, left and right | Interventricular Septum |
| | Ventricle, left | Ventricle, right |
| ORGANS & GI | Appendix | Cerebellum, left and right |
| | Bladder | Colon, Descending |
| | Lung | Colon, Transverse |
| | Lymph Nodes | Ileocecum |
| | Spleen | Ileum |
| | | Lung |
| | | Rectum |
| | | Stomach |
| FETAL | Heart | Lung |
| | Spleen | |
| | Placenta | |
| OTHER | Peripheral Blood Leukocytes | Bone Marrow |
| CELL LINES | Colorectal Adenocarcinoma (SW480) | Chronic Myelogenous Leukemia (K562) |
| | | Lung Carcinoma (A549) |
| | | Burkitt's Lymphoma (Raji and Daudi) |

## 5.2 ADAM19 Substrates and Cytoplasmic Interactions

Synthesized as a zymogen in the ER, hADAM19 is cleaved by furin to remove its prodomain, and is then processed at $E^{586}$-$S^{587}$ within the cysteine-rich domain by autolysis in the TGN (Kang *et al.* 2002b, 2002c). This produces an active N-terminal fragment consisting of the metalloproteinase and disintegrin domains, and a fraction of the cysteine-rich domain, which may then exert sheddase activity, such as the release of beta type-NRG in the TGN (Kurohara *et al.* 2004; Shirakabe *et al.* 2001) or TRANCE (Chesneau *et al.* 2003; Zhou *et al.* 2004). Alternatively, the active N-terminal fragment, associated with its C-terminal fragment by disulfide bonds, may then be transported to the cell surface, where it acts as a sheddase (Chesneau *et al.* 2003; Kurohara *et al.* 2004; Shirakabe *et al.* 2001; Zhou *et al.* 2004) or binds with components of the extracellular matrix (ECM), as proposed in Figure 2.

TNF-α may serve as one potential substrate of ADAM19, as soluble mADAM19 is able to cleave a peptide corresponding to the known cleavage sites of TNF-α (Chesneau *et al.* 2003). Also, KL-1 shedding is down-regulated by overexpression of mADAM19 in cell culture systems (Chesneau *et al.* 2003). Mutation data generated by this laboratory has revealed the $E^{586}$-$A^{587}$ shedding site to be optimal for the autolytic processing of hADAM19 (Kang *et al.* 2002c), indicating that ADAM19 might also serve as an aggrecanase, cleaving aggrecan at $E^{373}$-$A^{374}$, versican at $E^{441}$-$A^{442}$, and/or brevican at $E^{395}$-$A^{396}$, although direct evidence for such a role remains to be provided. Interestingly, a truncated form of murine meltrin β, lacking the prodomain, metalloproteinase and disintegrin domains, described as meltrin β mini, has been reported, and it was shown that this novel meltrin β isoform induced neurite outgrowth in neuronal cells (Kurisaki *et al.* 2002).

The potential functions for the processed C-terminal remnant of ADAM19, containing a fraction of the cysteine-rich domain, the transmembrane domain, and a cytoplasmic domain containing two SH3 binding sites, remain to be uncovered. Huang *et al.* showed that the cytoplasmic tail of ADAM19 interacts with ArgBP1, beta-cop, ubiquitin, and another unknown protein by a yeast two-hybrid screen (Huang *et al.* 2002). ADAM19 was also identified as a binding partner when isolated by a phage display screen with the fifth SH3 domain of Fish, which is a scaffolding protein and Src substrate (Abram *et al.* 2003), suggesting that ADAM19 may function in signaling pathways related to the Src family.

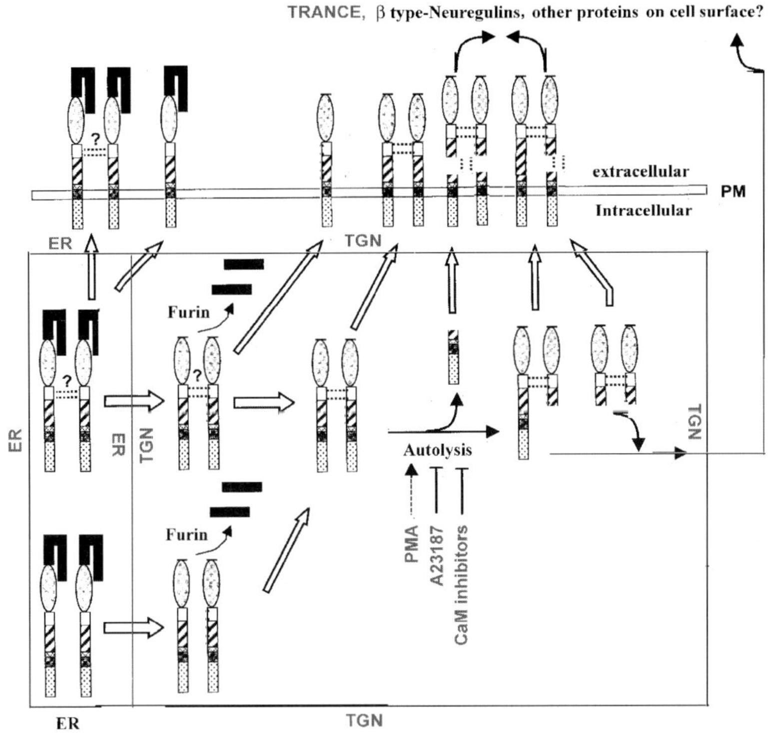


*Figure 2.* A proposed model for the maturation and function of ADAM19.

: Prodomain; : Catalytic domain; : Disintegrin domain; : Cysteine-rich domain; : EGF-like domain; : Transmembrane domain; : Cytoplasmic domain; : disulfide bonds; : transportation; : processing; : enhancement; : blocking. Synthesized as a zymogen (as a dimer or a single molecule) in the endoplamic reticulum (ER), ADAM19 is cleaved by furin to remove its prodomain, and then effects autolysis at $E^{586}$-$S^{587}$ within the cysteine-rich domain in the trans-Golgi networks (TGN). This processing produces an active N-terminal fragment containing the metalloproteinase and disintegrin domains and a fraction of the cysteine-rich domain, which is then capable of exerting sheddase activity, such as the release of TRANCE or beta type-neuregulin, either at the level of the TGN or later at the cell surface. This processing can be enhanced by PMA, but blocked by calcium ionophore A23187 and inhibitors of calmodulin (CaM). Alternatively, the active N-terminal fragment, associated with its C-terminal fragment by disulfide bonds, may then be transported to the plasma membrane (PM), where again, it may exert sheddase activity on the cell surface.

# 6. CONCLUSION

ADAM proteins constitute a large family, with at least 34 members known so far, and they play important roles in many biological processes involving cell-surface proteolysis and cell-cell or cell-matrix interactions (Kheradmand and Werb 2002; Primakoff and Myles 2000; Schlondorff and Blobel 1999; Seals and Courtneidge 2003; Wolfsberg and White 1996). Therefore, ADAMs have been implicated in many vital functions during development (Kheradmand and Werb 2002; Schlondorff and Blobel 1999; Wolfsberg and White 1996) and in the pathogenesis of cancer (Kheradmand and Werb 2002; Seals and Courtneidge 2003), rheumatoid arthritis (Sandy and Verscharen 2001), Alzheimer's disease (Anders *et al.* 2001; Buxbaum *et al.* 1998; Colciaghi *et al.* 2002; Hooper and Turner 2002; Koike *et al.* 1999), asthma and bronchial hyperresponsiveness (Seals and Courtneidge 2003; Van Eerdewegh *et al.* 2002), and inflammatory responses (Sandy *et al.* 2000, 2001).

Much progress has been made in the study of ADAM19 since it was first isolated in 1998, and ADAM19 has been clearly demonstrated to have essential roles in heart development, as shown in null-mice (Kurohara *et al.* 2004; Zhou *et al.* 2004), and to be a suitable marker for human monocyte-derived dendritic cells (Fritsche *et al.* 2000, 2003). However, it is difficult to explain these important roles because the physiological and pathological substrates and/or binding partners for ADAM19 have yet to be conclusively identified, even though its regulation at the post-translational level has been extensively investigated (Kang *et al.* 2002b, 2002c, 2004). For instance, β-type NRGs are normally shed in null-mice of ADAM19 (Kurohara *et al.* 2004; Zhou *et al.* 2004), yet β- type NRGs seem to be substrates of ADAM19 in cell culture systems (Shirakabe *et al.* 2001); the enzyme responsible for the endogenous shedding of TRANCE remains to be identified, although results from co-transfections revealed that the shedding of TRANCE is enhanced by ADAM19, and ADAM19 is capable of cleaving a peptide corresponding to the known cleavage site of TRANCE (Chesneau *et al.* 2003). We believe that molecules exhibiting specific interactions with ADAM19 will yet be identified, acting either as substrates or binding partners under physiological or pathological conditions, with critical functional roles in systems as diverse as heart development, dendritic differentiation, the nervous system, and inflammation.

# REFERENCES


Abe, E., Mocharla. H., Yamate, T., Taguchi, Y., Manolagas, S.C., 1999, Meltrin-alpha, a fusion protein involved in multinucleated giant cell and osteoclast formation. *Calcif Tissue Int.* **64:** 508-515.

Abram, C.L., Seals, D.F., Pass, I., Salinsky, D., Maurer, L., Roth, T.M., Courtneidge, S.A., 2003, The adaptor protein fish associates with members of the ADAMs family and localizes to podosomes of Src-transformed cells. *J. Biol. Chem.* **278:** 16844-16851.

Alfalah, M., Parkin, E. T., Jacob, R., Sturrock, E. D., Mentele, R., Turner, A. J., Hooper, N. M., Naim, H. Y., 2001, A point mutation in the juxtamembrane stalk of human angiotensin I-converting enzyme invokes the action of a distinct secretase. *J. Biol. Chem.* **276:** 21105-21109.

Amour, A., Knight, C. G., Webster, A., Slocombe, P. M., Stephens, P. E., Knauper, V., Docherty, A. J., Murphy, G., 2000, The in vitro activity of ADAM-10 is inhibited by TIMP-1 and TIMP-3. *FEBS Lett.* **473:** 275-279.

Amour, A., Slocombe, P. M., Webster, A., Butler, M., Knight, C. G., Smith, B. J., Stephens, P. E., Shelley, C., Hutton, M., Knauper, V., Docherty, A. J., Murphy, G., 1998, TNF-alpha converting enzyme (TACE) is inhibited by TIMP-3. *FEBS Lett.* **435:** 39-44.

Anders, A., Gilbert, S., Garten, W., Postina, R., Fahrenholz, F., 2001, Regulation of the alpha-secretase ADAM10 by its prodomain and proprotein convertases. *FASEB J.* **15**: 1837-1839.

Annabi, B., Pilorget, A., Bousquet-Gagnon, N., Gingras, D., Beliveau, R., 2001, Calmodulin inhibitors trigger the proteolytic processing of membrane type-1 matrix metalloproteinase, but not its shedding in glioblastoma cells. *Biochem. J.* **359:** 325-333.

Arribas, J., Lopez-Casillas, F., Massague, J., 1997, Role of the juxtamembrane domains of the transforming growth factor-alpha precursor and the beta-amyloid precursor protein in regulated ectodomain shedding. *J. Biol. Chem.* **272:** 17160-17165.

Arribas, J., Coodly, L., Vollmer, P., Kishimoto, T. K., Rose-John, S., Massague, J., 1996, Diverse cell surface protein ectodomains are shed by a system sensitive to metalloprotease inhibitors. *J. Biol. Chem.* **271:** 11376-11382.

Bailly, V., Zhang, Z., Meier, W., Cate, R., Sanicola, M., Bonventre, J. V., 2002, Shedding of kidney injury molecule-1, a putative adhesion protein involved in renal regeneration. *J. Biol. Chem.* **277:** 39739-39748.

Bergers, G., and Coussens, L.M., 2000, Extrinsic regulators of epithelial tumor progression: metalloproteinases. *Curr. Opin. Genet. Dev.* **10:** 120 127.

Blobel, C. P., 2000, Remarkable roles of proteolysis on and beyond the cell surface. *Curr. Opin. Cell Biol.* **12:** 606-612.

Borland, G., Murphy, G., Ager, A., 1999, Tissue inhibitor of metalloproteinases-3 inhibits shedding of L-selectin from leukocytes. *J. Biol. Chem.* **274:** 2810-2815.

Brew, K., Dinakarpandian, D., Nagase, H., 2000, Tissue inhibitors of metalloproteinases: evolution, structure and function. *Biochim. Biophys. Acta.* **1477:** 267-283.

Buxbaum, J. D., Liu, K. N., Luo, Y., Slack, J. L., Stocking, K. L., Peschon, J. J., Johnson, R. S., Castner, B. J., Cerretti, D. P., Black, R. A., 1998, Evidence that tumor necrosis factor alpha converting enzyme is involved in regulated alpha-secretase cleavage of the Alzheimer amyloid protein precursor. *J. Biol. Chem.* **273:** 27765-27767.

Cal, S., Arguees, J. M., Fernandez, P. L., Lopez-Otin, C., 2001, Identification, characterization, and intracellular processing of ADAM-TS12, a novel human disintegrin with a complex structural organization involving multiple thrombospondin-1 repeats. *J. Biol. Chem.* **276:** 17932-17940.

Cao, Y., Kang, Q., Zhao, Z., Zolkiewska, A., 2002, Intracellular processing of metalloprotease disintegrin ADAM12. *J. Biol. Chem.* **277:** 26403-26411.

Chesneau, V., Becherer, J.D., Zheng, Y., Erdjument-Bromage, H., Tempst, P., Blobel, C.P., 2003, Catalytic properties of ADAM19. *J. Biol. Chem.* **278:** 22331-22340.

Colciaghi, F., Borroni, B., Pastorino, L., Marcello, E., Zimmermann, M., Cattabeni, F., Padovani, A., Di Luca, M., 2002, [alpha]-Secretase ADAM10 as well as [alpha]APPs is reduced in platelets and CSF of Alzheimer disease patients. *Mol. Med.* **8:** 67-74.

Crowe, P. D., Walter, B. N., Mohler, K. M., Otten-Evans, C., Black, R. A., Ware, C. F., 1995, A metalloprotease inhibitor blocks shedding of the 80-kD TNF receptor and TNF processing in T lymphocytes. *J. Exp. Med.* **181:** 1205-1210.

Desdouits-Magnen, J., Desdouits, F., Takeda, S., Syu, L. J., Saltiel, A. R., Buxbaum, J. D., Czernik, A. J., Nairn, A.C., Greengard, P., 1998, Regulation of secretion of Alzheimer amyloid precursor protein by the mitogen-activated protein kinase cascade. *J. Neurochem.* **70:** 524-530.

Diaz-Rodriguez, E., Esparis-Ogando, A., Montero, J. C., Yuste, L., Pandiella, A., 2000, Stimulation of cleavage of membrane proteins by calmodulin inhibitors. *Biochem. J.* **346:** 359-367.

Fan, H. and Derynck, R., 1999, Ectodomain shedding of TGF-alpha and other transmembrane proteins is induced by receptor tyrosine kinase activation and MAP kinase signaling cascades. *EMBO J.* **18:** 6962-6972.

Fourie, A.M., Coles, F., Moreno, V., Karlsson, L., 2003, Catalytic activity of ADAM8, ADAM15, and MDC-L (ADAM28) on synthetic peptide substrates and in ectodomain cleavage of CD23. *J. Biol. Chem.* **278:** 30469-30477.

Fitzgerald, M. L., Wang, Z., Park, P. W., Murphy, G., Bernfield, M., 2000, Shedding of syndecan-1 and -4 ectodomains is regulated by multiple signaling pathways and mediated by a TIMP-3-sensitive metalloproteinase. *J. Cell Biol.* **148:** 811-824.

Fors, B. P., Goodarzi, K., von Andrian, U. H., 2001, L-selectin shedding is independent of its subsurface structures and topographic distribution. *J. Immunol.* **167:** 3642-3651.

Fritsche, J., Muller, A., Hausmann, M., Rogler, G., Andreesen, R., Kreutz, M., 2003, Inverse regulation of the ADAM-family members, decysin and MADDAM/ADAM19 during monocyte differentiation. *Immunology.* **110:** 450-457.

Fritsche, J., Moser, M., Faust, S., Peuker, A., Buttner, R., Andreesen, R., Kreutz, M., 2000, Molecular cloning and characterization of a human metalloprotease disintegrin--a novel marker for dendritic cell differentiation. *Blood,* **96:** 732-739.

Gao, G., Westling, J., Thompson, V. P., Howell, T. D., Gottschall, P. E., Sandy, J.D., 2002, Activation of the proteolytic activity of ADAMTS4 (aggrecanase-1) by C-terminal truncation. *J. Biol. Chem.* **277:** 11034-11041.

Gaultier, A., Cousin, H., Darribere, T., Alfandari D., 2002, ADAM13 disintegrin and cysteine-rich domains bind to the second heparin-binding domain of fibronectin. *J. Biol. Chem.* **277:** 23336-23344.

Gechtman, Z., Alonso, J. L., Raab, G., Ingber, D. E., Klagsbrun, M., 1999, The shedding of membrane-anchored heparin-binding epidermal-like growth factor is regulated by the Raf/mitogen-activated protein kinase cascade and by cell adhesion and spreading. *J. Biol. Chem.* **274:** 28828-28835.

Gutwein, P., Oleszewski, M., Mechtersheimer, S., Agmon-Levin, N., Krauss, K., Altevogt, P., 2000, Role of Src kinases in the ADAM-mediated release of L1 adhesion molecule from human tumor cells. *J. Biol. Chem.* **275:** 15490-15497.

Hargreaves, P. G., Wang, F., Antcliff, J., Murphy, G., Lawry, J., Russell, R. G., Croucher, P. I., 1998, Human myeloma cells shed the interleukin-6 receptor: inhibition by tissue inhibitor of metalloproteinase-3 and a hydroxamate-based metalloproteinase inhibitor. *Br. J. Haematol.* **101:** 694-702.

Hirohata, S., Seldin, M.F., Apte, S.S., 1998, Chromosomal assignment of two ADAM genes, TACE (ADAM17) and MLTNB (ADAM19), to human chromosomes 2 and 5, respectively, and of Mltnb to mouse chromosome 11. *Genomics.* **54:** 178-179.

Hooper, N. M. and Turner, A. J., 2002, The search for alpha-secretase and its potential as a therapeutic approach to Alzheimer s disease. *Curr. Med. Chem.* **9:** 1107-1119.

Hooper, N. M., Karran, E. H., Turner, A. J., 1997, Membrane protein secretases. *Biochem. J.* **321:** 265-279.

Howard, L., Maciewicz, R. A., Blobel, C. P., 2000, Cloning and characterization of ADAM28: evidence for autocatalytic pro-domain removal and for cell surface localization of mature ADAM28. *Biochem. J.* **348:** 21-27.

Howard, L., Nelson, K. K., Maciewicz, R. A., Blobel, C. P., 1999, Interaction of the metalloprotease disintegrins MDC9 and MDC15 with two SH3 domain-containing proteins, endophilin I and SH3PX1. *J. Biol. Chem.* **274:** 31693-31699.

Huang, L., Feng. L., Yang, L., Zhou, W., Zhao, S., Li, C., 2002, Screen and identification of proteins interacting with ADAM19 cytoplasmic tail. *Mol. Biol. Rep.* **29:** 317-323.

Ilan, N., Mohsenin, A., Cheung, L., Madri, J.A., 2001, PECAM-1 shedding during apoptosis generates a membrane-anchored truncated molecule with unique signaling characteristics. *FASEB J.* **15:** 362-372.

Inoue, D., Reid, M., Lum, L., Kratzschmar, J., Weskamp, G., Myung, Y. M., Baron, R., Blobel, C. P., 1998, Cloning and initial characterization of mouse meltrin beta and analysis of the expression of four metalloprotease-disintegrins in bone cells. *J. Biol. Chem.* **273:** 4180-4187.

Itai, T., Tanaka, M., Nagata, S., 2001, Processing of tumor necrosis factor by the membrane-bound TNF-alpha-converting enzyme, but not its truncated soluble form. *Eur. J. Biochem.* **268:** 2074-2082.

Izumi, Y., Hirata, M., Hasuwa, H., Iwamoto, R., Umata, T., Miyado, K., Tamai, Y., Kurisaki, T., Sehara-Fujisawa, A., Ohno, S., Mekada, E., 1998, A metalloprotease-disintegrin, MDC9/meltrin-gamma/ADAM9 and PKCdelta are involved in TPA-induced ectodomain shedding of membrane-anchored heparin-binding EGF-like growth factor. *EMBO J.* **17:** 7260-7272.

Kahn, J., Walcheck, B., Migaki, G. I., Jutila, M. A., Kishimoto, T. K., 1998, Calmodulin regulates L-selectin adhesion molecule expression and function through a protease-dependent mechanism. *Cell* **92:** 809-812.

Kang, Q., Cao, Y., Zolkiewska A., 2001, Direct interaction between the cytoplasmic tail of ADAM 12 and the Src homology 3 domain of p85alpha activates phosphatidylinositol 3-kinase in C2C12 cells. *J. Biol. Chem.* **276:** 24466-24472.

Kang, Q., Cao, Y., Zolkiewska, A., 2000, Metalloprotease-disintegrin ADAM 12 binds to the SH3 domain of Src and activates Src tyrosine kinase in C2C12 cells. *Biochem. J.* 2000 **352:** 883-892.

Kang, T., Nagase, H., Pei, D., 2002(a), Activation of membrane-type matrix metalloproteinase 3 zymogen by the proprotein convertase furin in the trans-Golgi network. *Cancer Res.* **62:** 675-681.

Kang, T., Zhao, Y. G., Pei, D., Sucic, J. F., Sang, Q. X., 2002(b), Intracellular activation of human Adamalysin 19/disintegrin and metalloproteinase 19 by furin occurs via one of the two consecutive recognition sites. *J. Biol. Chem.* **277:** 25583-25591.

Kang, T., Park, H. I., Suh, Y., Zhao, YG., Tschesche, H., Sang, Q. X., 2002(c), Autolytic processing at Glu(586)-Ser(587) within the cysteine-rich domain of human adamalysin 19/disintegrin-metalloproteinase 19 is necessary for its proteolytic activity. *J. Biol. Chem.* **277:** 48514-48522.

Kang, T., Tschesche, H., Sang, Q. X., 2004, Evidence for involvement of disulfides in the regulation of intramolecular autolytic processing of human adamalysin19/ADAM19. *Exp. Cell Res.* **298:** 285-295.

Kashiwagi, M., Tortorella, M., Nagase, H., Brew, K., 2001, TIMP-3 is a potent inhibitor of aggrecanase 1 (ADAM-TS4) and aggrecanase 2 (ADAM-TS5). *J. Biol. Chem.* **276:** 12501-12504.

Kheradmand, F. and Werb, Z., 2002, Shedding light on sheddases: role in growth and development. *BioEssay* **24:** 8-12.

Koike, H., Tomioka, S., Sorimachi, H., Saido, T. C., Maruyama, K., Okuyama, A., Fujisawa-Sehara, A., Ohno, S., Suzuki, K., Ishiura, S., 1999, Membrane-anchored metalloprotease MDC9 has an alpha-secretase activity responsible for processing the amyloid precursor protein. *Biochem. J.* **343:** 371-375.

Kuno, K., Terashima, Y., Matsushima, K., 1999, ADAMTS-1 is an active metalloproteinase associated with the extracellular matrix. *J. Biol. Chem.* **274:** 18821-18826.

Kurisaki, T., Wakatsuki, S., Sehara-Fujisawa, A., 2002, Meltrin beta mini, a new ADAM19 isoform lacking metalloprotease and disintegrin domains, induces morphological changes in neuronal cells. *FEBS Lett.* **532:** 419-422.

Kurisaki, T., Masuda, A., Osumi, N., Nabeshima, Y., Fujisawa-Sehara A., 1998, Spatially- and temporally-restricted expression of meltrin alpha (ADAM12) and beta (ADAM19) in mouse embryo. *Mech. Dev.* **73:** 211-215.

Kurohara, K., Masuda, Y., Nagabukuro, N., Tsuji, A., Amagasa, T., Fujisawa-Sehara, 2000, Meltrin beta (ADAM19) gene: cloning, mapping, and analysis of the regulatory region. *Biochem. Biophys. Res. Commun.* **270:** 522-527.

Kurohara, K., Komatsu, K., Kurisaki, T., Masuda, A., Irie, N., Asano, M., Sudo, K., Nabeshima, Y., Iwakura, Y., Sehara-Fujisawa, A., 2004, Essential roles of Meltrin beta (ADAM19) in heart development. *Dev. Biol.* **267:** 14-28.

Loechel, F., Fox, J. W., Murphy, G., Albrechtsen, R., Wewer, U. M., 2000, ADAM 12-S cleaves IGFBP-3 and IGFBP-5 and is inhibited by TIMP-3. *Biochem. Biophys. Res. Commun.* **278:** 511-515.

Loechel, F., Overgaad, M. T., Oxvig, C., Albrechtsen, R., Wewer, U. M., 1999, Regulation of human ADAM 12 protease by the prodomain. Evidence for a functional cysteine switch. *J. Biol. Chem.* **274:** 13427-13433.

Lum, L., Reid, M. S., Blobel, C. P., 1998, Intracellular maturation of the mouse metalloprotease disintegrin MDC15. *J. Biol. Chem.* **273:** 26236-26247.

Lum, L., and Blobel, C. P., 1997, Evidence for distinct serine protease activities with a potential role in processing the sperm protein fertilin. *Dev. Biol.* **191:** 131-145.

Migaki, G. I., Kahn, J., Kishimoto, T. K., 1995, Mutational analysis of the membrane-proximal cleavage site of L-selectin: relaxed sequence specificity surrounding the cleavage site. *J. Exp. Med.* **182:** 549-557.

Milla, M. E., Leesnitzer M. A., Moss. M. L., Clay, W. C. Carter, H. L., Miller, A. B., Su, J., Lambert, M. H., Willard, D.H., Sheeley, D. M., Kost, T. A., Burkhart, W., Moyer, M., Blackburn, R. K., Pahel, G. L., Mitchell, J. L., Hoffman, C. R., Becherer, J. D., 1999, Specific sequence elements are required for the expression of functional tumor necrosis factor-alpha-converting enzyme (TACE). *J. Biol. Chem.* **274:** 30563-30570.

Müllberg, J., Rauch, C. T., Wolfson, M. F., Castner, B., Fitzner, J. N., Otten-Evans, C., Mohler, K. M., Cosman, D., Black, R. A., 1997, Further evidence for a common mechanism for shedding of cell surface proteins. *FEBS Lett.* **401:** 235-238.

Nakayama, K., 1997, Furin: a mammalian subtilisin/Kex2p-like endoprotease involved in processing of a wide variety of precursor proteins. *Biochem. J.* **327:** 625-635.

Nath, D., Williamson, N. J., Jarvis, R., Murphy, G., 2001, Shedding of c-Met is regulated by crosstalk between a G-protein coupled receptor and the EGF receptor and is mediated by a TIMP-3 sensitive metalloproteinase. *J. Cell Sci.* **114:** 1213-1220.

Naus, S., Richter, M., Wildeboer, D., Moss, M., Schachner, M., Bartsch, J.W., 2004, Ectodomain shedding of the neural recognition molecule CHL1 by the metalloprotease-disintegrin ADAM8 promotes neurite outgrowth and suppresses neuronal cell death. *J. Biol. Chem.* **279:** 16083-16090.

Pandiella, A. and Massague, J., 1991, Multiple signals activate cleavage of the membrane transforming growth factor-alpha precursor. *J. Biol. Chem.* **266:** 5769-5773.

Peschon, J. J., Slack, J. L., Reddy, P., Stocking, K. L., Sunnarborg, S. W., Lee, D. C., Russell, W. E., Castner, B. J., Johnson, R. S., Fitzner, J. N., Boyce, R. W., Nelson, N., Kozlosky, C. J., Wolfson, M. F., Rauch, C. T., Cerretti, D. P., Paxton, R. J., March, C. J., Black, R. A., 1998, An essential role for ectodomain shedding in mammalian development. *Science* **282:** 1281-1284.

Poghosyan, Z., Robbins, S. M., Houslay, M.D., Webster, A., Murphy, G., Edwards, D. R., 2002, Phosphorylation-dependent interactions between ADAM15 cytoplasmic domain and Src family protein tyrosine kinase. *J. Biol. Chem.* **277:** 4999-5007.

Primakoff, P., and Myles, D. G., 2000, The ADAM gene family: surface proteins with adhesion and protease activity. *Trends. Genet.* **16:** 83-87.

Reddy, P., Slack, J. L., Davis, R., Cerretti, D. P., Kozlosky, C. J., Blanton, R. A., Shows, D., Peschon, J. J., Black, R. A., 2000, Functional analysis of the domain structure of tumor necrosis factor-alpha converting enzyme. *J. Biol. Chem.* **275:** 14608-14614.

Roghani, M., Becherer, J. D., Moss, M. L., Atherton, R. E., Erdjument-Bromage, H., Arribas, J., Blackburn, R. K., Weskamp, G., Tempst, P., Blobel CP., 1999, Metalloprotease-disintegrin MDC9: intracellular maturation and catalytic activity. *J. Biol. Chem.* **274:** 3531-3540.

Sandy, J. D., and Verscharen, C., 2001, Analysis of aggrecan in human knee cartilage and synovial fluid indicates that aggrecanase (ADAMTS) activity is responsible for the catabolic turnover and loss of whole aggrecan whereas other protease activity is required for C-terminal processing in vivo. *Biochem. J.* **358:** 615-626.

Sandy, J. D., Westling, J., Kenagy, R. D., Iruela-Arispe, M. L., Verscharen, C., Rodriguez-Mazaneque, J. C., Zimmermann, D. R., Lemire, J. M., Fischer, J. W., Wight, T. N., Clowes, A.W., 2001, Versican V1 proteolysis in human aorta in vivo occurs at the Glu441-Ala442 bond, a site that is cleaved by recombinant ADAMTS-1 and ADAMTS-4. *J. Biol. Chem.* **276:** 13372-13378.

Sandy, J. D., Thompson, V., Doege, K., Verscharen, C., 2000, The intermediates of aggrecanase-dependent cleavage of aggrecan in rat chondrosarcoma cells treated with interleukin-1. *Biochem. J.* **351:** 161-166.

Schlomann, U., Wildeboer, D., Webster, A., Antropova, O., Zeuschner, D., Knight, C. G., Docherty, A. J., Lambert, M., Skelton, L., Jockusch, H., Bartsch, J. W., 2002, The metalloprotease disintegrin ADAM8. Processing by autocatalysis is required for proteolytic activity and cell adhesion. *J. Biol. Chem.* **277:** 48210-48219.

Chapter 10

# ADAM28

Anne M. Fourie
*Johnson & Johnson Pharmaceutical Research and Development, San Diego, USA*

Abstract: The ADAM family of disintegrin metalloproteases plays important roles in proteolytic "ectodomain shedding" and adhesion functions. Currently, this family has 34 members, approximately half of which are predicted to be active proteases, including ADAM28. ADAM28 is expressed in human lymphocytes, murine thymic epithelial cells, the epididymis in multiple species, and is upregulated in certain cancer cells. Both membrane-bound and secreted isoforms of ADAM28 have been identified. ADAM28 is activated by autocatalytic removal of the pro-domain and the mature transmembrane protein is expressed on the cell surface. The active form of ADAM28 cleaves specific sites in numerous peptides and protein substrates, including the low affinity IgE receptor, CD23, and IGFBP-3. The substrate selectivity of ADAM28 is very similar to that of the closely-related ADAMs, -8 and -15, but distinct from that of the more distantly-related ADAM17. An extended region of the disintegrin domain of ADAM28 specifically recognizes the leukocyte integrin, $\alpha4\beta1$, in an activation-dependent conformation. The physiological functions of ADAM28 are not known, but its expression pattern, together with its substrate and integrin binding selectivity, suggest potential roles in spermatogenesis, lymphocyte maturation and function, inflammation and cancer.

Key words: metalloprotease, lymphocyte, thymic epithelium, epididymis, MDC-L, eMDC II, autocatalytic, CD23, IGFBP-3, disintegrin, $\alpha4\beta1$

## 1. INTRODUCTION

The ADAM (A Disintegrin And Metalloprotease) family comprises multi-domain proteins containing pro-, metalloprotease, disintegrin, cysteine-rich, epidermal growth factor (EGF)-like, transmembrane and cytoplasmic domains. These diverse domains are believed to play important



*N. M. Hooper and U. Lendeckel (eds.), The ADAM Family of Proteases,* 223-238.

roles in the proteolytic "ectodomain shedding" and adhesion functions of ADAMs (recently reviewed in Becherer and Blobel, 2003; Duffy *et al.*, 2003; Moss and Bartsch, 2004; Seals and Courtneidge, 2003; White, 2003). To date, this protein family has 34 known members, approximately half of which have consensus sequences for active metalloproteases. ADAM28 is a catalytically active member, originally identified in a search for novel ADAMs expressed in human lymphocytes (Roberts *et al.*, 1999). Two isoforms were cloned, a membrane-bound form, named MDC-Lm, and a secreted form, named MDC-Ls. In the same year, another group (Jury *et al.*, 1999) cloned the macaque homologue, eMDC II, from epididymis. Murine ADAM28 was identified in mouse lung the following year (Howard *et al.*, 2000) and shown to be highly expressed in epididymis, and later in thymic epithelial cells (Haidl *et al.*, 2002). Recently, the rat sequence was also cloned from epididymis and deposited in the Genbank nucleotide database (Oh and Cho, unpublished 2003). While the initial cloning of human ADAM28 from lymph nodes, and the thymic expression of the mouse orthologue, suggests an immunological function for this protease, the strong expression in male reproductive tissues suggests a possible role in sperm maturation. Various studies reviewed here have identified proteolytic substrates and adhesion ligands for ADAM28, but the physiological functions of the membrane-bound and secreted forms of this member of the ADAM family are currently unknown.

## 2. STRUCTURE

Two alternative forms of human ADAM28 have been identified (Roberts *et al.*, 1999). One is a prototypical membrane-bound form, named MDC-Lm, containing an N-terminal signal sequence and a prodomain, followed by metalloprotease, disintegrin-like, cysteine-rich, EGF-like, transmembrane and cytoplasmic domains. The second, secreted form, named MDC-Ls, lacks the transmembrane and cytoplasmic domains because of a stop codon in the cysteine-rich domain. Secreted forms of ADAM28 have not been unequivocally identified in species other than human, although differential recognition of murine transcripts by specific hybridization probes suggested that the 2.3 kb transcript in epididymis may code for a secreted form (Howard *et al.*, 2000). The sequence identified for rat ADAM28 does not appear to contain a transmembrane domain and may thus represent a secreted isoform. Transmembrane forms have been cloned in human (Howard *et al.*, 2000; Roberts *et al.*, 1999), macaque (Jury *et al.*, 1999) and mouse (Howard *et al.*, 2000). The amino acid sequence of the human protein is 91% identical to that of the macaque, and 70% identical to the mouse sequence, while the mouse and macaque sequences are 73% identical (Howard *et al.*, 2000). Figures 1 and 2 show an alignment of the amino acid sequences of human, macaque, mouse and rat ADAM28. ADAM28 is phylogenetically most closely related to the non-protease ADAM7, followed by the active protease, ADAM8 (Cerretti *et al.*, 1999). A summary of ADAM28 nucleotide and protein sequence accession numbers can be found in Table 1.

The metalloprotease domain of ADAM28 contains a consensus zinc binding sequence ($H^{342}$EXGHXXGXXHD) (residue numbering according to sequence alignment in Figs. 1 and 2), followed by a methionine turn ($C^{364}$VMD/E), typical of the reprolysin subfamily of the metzincins (Bode *et al.*, 1993). The prodomain contains a sequence ($S^{169}$T/NCGM/T/V), similar to the "cysteine switch" found in matrix metalloproteases, believed to bind to the catalytic zinc to maintain the inactive form of the protease during biosynthesis (Roberts *et al.*, 1999). The sequence at the predicted boundary between the pro- and metalloprotease domains (K/$N^{197}$DR/GK/M) is not a typical furin cleavage site, and it has been shown that ADAM28 is activated by autocatalytic removal of the prodomain by cleavage at three other sites (Howard *et al.*, 2000). Transiently transfected COS-7 cells produce immunoreactive murine ADAM28 as a precursor form with an apparent molecular weight of 115 kDa, which is processed to a mature and presumably active form of 88 kDa. Trypsin susceptibility suggests that most or all of the mature form is on the cell surface (Howard et al., 2000). Immunoblotting with antiserum against the disintegrin domain of human MDC-L showed two bands of 87 and 67 kDa in peripheral blood leukocytes. The 67 kDa form may represent a processed form of MDC-Lm or the truncated, secreted form, MDC-Ls (Roberts *et al.*, 1999).

Numerous members of the ADAM family have been shown to interact specifically with particular integrins (White, 2003), although ADAM15 is the only member of the family that contains an RGD sequence, similar to snake venom disintegrins (Krätzschmar *et al.*, 1996). The disintegrin domain of ADAM28 contains the sequence $K^{472}$DE, within a putative disintegrin loop ($C^{468}$-$L^{477}$) (Roberts *et al.*, 1999). The ADAM28 disintegrin domain has been shown to bind the leukocyte integrin, $\alpha 4\beta 1$, in a concentration- and activation-dependent manner (Bridges *et al.*, 2002). ADAM28 does not contain sequences similar to those found in other $\alpha 4\beta 1$ ligands such as fibronectin (CS-1 domain) or vascular adhesion molecule-1 (VCAM-1), respectively. A recent study (Bridges *et al.*, 2003) has shown that multiple charged residues outside the disintegrin loop are essential for $\alpha 4\beta 1$-dependent binding of cells to the ADAM28 disintegrin domain.

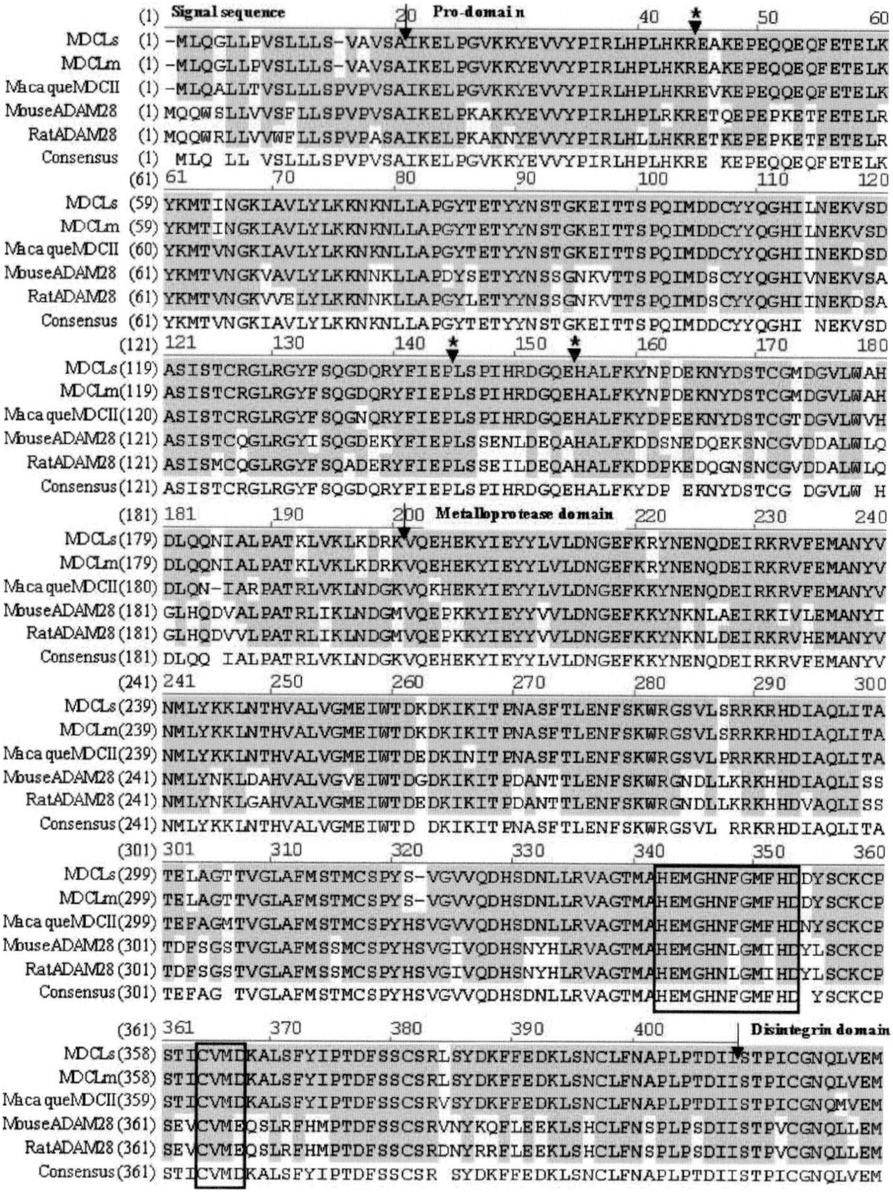


*Figure 1.* Alignment (Part1) of protein sequences of ADAM28 from human (secreted and membrane-bound), macaque, mouse and rat. Human MDC-Ls (NP_068547), human MDC-Lm (NP_068548), macaque eMDC II (CAB42090), murine ADAM28 (NP_034212), and rat ADAM28 (NM_859044) were aligned using Vector NTI AlignX. Identical residues are shown in gray shaded boxes. The catalytic zinc-binding site and methionine turn are outlined by black boxes. Arrows indicate approximate domain boundaries. Sites of autocatalytic cleavage are indicated by arrowheads with asterisks.

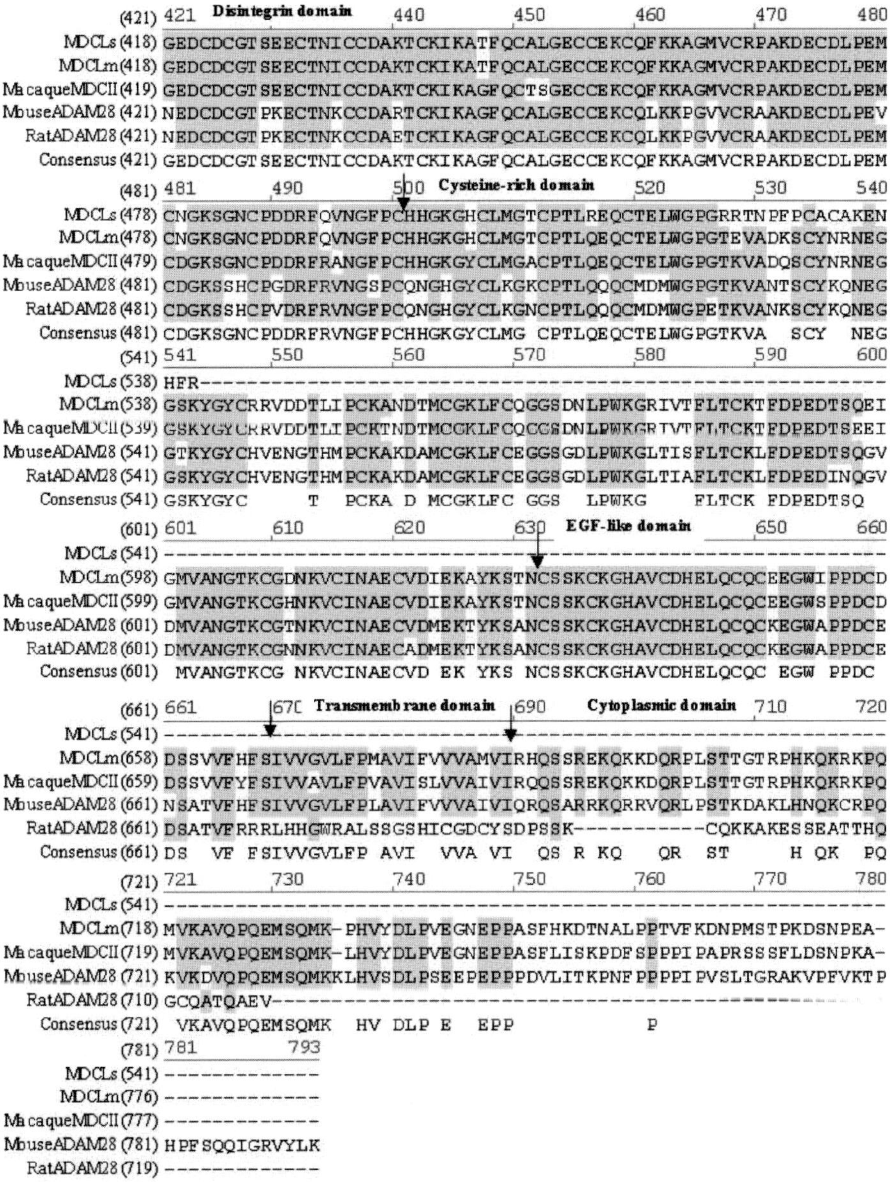


*Figure 2* Alignment (Part2) of protein sequences of ADAM28 from human (secreted and membrane-bound), macaque, rat and mouse (see legend for Fig. 1, Part 1 of alignment, for details).

Molecular modeling studies suggest that these residues ($K^{440, 445, 458, 462, 463, 472}$ and $E^{479}$) are located on one face of the disintegrin domain in the

ADAM28 molecule. The cytoplasmic tail of ADAM28 contains numerous proline-rich regions which may interact with Src homology 3 (SH3) domains in other proteins (Howard *et al.*, 2000), as has been shown for ADAM9, ADAM15 and ADAM13 (Becherer and Blobel, 2003).

*Table 1.* Human, macaque, murine and rat ADAM28 sequences and expression patterns.

| Species | Other names | Nucleotide Accession numbers | Protein accession numbers | mRNA expression | Protein expression |
|---|---|---|---|---|---|
| Human *(Roberts et al. 1999, Jury et al. 1999, Howard et al. 2000)* | MDC-Lm, MDC-Ls | NM_021778<br>NM_021777<br>NM_014265 | NP_068548<br>NP_068547<br>NP_055080 | Epididymis, lymph nodes, spleen, small intestine, stomach, colon, appendix, trachea, small intestine, colon | Lymphocytes, monocytes, neutrophils |
| Macaque *(Jury et al. 1999)* | eMDC II | AJ242014 | CAB42090 | Epididymis, lymph node, pancreas, ovary, uterus | |
| Mouse *(Howard et al. 2000)* | | NM_010082<br>NM_176991<br>NM_183366 | NP_034212<br>NP_795965<br>NP_899222 | Epididymis, thymic epithelium, trachea, thyroid, stomach, lung | |
| Rat *(Oh and Cho, unpublished 2003)* | | NM_181693 | NP_859044 | Epididymis | |

At the RNA level, three transcript variants have been identified for human ADAM28. Transcript variants 1 and 2 encode the membrane-bound forms originally named eMDC II (Jury *et al.*, 1999), or MDC-Lm (Roberts *et al.*, 1999), respectively. Variant 1 has a 3' untranslated region 272 base pairs longer than variant 2, and the next to last amino acid in variant 1 is lysine, versus glutamic acid in variant 2. Transcript variant 3 is the result of alternative splicing to code for the secreted form, MDC-Ls (Roberts *et al.*, 1999). Initially, three different transcripts were identified for murine ADAM28: two larger transcripts of 4.3 and 2.9 kb, expressed in both epididymis and lung, and a smaller, 2.3 kb transcript, expressed only in epididymis (Howard et al., 2000). Later work provided evidence for a total of 6 isoforms, the result of alternative splicing in the cytoplasmic domain, as well as variation in the 3' untranslated region (Haidl *et al.*, 2002). A recent study also provides evidence for variation and positive selection in the cytoplasmic domains of human, murine and macaque ADAM28 (Glassey and Civetta, 2004).

The ADAM28 gene is part of a metalloprotease gene cluster in both human and mouse, on chromosomes 8p21 and 14D1, respectively. This cluster comprises the genes for ADAM28, decysin (ADAMDEC1), and ADAM7 (Bates *et al.*, 2002). The gene of murine ADAM28 spans 55 kilobases and contains 23 exons (Haidl *et al.*, 2002). The exon-intron boundaries in the ADAM28 gene do not correspond to protein domains, and are very similar to those determined for ADAM2, even though these two ADAMs are phylogenetically quite distinct (Cerretti *et al.*, 1999).

## 3. EXPRESSION AND REGULATION

Human ADAM28 or MDC-L was originally cloned from lymph node and Northern blot analysis showed expression of the membrane-bound form, MDC-Lm, predominantly in lymph node, and the secreted form, MDC-Ls mainly in spleen (Roberts *et al.*, 1999). Further dot blot analysis of polyadenylated RNA from human tissues showed the highest expression of human ADAM28 in secondary lymphoid tissues such as stomach, lymph node, appendix, spleen, trachea, small intestine and colon. The macaque homologue, eMDC II, also showed low expression in lymph node by RT-PCR. Further analysis of human ADAM28 in leukocytes (Roberts *et al.*, 1999) by nested RT-PCR showed expression of MDC-Lm (and lower levels of MDC-Ls) in CD2-positive T lymphocytes and CD19-positive B lymphocytes. In addition, MDC-Lm expression was detected by RT-PCR in CD14 and CD16 positive leukocyte populations, representing predominantly monocytes and neutrophils, respectively (Roberts *et al.*, 1999). RT-PCR of T cell lines (HuT78, A139.1) and a B cell line (RD105U) also indicated that they expressed MDC-L. Immunoblotting with antiserum to the C-terminus of MDC-Lm has shown expression in the B cell lines, JY and Raji, as well as U937 and THP-1 monocyte/macrophage lines (Fourie, AM unpublished). Expression of ADAM28 RNA in THP-1 cells has recently been demonstrated by reverse transcription and real-time PCR (Worley *et al.*, 2003).

Apart from RT-PCR and immunoblotting evidence for expression of human ADAM28 in lymphocytes, immunohistochemistry on various lymphoid tissues with antiserum against the MDC-L disintegrin domain showed MDC-L positive cells with typical lymphocyte morphology in all tissues examined (Roberts *et al.*, 1999). However, the murine counterpart showed a different expression pattern in lymphoid tissues. Murine ADAM28 expression was observed in thymic epithelial cells, and developmentally related tissues including trachea, thyroid, stomach and lung (Haidl *et al.*, 2002). ADAM28 was not detected by RT-PCR in murine spleen, lymph node, bone marrow or thymocytes. Furthermore, analysis of ADAM28 expression in RAG-2 knockout mice which do not contain mature B or T lymphocytes showed an identical expression pattern to wild-type mice for ADAM28, suggesting that expression observed in lymphoid tissues is not lymphocyte-derived. The functional relevance of the difference between human and murine ADAM28 expression in lymphocytes remains to be determined.

While ADAM28 expression has been detected in lymphoid tissues from multiple species, the highest expression in human (Howard *et al.*, 2000), macaque (Jury *et al.*, 1999) and mouse (Howard *et al.*, 2000) was in the epididymis, and the rat homologue was also cloned from epididymis (Oh and Cho, 2003). Immunoblotting of a number of murine tissues for membrane-bound ADAM28 detected expression only in the epididymis, although expression of ADAM28 RNA had also been detected in lung tissue (Howard *et al.*, 2000). RT-PCR analysis of macaque eMDC II transcripts showed the highest expression in the caput region of the epididymis, and lower, but very significant, expression in the cauda epididymis (Jury *et al.*, 1999). The predominant expression of ADAM28 in the epididymis suggests a function for this protease in sperm maturation. A summary of ADAM28 expression patterns can be found in Table 1.

Limited information is available on regulation of ADAM28 expression. No promoter analysis data has been reported to date. In a recent study in phorbol ester-simulated THP-1 cells, regulation of ADAM28 mRNA expression by agonists of peroxisome proliferator-activated receptor-$\gamma$ (PPAR$\gamma$) and retinoid X receptor was compared to that of MMP9 and other ADAM proteases (Worley *et al.*, 2003). Phorbol ester (PMA) treatment caused significant upregulation of MMP9, ADAM8, -9 and -17 at 6h and 9h and repression of ADAM15 mRNA, as measured by real-time PCR.

ADAM28 and ADAMTS4 (Aggrecanase-1) transcription was unaffected by PMA at these time points, but longer incubations up to 24h induced ADAM28 and ADAMTS4 by 5- and 16-fold, respectively. The longer time required for these effects suggests that these genes are not directly induced by PMA, but by secondary signaling events. In addition, ADAM28 transcription was induced a further 2-fold by the retinoid X receptor agonist, 9-cis-retinoic acid, alone or in combination with PPARγ agonists, GW7845 or $PGJ_2$. PPARγ agonists alone had no effect on ADAM28, but pre-treatment with a PPARγ antagonist suppressed the retinoic acid-induced transcription, suggesting some role for PPARγ regulation. In contrast, ADAMTS4 induction by PMA was repressed by the retinoid receptor and PPARγ agonists, particularly in combination, similarly to the repression of PMA induction of MMP9 RNA at earlier time-points (Worley *et al.*, 2003). The effects of PMA, PPARγ and 9-*cis*-retinoic acid on ADAM28 in THP-1 cells are thus complex, but unique compared to regulation of other MMPs and ADAMs. The functional relevance of this differentiated regulation of ADAM28 transcription is not clear.

Differential expression of ADAM28 mRNA has been observed in a breast carcinoma cell line. ADAM28 was one of 28 genes more highly expressed (3-fold) in the breast cancer cell line MDA-MB-231, compared to the normal breast cell line MCF12A (Xie *et al.*, 2002). A recent study investigated expression of ADAMs in renal cell carcinoma, in both cancerous and non-cancerous regions of patient tissues (Roemer *et al.*, 2004). ADAM28 and other ADAMs (ADAM-8, -17, -19 and –TS2) showed increased expression in the non-cancerous tissue, as the grade of the tumors increased from pT1 through pT3. In patients with pT1 tumors without distant metastases, ADAM28 expression was 3-fold higher in cancerous versus non-cancerous tissue. A recent proteomic study showed upregulation of ADAM28 as one of 60 proteins whose expression was affected by the carcinogenic alkylating agent, N-methyl-N'-nitro-nitrosoguanidine (MNNG) (Jin *et al.*, 2003).

The catalytic function of ADAM28 has been shown to be activated by autocatalytic removal of the pro-domain in a late Golgi compartment in transfected cells (Howard *et al.*, 2000). ADAM28 can also be activated *in vitro* by MMP7 to cleave insulin-like growth factor binding protein-3 (IGFBP-3) (Mochizuki *et al.*, 2004). This activity of ADAM28 can be inhibited by endogenous metalloprotease inhibitors, TIMP-3 and TIMP-4, but not TIMP-1 and –2. This pattern of selective inhibition by TIMP-3, versus TIMP-1 and –2, has also been shown for proteolytic activity of ADAM12, ADAM17, ADAMTS4 and ADAMTS5, whereas ADAM10 activity is susceptible to inhibition by TIMP1 and TIMP-3 (Becherer and

Blobel, 2003; Kashiwagi *et al.*, 2001). Recently ADAM33 activity has also been shown to be inhibited by TIMP-3 and TIMP-4 (Zou *et al.*, 2004).

## 4. FUNCTION

The physiological function of ADAM28 is currently unknown, but various hypotheses have been put forward based on the functional domains within the structure of ADAM28 and other members of the family, the cell- and tissue-specific expression of ADAM28, and information about its proteolytic substrate and integrin binding selectivity.

Although the deduced amino acid sequence of ADAM28 predicted that it would be an active metalloprotease (Roberts *et al.*, 1999), and there was evidence for autocatalytic activation of ADAM28 (Howard *et al.*, 2000), the first formal demonstration of proteolytic activity was provided by cleavage of myelin basic protein (Howard *et al.*, 2001). Analysis of the cleavage sites for ADAM28 (summarized in Table 2) and ADAM10 within MBP showed both common and different selectivity for cleavage sites between these two ADAM proteases. The autocatalytic cleavage sites within the prodomain of ADAM28 were also identified in the same study and are listed in Table 2. ADAM 28 catalytic selectivity was recently compared to that of ADAM8, -15 and -17 (Fourie *et al.*, 2003). By screening a library of synthetic peptides, it was shown that recombinant metalloprotease domains of ADAM8, -15 and -28 had almost identical selectivity for cleavage of a subset of peptides, out of 49 tested. The peptide sequences that were most efficiently cleaved by ADAM28 are shown in Table 2. The selectivity for peptide cleavage by ADAM17 was clearly distinct from that for ADAM8, -15 and -28. The proteolytic selectivity of these four enzymes is consistent with their phylogenetic relationships. ADAM28 is most closely related to ADAM8, followed by ADAM15, while ADAM17 is more distantly related (Cerretti *et al.*, 1999). ADAM17 is very closely related to ADAM10, which also showed differences for preferential cleavage sites in MBP compared to ADAM28 (Howard *et al.*, 2001). Specific cleavage sites for some peptides were determined for ADAM8 (Fourie *et al.*, 2003). Based on the close correlation of their peptide cleavage selectivity and phylogenetic relationship, similar cleavage sites would be predicted for ADAM28, and these are shown for reference in Table 2.

Consistent with their similar selectivity for peptide cleavage, soluble recombinant metalloprotease domains of ADAM8, -15 and -28 all catalyzed ectodomain cleavage of membrane-bound CD23 (the low affinity IgE receptor) in transfected cells, releasing soluble CD23 (Fourie *et al.*, 2003). ADAM17 was unable to cleave membrane-bound CD23, once again demonstrating its distinct specificity for protein as well as peptide substrates. CD23 is expressed on B cells, monocytes, macrophages, eosinophils (Delespesse *et al.*, 1992), T cells from allergic individuals (Gagro and Rabatic, 1994), and neutrophils from individuals with rheumatoid arthritis (Vella *et al.*, 1999). Both transmembrane and soluble forms of CD23 are elevated in these and other inflammatory diseases (Bonnefoy *et al.*, 1996). Soluble CD23 up-regulates IgE production through interaction with B cells (Wheeler *et al.*, 1998) and promotes induction of inflammatory cytokines in macrophages (Bonnefoy *et al.*, 1996). The identity of the metalloprotease which cleaves membrane-bound CD23 to release soluble forms is currently unknown, but based on the recent data (Fourie *et al.*, 2003), ADAM28 is a potential candidate, along with ADAM8 and -15. In this regard, it is of interest that ADAM28 expression has been detected on lymphocytes, monocytes and neutrophils (Roberts *et al.*, 1999), similar to CD23 expression.

*Table 2.* Peptide and protein substrates for ADAM28.

| Protein Or Peptide Name | Sequence And/Or Cleavage Site | Reference |
|---|---|---|
| MBP | ....SKYLASA ^ STMDHAR..... | Howard *et al.* 2001 |
| MBP | ....THYGSLP ^ QKAQGHR.... | Howard *et al.* 2001 |
| MBP | ...PSQGKGR ^ GLSLSRF..... | Howard *et al.* 2001 |
| ADAM28 | ...LHPLRKR ^ ETQEPEP... | Howard *et al.* 2001 |
| ADAM28 | .....EKYFIEP ^ LSSENLD.... | Howard *et al.* 2001 |
| ADAM28 | ...ENLDEQA ^ HALFKDD.... | Howard *et al.* 2001 |
| CatE1 | *X*KPAKF ^ FRLZ * | Fourie *et al.* 2003 |
| CatE | *X*KPAAF ^ FRLZ * | Fourie *et al.* 2003 |
| CD27L | *X*RFAQA ^ QQQLPZ * | Fourie *et al.* 2003 |
| TNFα | *X*PLAQAVRS ^ SSZ * | Fourie *et al.* 2003 |
| TNFB2 | *X*PSAAQTARQHPZ | Fourie *et al.* 2003 |
| KL1 | *X*PPVAASSLRNZ | Fourie *et al.* 2003 |
| CD23 | Cleavage site(s) unkown | Fourie *et al.* 2003 |
| IGFBP-3 | ...SAVSRLR ^ AYLLPAP.... | Mochizuki *et al.*2004 |

*cleavage site for ADAM8. MBP = myelin basic protein, *X* = Aedans-E, *Z* = Dabcyl-K

Autocatalytic activation has been observed for ADAM28 in transfected cells (Howard *et al.*, 2000) and *in vitro* (Fourie *et al.*, 2003). ADAM28 can also be activated by MMP7 *in vitro* (Mochizuki *et al.*, 2004) to cleave IGFBP-3 at a specific site, $R^{97}$-$A^{98}$ in the central domain (see Table 2). ADAM28 cleaves IGFBP-3 when in complex with IGF-I or IGF-II without digestion of the IGFs, which can then be released intact. It is thus possible that ADAM28 may reactivate IGFs in complex with IGFBP-3 by selective proteolysis of the binding protein. These *in vitro* data suggest that MMP7 may activate ADAM28 in tissues or cells where they are co-expressed, such as breast carcinoma (Mochizuki *et al.*, 2004; Xie *et al.*, 2002). Insulin-like growth factors are potent mitogens for breast cancer cells and are regulated by binding proteins such as IGFBP-3. If ADAM28 releases active IGFs by selective digestion of IGFBP-3 *in vivo*, the observation that ADAM28 is upregulated in a breast cancer cell line (Xie *et al.*, 2002) may have clinical significance.

The disintegrin-like domain of ADAM28 is recognized by the integrin α4β1 (or VLA-4) in an activation-dependent manner. The recognition requires multiple charged residues outside the disintegrin loop, predicted to be located on one face of the ADAM28 disintegrin domain (Bridges *et al.*, 2003; Bridges *et al.*, 2002). The interaction between ADAM28 and α4β1 mediates cellular adhesion of Jurkat T cell leukemia and Daudi B lymphoma lines. The α4β1 integrin is known to be involved in interactions of lymphocytes, dendritic cells and stem cells with the extracellular matrix and endothelial cells, such as rolling and firm adhesion of leukocytes to inflamed tissues through ligation of VCAM-1 or fibronectin (Chigaev *et al.*, 2004). This integrin also plays a role in T cell development in the thymus (Salomon *et al.*, 1997). The interaction between ADAM28 and α4β1 may target the protease to other ligands of the integrin such as VCAM-1 or fibronectin, or may serve to sequester ADAM28 protease away from other potential target proteins. The leukocyte-specific expression of human ADAM28 (Roberts *et al.*, 1999), and thymic epithelial expression of the murine homologue (Haidl *et al.*, 2002), suggest roles for the disintegrin- α4β1 recognition in targeting or sequestering ADAM28 protease activity, to modulate leukocyte-endothelium interactions, or lymphocyte development in the thymus, respectively. It is interesting to note that thymic epithelial cells release soluble CD23, which contributes to the maturation of prothymocytes (Dalloul *et al.*, 1991), suggesting a possible role for the proteolytic activity of ADAM28 in this process.

Many ADAM family members are predominantly expressed in male reproductive tissues, and are believed to play roles in fertilization or spermatogenesis (Cerretti *et al.*, 1999). The strong expression of ADAM28 observed in male reproductive tissues from numerous species (Howard *et al.*,

2000; Jury *et al.*, 1999) thus suggests possible function(s) in these processes. In this regard, it is interesting to note that the ADAM sequence most closely related to ADAM28 is that of the non-protease ADAM7, a gene expressed specifically in the caput epididymis (Cornwall and Hsia, 1997), similarly to ADAM28 (Jury *et al.*, 1999). These two epididymal ADAM proteins presumably play roles in sperm maturation involving proteolytic and/or adhesive functions, but the details of this function are currently unclear.

ADAM28 expression is upregulated by carcinogens and in certain cancers (Jin *et al.*, 2003; Mochizuki *et al.*, 2004; Roemer *et al.*, 2004; Xie *et al.*, 2002). ADAM28 could conceivably play a role in carcinogenesis through the release of active growth factors, degradation of the extracellular matrix to facilitate invasion, or effects on cell-cell contact and signaling through integrins. As discussed previously, ADAM28 has been shown to release the mitogen, IGF-I from IGFBP-3 by selective proteolysis of the binding protein (Mochizuki et al., 2004), but did not degrade extracellular matrix proteins when tested *in vitro* (Howard et al., 2001). The specific recognition and interaction of ADAM28 with the integrin, $\alpha4\beta1$, may have significance for potential roles in cancer. Studies have shown that $\alpha4\beta1$ is important in invasion and metastasis of renal cell carcinoma (Gilcrease et al., 1996; Neeson et al., 2003), a disease where ADAM28 has been shown to be upregulated (Roemer et al., 2004). The $\alpha4\beta1$ integrin also plays a role in lymphocyte-facilitated adhesion of tumor cells to endothelial cells (Neeson *et al.*, 2003). A number of other ADAMs have been implicated in cancer (Duffy *et al.*, 2003), but the potential role of ADAM28 in cancer remains to be elucidated.

## 5. CONCLUSION

Since the original discovery of ADAM28 in 1999, much has been learnt about its expression pattern in different species, cell types and tissues, the selectivity of its metalloprotease activity for peptide and protein substrates, and the specific recognition by the disintegrin domain for the integrin, $\alpha4\beta1$. However, the physiological functions of this member of the ADAMs family are still unknown and a number of questions remain unanswered, including why human, but not murine, ADAM28 is expressed in lymphocytes, and what the functional significance is for the expression of both membrane-bound and secreted isoforms of this protein. The answers to these and other questions will require further functional studies, including gene silencing of ADAM28 in cells by siRNA techniques, and targeted deletion in mice. These approaches will help elucidate the potential roles for ADAM28 in development, reproduction, adult homeostasis, host defense, and cancer.

## REFERENCES


Bates, E.E., W.H. Fridman, and C.G. Mueller. 2002. The ADAMDEC1 (decysin) gene structure: evolution by duplication in a metalloprotease gene cluster on chromosome 8p12. *Immunogenetics*. 54:96-105.

Becherer, J.D., and C.P. Blobel. 2003. Biochemical properties and functions of membrane-anchored metalloprotease-disintegrin proteins (ADAMs). *Curr Top Dev Biol*. 54:101-23.

Bode, W., F.X. Gomis-Ruth, and W. Stockler. 1993. Astacins, serralysins, snake venom and matrix metalloproteinases exhibit identical zinc-binding environments (HEXXHXXGXXH and Met-turn) and topologies and should be grouped into a common family, the 'metzincins'. *FEBS Lett*. 331:134-40.

Bonnefoy, J.Y., C. Plater-Zyberk, S. Lecoanet-Henchoz, J.F. Gauchat, J.P. Aubry, and P. Graber. 1996. A new role for CD23 in inflammation. *Immunol Today*. 17:418-20.

Bridges, L.C., K.R. Hanson, P.H. Tani, T. Mather, and R.D. Bowditch. 2003. Integrin alpha4beta1-dependent adhesion to ADAM 28 (MDC-L) requires an extended surface of the disintegrin domain. *Biochemistry*. 42:3734-41.

Bridges, L.C., P.H. Tani, K.R. Hanson, C.M. Roberts, M.B. Judkins, and R.D. Bowditch. 2002. The lymphocyte metalloprotease MDC-L (ADAM 28) is a ligand for the integrin alpha4beta1. *J Biol Chem*. 277:3784-92.

Cerretti, D.P., R.F. DuBose, R.A. Black, and N. Nelson. 1999. Isolation of two novel metalloproteinase-disintegrin (ADAM) cDNAs that show testis-specific gene expression. *Biochem Biophys Res Commun*. 263:810-5.

Chigaev, A., G.J. Zwartz, T. Buranda, B.S. Edwards, E.R. Prossnitz, and L.A. Sklar. 2004. Conformational Regulation of {alpha}4{beta}1-Integrin Affinity by Reducing Agents: "Inside-Out" Signaling Is Independent Of And Additive To Reduction-Regulated Integrin Activation. *J. Biol. Chem.* 279:32435-32443.

Cornwall, G.A., and N. Hsia. 1997. ADAM7, a member of the ADAM (a disintegrin and metalloprotease) gene family is specifically expressed in the mouse anterior pituitary and epididymis. *Endocrinology*. 138:4262-72.

Dalloul, A.H., C. Fourcade, P. Debre, and M.D. Mossalayi. 1991. Thymic epithelial cell-derived supernatants sustain the maturation of human prothymocytes: involvement of interleukin 1 and CD23. *Eur J Immunol*. 21:2633-6.

Delespesse, G., M. Sarfati, C.Y. Wu, S. Fournier, and M. Letellier. 1992. The low-affinity receptor for IgE. *Immunol Rev*. 125:77-97.

Duffy, M.J., D.J. Lynn, A.T. Lloyd, and C.M. O'Shea. 2003. The ADAMs family of proteins: from basic studies to potential clinical applications. *Thromb Haemost*. 89:622-31.

Fourie, A.M., F. Coles, V. Moreno, and L. Karlsson. 2003. Catalytic activity of ADAM8, ADAM15, and MDC-L (ADAM28) on synthetic peptide substrates and in ectodomain cleavage of CD23. *J Biol Chem*. 278:30469-77.

Gagro, A., and S. Rabatic. 1994. Allergen-induced CD23 on CD4+ T lymphocytes and CD21 on B lymphocytes in patients with allergic asthma: evidence and regulation. *Eur J Immunol*. 24:1109-14.

Gilcrease, M.Z., L. Truong, and R.W. Brown. 1996. Correlation of very late activation integrin and CD44 expression with extrarenal invasion and metastasis of renal cell carcinomas. *Hum Pathol*. 27:1355-60.

Glassey, B., and A. Civetta. 2004. Positive selection at reproductive ADAM genes with potential intercellular binding activity. *Mol Biol Evol*. 21:851-9.

Haidl, I.D., G. Huber, and K. Eichmann. 2002. An ADAM family member with expression in thymic epithelial cells and related tissues. *Gene*. 283:163-70.

Howard, L., R.A. Maciewicz, and C.P. Blobel. 2000. Cloning and characterization of ADAM28: evidence for autocatalytic pro-domain removal and for cell surface localization of mature ADAM28. *Biochem J.* 348 Pt 1:21-7.

Howard, L., Y. Zheng, M. Horrocks, R.A. Maciewicz, and C. Blobel. 2001. Catalytic activity of ADAM28. *FEBS Lett.* 498:82-6.

Jin, J., Z. Gao, L. Guo, J. Yang, and Y. Yu. 2003. Altered expression of zinc finger proteins, ADAMs, and integrin-related proteins following treatment of cultured human cells with a low concentration of N-methyl-N'-nitro-N-nitrosoguanidine. *Environ Mol Mutagen.* 41:344-52.

Jury, J.A., A.C. Perry, and L. Hall. 1999. Identification, sequence analysis and expression of transcripts encoding a putative metalloproteinase, eMDC II, in human and macaque epididymis. *Mol Hum Reprod.* 5:1127-34.

Kashiwagi, M., M. Tortorella, H. Nagase, and K. Brew. 2001. TIMP-3 is a potent inhibitor of aggrecanase 1 (ADAM-TS4) and aggrecanase 2 (ADAM-TS5). *J Biol Chem.* 276:12501-4.

Krätzschmar, J., L. Lum, and C.P. Blobel. 1996. Metargidin, a Membrane-anchored Metalloprotease-Disintegrin Protein with an RGD Integrin Binding Sequence. *J. Biol. Chem.* 271:4593-4596.

Mochizuki, S., M. Shimoda, T. Shiomi, Y. Fujii, and Y. Okada. 2004. ADAM28 is activated by MMP-7 (matrilysin-1) and cleaves insulin-like growth factor binding protein-3. *Biochem Biophys Res Commun.* 315:79-84.

Moss, M.L., and J.W. Bartsch. 2004. Therapeutic benefits from targeting of ADAM family members. *Biochemistry.* 43:7227-35.

Neeson, P.J., P.J. Thurlow, G.P. Jamieson, and C. Bradley. 2003. Lymphocyte-facilitated tumour cell adhesion to endothelial cells: the role of high affinity leucocyte integrins. *Pathology.* 35:50-5.

Oh, J., and C. Cho. 2003. Rattus norvegicus ADAM28 isoform-1 mRNA. *unpublished.* NM_181693.

Roberts, C.M., P.H. Tani, L.C. Bridges, Z. Laszik, and R.D. Bowditch. 1999. MDC-L, a novel metalloprotease disintegrin cysteine-rich protein family member expressed by human lymphocytes. *J Biol Chem.* 274:29251-9.

Roemer, A., L. Schwettmann, M. Jung, J. Roigas, G. Kristiansen, D. Schnorr, S.A. Loening, K. Jung, and R. Lichtinghagen. 2004. Increased mRNA expression of ADAMs in renal cell carcinoma and their association with clinical outcome. *Oncol Rep.* 11:529-36.

Salomon, D.R., L. Crisa, C.F. Mojcik, J.K. Ishii, G. Klier, and E.M. Shevach. 1997. Vascular cell adhesion molecule-1 is expressed by cortical thymic epithelial cells and mediates thymocyte adhesion. Implications for the function of alpha4beta1 (VLA4) integrin in T-cell development. *Blood.* 89:2461-71.

Seals, D.F., and S.A. Courtneidge. 2003. The ADAMs family of metalloproteases: multidomain proteins with multiple functions. *Genes Dev.* 17:7-30.

Vella, A., P. Bellavite, A. Adami, R. Ortolani, G. Benoni, A. Carletto, D. Biasi, P. Caramaschi, and G. Tridente. 1999. Expression of FCepsilonII/CD23 on human neutrophils isolated from rheumatoid arthritis patients. *Inflammation.* 23:471-9.

Wheeler, D.J., S. Parveen, K. Pollock, and R.J. Williams. 1998. Inhibition of sCD23 and immunoglobulin E release from human B cells by a metalloproteinase inhibitor, GI 129471. *Immunology.* 95:105-10.

White, J.M. 2003. ADAMs: modulators of cell-cell and cell-matrix interactions. *Curr Opin Cell Biol.* 15:598-606.

Worley, J.R., M.D. Baugh, D.A. Hughes, D.R. Edwards, A. Hogan, M.J. Sampson, and J. Gavrilovic. 2003. Metalloproteinase expression in PMA-stimulated THP-1 cells. Effects of peroxisome proliferator-activated receptor-gamma (PPAR gamma) agonists and 9-cis-retinoic acid. *J Biol Chem*. 278:51340-6.

Xie, D., A. Jauch, C.W. Miller, C.R. Bartram, and H.P. Koeffler. 2002. Discovery of over-expressed genes and genetic alterations in breast cancer cells using a combination of suppression subtractive hybridization, multiplex FISH and comparative genomic hybridization. *Int J Oncol*. 21:499-507.

Zou, J., F. Zhu, J. Liu, W. Wang, R. Zhang, C.G. Garlisi, Y.H. Liu, S. Wang, H. Shah, Y. Wan, and S.P. Umland. 2004. Catalytic activity of human ADAM33. *J Biol Chem*. 279:9818-30.

Chapter 11

# MAMMALIAN ADAMS WITH TESTIS-SPECIFIC OR -PREDOMINANT EXPRESSION

*Testicular ADAMs*

Chunghee Cho
*Department of Life Science, Gwangju Institute of Science and Technology, Gwangju 500-712, Republic of Korea*

Abstract: A family of multidomain membrane proteins, the ADAM family (a disintegrin and metalloprotease), comprises at least 32 members. Based on an ADAM phylogenetic tree, mammalian ADAMs with testis-specific or -predominant expression are divided into two major groups: ADAMs 1, 4, 6, 20, 21, 24, 25, 26, 29, 30 and 34 (the first group), and ADAMs 2, 3, 5, 27 and 32 (the second group). All of the mammalian, testicular ADAMs predicted as active metalloproteases (ADAMs 1, 20, 21, 24, 25, 26, 30 and 34) belong to the first phylogenetic group and are unique in several aspects. All of these ADAM genes lack introns in their coding sequences and many of them are present as multiple copies in the mouse genome, resulting in total of 11 functional genes (ADAMs 1a, 1b, 21, 24, 25a, 25b, 25c, 26a, 26b, 30 and 34) predicted to encode active proteases. These genes are transcribed by both somatic and germ cells with higher expression level in post-meiotic germ cells in the testis. The ADAM 1 protein expressed in testicular germ cells is complexed with ADAM 2 to form a heterodimer and processed during spermatogenesis. Mouse knockout studies indicate that ADAM 1a/2 heterodimer in testicular germ cells is implicated in the regulation or localization of sperm proteins involved in sperm progression in the female reproductive tract, sperm penetration into the cumulus cell layer and sperm-egg zona pellucida binding, thus playing a central role in fertilization. ADAM 24 is a potential sperm protease implicated in sperm function during sperm maturation or fertilization. A number of the 11 mouse ADAMs do not have orthologues in human and, even if they exist, some orthologues are pseudogenes in human. As a result, only 3 human ADAMs (ADAMs 20, 21 and 30) are functional genes encoding potential metalloproteases. Uncovering the in vivo functions of the testicular ADAM proteases present in both species should provide insights into the mammalian reproductive system involving protease-mediated events.

Key words: Disintegrin; fertilization; metalloprotease; orthologue; sperm; testis



*N. M. Hooper and U. Lendeckel (eds.), The ADAM Family of Proteases,* 239-259.

# 1. INTRODUCTION

ADAMs are a family of transmembrane proteins that contain A Disintegrin And Metalloprotease domain. The ADAM family members have been discovered in a variety of tissues and species, and the family currently has at least 32 members. The ADAM family proteins share conserved multidomain structure: pro-, metalloprotease, disintegrin, cysteine-rich, epidermal growth factor (EGF)-like, transmembrane and cytoplasmic domains (Fig. 1). They show sequence homology with soluble snake venom peptides and proteins that contain one or more of pro-, metalloprotease, disintegrin, and cysteine-rich domains. A primordial ADAM gene, having arisen by the assembly of the separate domains, may have been amplified to generate new genes. The ADAM proteins have at least two potential functions. They function as cell adhesion molecules. In many snake venom disintegrin peptides, the active site of the molecules is defined by a tripeptide sequence motif RGD. The RGD-containing disintegrins bind to platelet integrin gpIIb/IIIa and prevent fibrinogen binding and platelet aggregation. In the disintegrin domains of the ADAM proteins, the RGD sequence is replaced by other amino acids. The predicted binding partners of the ADAM proteins are integrins, but non-integrin adhesion partners could bind to particular ADAMs. The ADAM proteins with a consensus sequence for an active metalloprotease site potentially have protease activity. The consensus sequence to be an active protease is HEXGHXXGXXHD in which the histidines bind zinc and the glutamic acid is the catalytic residue. ADAMs with the consensus sequence contain a conserved, free cysteine residue responsible for the cysteine-switch mechanism in the prodomain. In addition, these ADAM proteins contain a 'methionine turn' downstream to the catalytic site and thus belong to the 'metzincin' class together with the ADAM-TS proteins and matrix metalloproteases. The ADAM proteins having cell adhesion activity or protease activity have been found to be active in various biological processes (reviewed in Wolfsberg *et al.* 1995a; Blobel 1997; Black and White 1998; Primakoff and Myles 2000; White 2003).

Of the 32 known ADAMs, 16 are expressed exclusively or predominantly in the mammalian testis. The 16 ADAMs are divided phylogenetically into two major groups: ADAMs 1, 4, 6, 20, 21, 24, 25, 26, 29, 30 and 34 (the first group); and ADAMs 2, 3, 5, 27 and 32 (the second group) (Fig. 1). All of the mammalian, testicular ADAMs predicted as active metalloproteases belong to the first phylogenetic group. They are ADAMs 1, 20, 21, 24, 25, 26, 30 and 34 (Fig. 1). This chapter focuses mainly on ADAM 1, one of the best-studied ADAMs, and also the other testicular

ADAMs of which functions are largely unknown, discussing their genomic nature, protein characteristics and roles in reproduction.

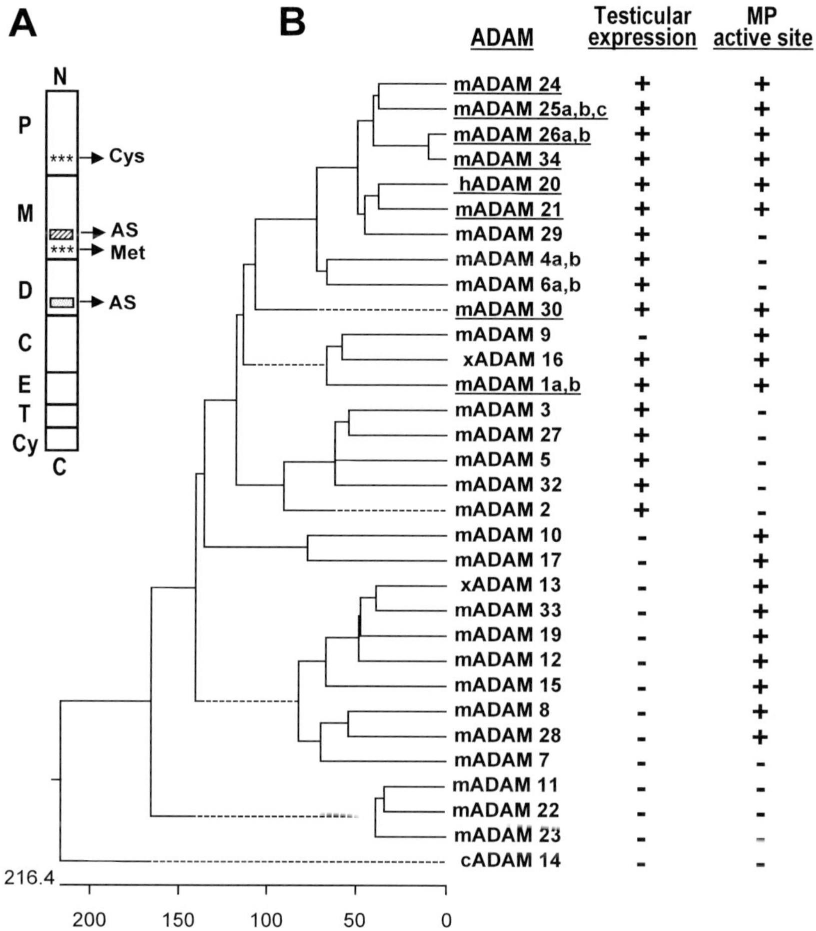


*Figure 1.* ADAM domain structure (A) and the phylogenetic tree of ADAM family members (B). Residues responsible for cysteine-switch (Cys), active sites (AS) and methionine-turn (Met) are indicated at the domain structure. Domain abbreviations are as follows: P, prodomain; M, metalloprotease; D, disintegrin; C, cysteine-rich; E, EGF-like; T, transmembrane; Cy, cytoplasmic. The length of the branch in the phylogenetic tree represents the level of divergence between ADAMs. Mammalian ADAM members with testis-specific or -predominant gene expression, and the active metalloprotease (MP) site are underlined. GenBank accession numbers for the sequences are AB048844 (ADAM 1a), AB048843 (ADAM 1b), NM_009618 (ADAM 2), X64227 (ADAM 3), AY158688 (ADAM 4a), BK001466 (ADAM 4b), U22059 (ADAM 5), AY158689 (ADAM 6a), BK001467 (ADAM 6b), NM_007402 (ADAM 7), NM_007403 (ADAM 8), U41765 (ADAM 9), NM_007399

(ADAM 10), AB009676 (ADAM 11), NM_007400 (ADAM 12), U66003 (ADAM 13), U68185 (ADAM 14), AF006196 (ADAM 15), U78185 (ADAM 16), AB021709 (ADAM 17), AF019887 (ADAM 19), AF158643 (ADAM 20), AF251559 (ADAM 21), AB009674 (ADAM 22), AB009673 (ADAM 23), AF167402 (ADAM 24), AF167403 (ADAM 25a), BN000121 (ADAM 25b), BN000119 (ADAM 25c), AF167404 (ADAM 26a), BK001468 (ADAM 26b), AF167405 (ADAM 27), AF153350 (ADAM 28), AY190759 (ADAM 29), AY190760 (ADAM 30), AF513715 (ADAM 32), AF472524 (ADAM 33) and AF373288 (ADAM 34). Species abbreviations: m, *Mus musculus*; h, *Homo sapiens*; x, *Xenopus lavies*; c, *Caenorhabditis elegans*.

# 2. ADAM 1

## 2.1 Organization and Expression of the ADAM 1 Gene

ADAM 1 was one of the first-identified and best-studied ADAMs. Complementary DNA for ADAM 1 was first cloned in guinea pig (Blobel *et al.* 1992; Wolfsberg *et al.* 1993) and subsequently, ADAM 1 cDNA sequences from a number of mammalian species have been reported, including mouse, rat, rabbit, bull and monkey (Perry *et al.* 1995; Wolfsberg *et al.* 1995b; Hardy and Holland 1996; McLaughlin *et al.* 1997; Waters and White 1997; Nishimura *et al.* 2002).

One of the notable features of the ADAM 1 gene is the presence of two copies for the gene in a range of mammalian species. In mouse, two genes for ADAM 1 were identified, encoding different isoforms named ADAM 1a (791 residues) and ADAM 1b (806 residues). Both genes for ADAM 1 were mapped to one site of a single chromosome in mouse (distal region of chromosome 5). Consistently, a genome database search revealed that the two genes are adjacent in the mouse chromosome, suggesting that the primordial ADAM 1 gene has undergone duplication during evolution (Cho *et al.* 1996, 1997; Nishimura *et al.* 2002). The amino acid sequences of mouse ADAM 1a and ADAM 1b are highly homologous (99%) to each other in the N-terminal region such as the pro- and metalloprotease domains. The metalloprotease active site sequence is the same between the two isoforms: HELGHNLGIQHD. By contrast, the two isoforms share a low degree of identity (37%) in the C-terminal region including the disintegrin and cysteine-rich domains (Nishimura *et al.* 2002). Similarly, cDNAs for two ADAM 1 isoforms, fertilin αI and fertilin αII, were identified in monkey (*Macaca fascicularis*), encoding proteins of 905 and 825 residues, respectively. The two monkey isoforms share extensive identity, but differ significantly towards their N- and C-termini (Perry *et al.* 1995). A study with a range of additional primate species revealed the presence of two copies for the ADAM 1 gene in monkey species such as baboon and tamarin (Jury *et al.* 1998). On the contrary, higher primates such as orang-utan, gorilla and human were found to have only a single gene for ADAM 1. Nucleotide sequence comparison showed that the 5' end of the ADAM 1 gene in these three species is highly homologous to the 5' end of monkey fertilin αI (but not fertilin αII), the central region shows a high degree of identity to the central region common to both monkey isoforms and the 3' end of the genes are significantly homologous to the 3' region of monkey fertilin αII (but not fertilin αI). This suggests that the ADAM 1 gene in orang-utan, gorilla and human has been derived by genetic recombination between the two ADAM 1 genes maintained presumably from mouse to monkeys during evolution. This recombination event may have occurred during the divergence of monkeys and great apes.

Another distinctive feature of the ADAM 1 gene is that all the mammalian ADAM 1 genes lack introns, and the gorilla and human genes are inactive (Jury *et al.* 1997, 1998; Nishimura *et al.* 2002). This suggests that the ADAM 1 gene may have arisen through retrotransposition of spliced mRNA transcribed from a primordial ADAM gene into the mammalian genome, resulting in the loss of introns. The gorilla and human ADAM 1 genes contain a number of nucleotide insertions and deletions in the coding regions, disrupting the reading frames. This suggests that the processed, intronless ADAM 1 genes may have accumulated mutations in the mammalian lineage over time, due to the absence of selective pressure during evolution.

The ADAM 1 gene is expressed exclusively or predominantly in the mammalian testis. Northern blot analyses demonstrated the testis-specific expression of the mouse and guinea pig ADAM 1 genes, whereas RT-PCR analyses, a more sensitive assay, showed that the mouse and rat ADAM 1 genes are expressed abundantly in the testis and at low levels in other tissues (Wolfsberg *et al.* 1993, 1995b; Frayne *et al.* 1997; Nishimura *et al.* 2002). When a certain gene is found to be expressed in the testis, it is of particular interest whether the gene is expressed in germ cells or somatic cells in the testis. Male germ cell development occurs in seminiferous tubules containing a mixture of spermatogenic cells and somatic cells such as Sertoli cells. In the first round of spermatogenesis in prepubertal animals, stem cells proliferate and differentiate gradually to yield the sequence of spermatogonia, spermatocytes, spermatids and sperm. This process takes about 35 days in the mouse with three different phases such as mitotic (11 days), meiotic (10 days) and postmeiotic (14 days) phases (Eddy 1998). If a particular gene is expressed in germ cells during spermatogenesis, a transcript for that gene will appear in the testis at a certain post-partum time point corresponding to a specific stage of spermatogenesis. Northern blot analysis using prepubertal and mature mouse testes showed that the ADAM 1a and ADAM 1b genes are first expressed from days 24 and 20, respectively, suggesting germ cell-specific expression of the genes (Nishimura *et al.* 2002). By contrast, separate studies using PCR assay demonstrated that the rat ADAM 1 gene is expressed at all stages of development. This suggests that the rat ADAM 1 gene is not exclusively expressed in germ cells (Fryne *et al.* 1997; McLaughlin *et al.* 1997). Considering a difference between Northern blot and PCR assay in sensitivity and quantification, it is likely that the ADAM 1 gene is transcribed by both somatic and germ cells in the testis with higher level expression in germ cells. Alternatively, the expression of the ADAM 1 gene may be differentially regulated between mouse and rat testes.

## 2.2 Biochemical and Cellular Properties of ADAM 1

ADAM 1 (also known as PH30 α or fertilin α), an integral membrane glycoprotein, was identified originally in guinea pig sperm on which the protein is dimerized with the ADAM 2 protein (PH30 β or fertilin β) (Primakoff *et al.* 1987; Blobel *et al.* 1990). Although ADAM 2, unlike ADAM 1, does not have the correct amino acid sequence at the position of the metalloprotease active site to be a functional protease, it is closely related with ADAM 1 in protein interaction and function (reviewed in Myles 1993; Wolfsberg *et al.* 1995a; Snell and White 1996; Wolfsberg and White 1996; Myles and Primakoff 1997; Wassarman 1999; Blobel 2000; Primakoff and Myles 2000, 2002; Evans 2001; Talbot *et al.* 2003). A comprehensive, biochemical analysis using testicular cells, testicular sperm and epididymal sperm from guinea pig revealed the complicated processing pattern of ADAM 1/2 heterodimeric complex during sperm development and maturation. The guinea pig ADAM 1 and ADAM 2 proteins are synthesized as precursors in spermatogenic cells and undergo subsequent proteolytic cleavages during spermatogenesis (ADAM 1) and during passage through the epididymis (ADAM 2), a highly convoluted tubule in which sperm migrate from the testis to the vas deferens (Blobel *et al.* 1990). A further analysis of guinea pig ADAM 1 showed that only the processed form of ADAM 1, but not the ADAM 1 precursor, can be cell-surface biotinylated and some processed ADAM 1 is sensitive to endoglycosidase H. The proteolytic cleavage site was found to be next to four arginine residues (RRRR) between the metalloprotease domain and the disintegrin domain. This tetrabasic sequence is consistent with a consensus cleavage site, RX(K/R)R, as a target for a pro-protein convertase such as furin. These data suggest that ADAM 1 is processed by a pro-protein convertase in the secretory pathway of testicular cells before emerging on the cell surface (Lum and Blobel 1997). The proteolytic processing of the ADAM 2 precursor also removes the pro- and metalloprotease domains, leaving an N-terminal disintegrin domain on the processed form. The ADAM 1/2 complex is distributed throughout the testicular sperm head but becomes concentrated in the posterior head region simultaneously with ADAM 2 proteolytic processing during epididymal maturation of the sperm (Blobel *et al.* 1990; Phelps *et al.* 1990; Hunnicutt *et al.* 1997; Lum and Blobel 1997).

Similar ADAM 1 features with additional information were found in other mammalian species. Monkey fertilin αI, one of the ADAM 1 isoforms, is processed during spermatogenesis in testicular cells (Frayne *et al.* 1998). In bovine sperm, ADAM 1 and ADAM 2 are complexed to form a heterodimer and this complex appears to exist as a higher-order oligomer of the basic heterodimer (Waters and White 1997). As described above (section 2.1), there are two intronless mouse genes encoding different ADAM 1 isoforms, ADAM 1a and ADAM 1b (Nishimura *et al.* 2002). Both ADAM 1 isoforms were found to be present in testicular germ cells (Kim *et al.* 2003). The ADAM 1a precursor (100 kDa) is partially converted into a processed form (48 kDa) in round and elongating spermatids. Unlike ADAM 1 in other species, mouse ADAM 1a is missing in sperm from the epididymis and vas deferens. Both the ADAM 1a precursor and the processed form of ADAM 1a are sensitive to endoglycosidase H, suggesting that they reside in the endoplasmic reticulum (ER). ADAM 1a does not contain the motifs required for protein retention in the ER. Thus it is unknown how the ADAM 1a proteins are retained within the ER. The ADAM 1a precursor forms a heterodimeric complex with the ADAM 2 precursor in spermatogenic cells. On the other hand, ADAM 1b is made as a precursor (120 kDa) in pachytene spermatocytes, converted to an intermediate form (63 kDa) during spermiogenesis and further processed to a mature form (60 kDa) during the sperm transit in the epididymis. Both of the precursor and mature forms of ADAM 1b are capable of forming heterodimers with the ADAM 2 precursor and processed ADAM2, respectively. Most of the ADAM 1b precursor is associated with ADAM 2 in testicular cells, but only a part of the processed form of ADAM 1b contributes to the complex formation with ADAM 2 in mature sperm. Both of the heterodimeric and monomeric forms of ADAM 1b are localized on the cell surface of mature sperm. Taken together, the two ADAM 1 isoforms are distinguishable from each other in processing pattern and subcellular localization (Kim *et al.* 2003). Mouse ADAM 1b, but not ADAM 1a, is similar to guinea pig, bovine and monkey ADAM 1. Despite the existence of the two functional genes for ADAM 1 in several mammalian species lower than great apes (see section 2.1) besides mouse, it is unknown whether the ADAM 1a protein exists in these species. On the contrary, the human ADAM 1 gene is nonfunctional. It is possible that an ADAM 1a-like protein may be present and active in human. Resolution of these issues, requiring further investigation, is important because knockout studies imply that each of the two mouse ADAM 1 isoforms has a differential function in reproduction (see below; section 2.3).

ADAM 1 contains the pro-protein convertase consensus sequence between the metalloprotease domain and the disintegrin domain, directing removal of the pro- and metalloprotease domains from the membrane-anchored protein. This presents an apparent contrast to other catalytic ADAM members of which activation is dependent upon removal of the prodomain, leaving the N-terminal metalloprotease domain on the membrane-anchored, processed form (Blobel 2000). To answer if prodomain removal of ADAM 1 occurs and whether this processing event is necessary for protease activity will require further investigation.

## 2.3 Functions of ADAM 1 in Fertilization

Mammalian fertilization is a multi-step process. After sperm enter the uterus, they migrate into the oviduct where sperm fertilize eggs. The utero-tubal junction, a barrier for sperm ascent to the ampulla, permits migration of only a small fraction of uterine sperm into the oviduct (Yanagimachi, 1994). When sperm reach the egg, they first encounter a mass of cells, called the cumulus oophorus, that surrounds the egg. Sperm swim between these cells, dissolving the extracellular matrix rich in hyaluronic acid. When sperm reach the egg extracellular coat, the zona pellucida (ZP), they recognize it and bind to it. Subsequently, sperm penetrate the ZP and come to lie in the perivitelline space. Finally, the sperm bind to the egg plasma membrane and then fuse with it.

Mouse knockouts of ADAM 1 and related ADAMs revealed direct or indirect involvement of the ADAMs in the fertilization process. Recently, mice were produced that are homozygous nulls for ADAM 1a (Nishimura *et al.* 2004). The mutant male mice are infertile because of the severely impaired ability of sperm to migrate from the uterus into the oviduct. ADAM 1a-deficient sperm are also impaired in the dispersal of cumulus cells and binding to the egg ZP. The analysis of protein expression phenotype revealed that ADAM 1b and ADAM 2 are normally present in both testicular germ cells and sperm from ADAM 1a-deficient mice. However, the loss of ADAM 1a resulted in the severe reduction of ADAM 3 in epididymal sperm (Table 1). ADAM 3, also known as cyritestin, is a testicular ADAM member similar to ADAM 2 in structure, processing and subcellular localization: it is predicted as an inactive protease, processed during sperm maturation and present on mature sperm (Yuan *et al.* 1997). Mice lacking ADAM 2 were also produced and the mutant males were found to be infertile (Cho *et al.* 1998). Like ADAM 1a knockout, sperm from ADAM 2-deficient mice are defective in migration from the uterus into the oviduct and adhesion to the egg ZP. However, the protein expression phenotype of ADAM 2-deficient mice is dissimilar to that of ADAM 1a-deficient mice. In ADAM 2-deficient testicular germ cells, the amounts of ADAM 1a and ADAM 3 are normal, while the level of ADAM 1b is significantly reduced. Furthermore, ADAM 2-deficient sperm completely lack ADAM 1b and barely contain ADAM 3 (Table 1) (Cho *et al.* 1998, 2000; Nishimura *et al.* 2004). Finally, male mice with a deletion in the ADAM 3 gene were reported to be also infertile (Shamsadin *et al.* 1999; Nishimura *et al.* 2001). Comparing to ADAM 1a- and ADAM 2-deficient sperm, ADAM 3 knockout resulted in similar, but more restricted, effects on fertilization. ADAM 3 null sperm are deficient in adhesion to the egg ZP, but show normal transport from the uterus into the oviduct. The levels of ADAM 1a, ADAM 1b and ADAM 2 are unchanged in testicular germ cells and mature sperm from ADAM 3-deficient mice (Table 1) (Nishimura *et al.* 2004).

*Table 1.* Protein expression phenotype in ADAM knockouts.

| Knockout | Cell stage* | Expression level or integrity of ADAM proteins** | | | | | |
|---|---|---|---|---|---|---|---|
| | | 1a | 1b | 2 | 1a/2 | 1b/2 | 3 |
| ADAM 1a | TGC | - | + | + | - | + | + |
| | ES | - | + | + | - | + | - |
| ADAM 2 | TGC | + | - | - | - | - | + |
| | ES | - | - | - | - | - | - |
| ADAM 3 | TGC | + | + | + | + | + | - |
| | ES | - | + | + | + | + | - |

*Cell stage: TGC, testicular germ cell; ES, epididymal sperm.

**The normal expression level, and the absence or severely reduced expression level of ADAMs are indicated as + and -, respectively. 1a/2 or 1b/2 indicates ADAM 1/ADAM2 heterodimer. Lack of ADAM 1a and ADAM 1a/ADAM 2 in epididymal sperm is not due to gene knockouts since ADAM 1a is missing in wild-type sperm as described in section 2.2.

The intriguing feature of the ADAM knockouts is the intricate relationships between the protein expression phenotypes and the fertilization phenotypes (Table 1). Both of ADAM 1a- and ADAM 2-decificent sperm, but not ADAM 3 null sperm, are defective in migration from the uterus to the oviduct. A common protein expression phenotype between ADAM 1a and ADAM 2 knockouts is the absence of ADAM 1a/2 heterodimer and ADAM 3 in testicular germ cells and epididymal sperm, respectively (Table 1). Because ADAM 3 null sperm are normal in progression into the oviduct,

ADAM 1a/2 heterodimer, but not ADAM 3, is important for sperm migration. Considering lack of ADAM 1a/2 complex in wild-type sperm, this complex is indirectly involved in sperm transport from the uterus into the oviduct. During spermatogenesis, the ADAM 1a metalloprotease, of which activity presumably requires interaction with ADAM 2, might process and thus regulate sperm molecules directly involved in sperm transport (Fig. 2).

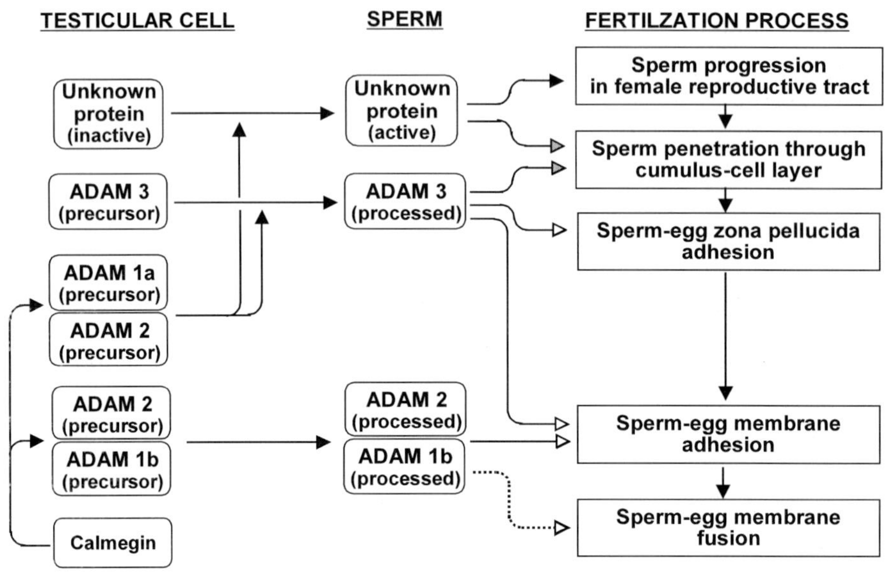


*Figure 2.* Schematic diagram showing the relationships between ADAMs and fertilization. Calmegin is necessary for ADAM 1/2 heterodimerization. ADAM 1a/2 regulates the activity or localization of sperm proteins including ADAM 3. An unknown sperm protein(s) regulated by ADAM 1a/2 is directly involved in sperm transport from the uterus to the oviduct. Sperm penetration through the layer of cumulus cells is partially dependent on the unknown protein and/or ADAM 3 (shaded arrowhead). The fertilization steps of sperm-egg zona pellucida binding involving ADAM 3, and sperm-egg plasma membrane binding involving ADAM 3, ADAM 2 or ADAM 1b are not physiologically essential for successful fertilization (open arrowhead) as found for the mutant sperm showing normal fertilization in the subsequent steps. ADAM 1b is not an absolute requirement for sperm-egg plasma membrane fusion (dotted arrow).

All ADAM 1a-, ADAM 2- and ADAM 3-deficient sperm are defective in binding to the egg ZP and all commonly lack ADAM 3 in mature sperm (Table 1). This suggests that the formation of the complex between ADAM1a and ADAM 2 in testicular germ cells is responsible for the appearance of ADAM 3 on the sperm surface, which plays a critical role in sperm-egg ZP binding (Fig. 2). Despite the impaired ability to bind the ZP, ADAM 1a-deficient sperm lacking ADAM 3 are capable of fertilizing cumulus-intact, ZP-intact eggs normally (Nishimura *et al.* 2004). This indicates that the step of sperm-egg ZP adhesion involving ADAM 3 is not a prerequisite for successful fertilization.

*In vitro* sperm-egg plasma membrane interaction analysis and amino acid sequence analysis suggested that ADAM 1 plays roles in sperm-egg plasma membrane adhesion and fusion. It was found that a recombinant protein and a peptide mimetic corresponding to the mouse ADAM 1a disintegrin domain inhibit sperm adhesion to the plasma membrane of ZP-free eggs (Evans *et al.* 1998; Wong *et al.* 2001). Because ADAM 1a is absent from mature sperm, however, assessment of ADAM 1a in sperm-egg binding is physiologically irrelevant. By contrast, mouse ADAM 1b is present on the surface of mature sperm and contains the RGD sequence, the integrin-binding motif, in the cysteine-rich domain (Nishimura et al. 2002; Kim et al. 2003). Consistent with this, ADAM 2 knockout sperm lacking ADAM 1b are defective in adhesion to the plasma membrane of ZP-free eggs (Table 1) (Cho *et al.* 1998). It should be noted that ADAM 2 and ADAM 3 also play a role in sperm-egg membrane binding because they are present on sperm surface and sperm lacking either or both proteins are deficient in adhesion to the egg membrane (Cho *et al.* 1998; Nishimura *et al.* 2001). The adhesion role of ADAM 2 and ADAM 3 is also supported by a number of *in vitro* studies in which peptide mimetics of the ADAM 2 and ADAM 3 disintegrin domains inhibit sperm-egg membrane binding (Myles *et al.* 1994; Almeida *et al.* 1995; Evans *et al.* 1995; Yuan *et al.* 1997). However, sperm lacking ADAM 1b, ADAM 2 or ADAM 3 were found to have little defect in fusion with the egg plasma membrane. Thus the adhesion step involving ADAM 1b, ADAM 2 or ADAM 3 is not an essential, physiological process that leads to fusion (Fig. 2).

A previously proposed model for the mechanism of gamete fusion suggests that a hydrophobic fusion peptide present in the ADAM 1 cysteine-rich domain promotes sperm-egg fusion (Blobel *et al.* 1992; Wolfsberg *et al.* 1995a; Snell and White 1996; Waters and White; 1997). A number of investigators have searched for fusion-promoting properties of peptides corresponding to the putative fusion peptide of ADAM 1 (Muga *et al.* 1994; Niidome *et al.* 1997; Martin *et al.* 1998; Wolfe *et al.* 1999). Fusion of mouse sperm lacking ADAM 2 with the egg plasma membrane occurs at about 50% of the wild-type rate (Cho *et al.* 1998). Because ADAM 1b is absent from the mutant sperm which can fuse, this indicates that the ADAM 1b fusion peptide is not an absolute requirement for sperm-egg membrane fusion (Fig. 2) (Cho *et al.* 2001; Nishimura *et al.* 2004).

The ADAM knockouts provide an insight into the molecular basis of mammalian fertilization. Figure 2 summarizes the relationships between the ADAM proteins and the fertilization process. Each of the two ADAM 1 isoforms interacts with ADAM 2 to form a heterodimer and differentially participates in fertilization. Spermatogenesis-specific chaperone calmegin was reported to be required for ADAM 1/2 heterodimerization (Ikawa *et al.* 1997; 2001). ADAM 1a is implicated in the regulation of sperm proteins, including ADAM 3, that function in sperm progression from the uterus into the oviduct. Identification of target proteins for ADAM 1a/2 complex should provide more information about the molecular network of fertilization.

## 3. ADAMS 20, 21, 24, 25, 26, 30 AND 34

### 3.1 Molecular and Genomic Nature of ADAMs

According to the ADAM phylogenetic tree (Fig. 1), the mammalian ADAMs with testis-specific or -predominant gene expression are divided into two major groups: ADAMs 1, 4, 6, 20, 21, 24, 25, 26, 29, 30 and 34 (group I); and ADAMs 2, 3, 5, 27 and 32 (group II). The first-group ADAMs can be further divided phylogenetically into two subgroups: ADAM 1 (group I-1) and the rest of the ADAMs (group I-2) (Fig. 1 and Table 2). Among the members in the group I-2, ADAMs with the metalloprotease active site are ADAMs 20, 21, 24 (testase 1), 25 (testase 2), 26 (testase 3), 30 and 34 (testase 4). Full-length cDNAs for these ADAMs were cloned from mouse and/or human (Hooft van Huijsduijnen 1998; Cerretti *et al.* 1999; Brachvogel *et al.* 2002; Poindexter *et al.* 1999; Zhu *et al.* 1999; Liu and Smith 2000; Choi *et al.* 2004). All of these ADAMs possess the consensus zinc-binding and catalytic sites HEXGHXXGXXHD and the 'methionine turn' downstream (Table 2). It is noteworthy that the group-I-2 ADAMs without the correct amino acid sequence to be a functional protease (especially ADAMs 4 and 29) show high similarity to the consensus sequence at the position of the metalloprotease active site, presenting a contrast to the group-II ADAMs which also do not have the consensus sequence (Table 2). It is likely that the metalloprotease active site arose in a common ancestor and subsequently, maintained or lost in the lineages of the group-I-2 ADAMs. Thus the testicular ADAMs with protease activity appear to be closely related by evolution (Cerretti *et al.* 1999; Choi *et al.* 2004). In addition to the conserved zinc-dependent metalloprotease active site, the prodomain of all of the group-I-2 ADAMs having the consensus sequence to be a functional protease contains an unpaired cysteine involved in the inhibition of the metalloprotease activity by interacting with zinc (Table 2).

Convertase-induced cleavage of the prodomain, removing the free cysteine, is expected to activate the metalloprotease activity. Unlike other ADAM proteases that have the consensus target sequence for a furin-like convertase, the group-I-2 ADAMs predicted as active proteases do not have this cleavage site (Table 2), suggesting a different processing mechanism (see section 3.2).

*Table 2.* Comparison of metalloprotease characteristics among mammalian, testicular ADAMs.

| Phylo-genetic Group | ADAM* | Prediction of active protease | Free cys. | Target site for furin | Active site sequence** | Met-turn |
|---|---|---|---|---|---|---|
| I-1 | 1 | + | + | + | **HE**L**GH**NL**G**IQ**HD** | + |
| I-2 | 20 | + | + | - | **HE**L**GH**NL**G**MQ**HD** | + |
| | 21 | + | + | - | **HE**L**GH**TF**G**MK**HD** | + |
| | 24 | + | + | - | **HE**I**GH**NL**G**MS**HD** | + |
| | 25 | + | + | - | **HE**M**GH**NL**G**ME**HD** | + |
| | 26 | + | + | - | **HE**M**GH**NF**G**MK**HD** | + |
| | 30 | + | + | - | **HE**L**GH**CV**G**MI**HD** | + |
| | 34 | + | + | - | **HE**M**GH**NL**G**MM**HD** | + |
| | 4 | - | + | - | **H**AV**GH**LLDVS**HD** | + |
| | 6 | - | - | - | NRGVRSL**G**LK**HD** | + |
| | 29 | - | - | - | **H**HL**GH**NL**G**MK**HD** | + |
| II | 2 | - | - | - | QLLSLSM**G**LAY**D** | + |
| | 3 | - | - | - | QLL**G**INL**G**LAY**D** | + |
| | 5 | - | - | - | QLLSIGM**G**LTY**D** | + |
| | 27 | - | - | - | QLI**G**LHI**G**LTY**D** | + |
| | 32 | - | - | - | QML**G**LSL**G**ISY**D** | + |
| | | | | Consensus | **HE**X**GH**XX**G**XX**HD** | |

*Except for ADAM 21 (human), all of the listed ADAMs are from mouse. In some ADAMs having isoforms, only the first isoforms (25a, 26a, 4a, 6a and 1a) are listed. The second or third isoform is similar to the first isoform in the characteristics of each ADAM.
** Amino acids identical to the conserved residues in the consensus sequence are bolded.

In genomic organization, ADAMs in the phylogenetic group I-2, containing most of testicular members predicted as active proteases, are similar to ADAM 1, but strikingly different from the group-II ADAMs without protease activity. First, all of the group-I-2 ADAM genes lack introns that interrupt protein-coding exons (Poindexter *et al.* 1999; Choi *et al.* 2004). In PCR analysis using genomic DNA and testis cDNA with primers amplifying the complete coding sequence of each mouse ADAM, PCR products with the identical size were generated in each ADAM (Choi *et al.* 2004). Similarly, genome database searches revealed lack of introns in the open reading frames in all of the ADAMs belonging to the group I-2. This presents a great contrast to the group-II ADAMs containing multiple

(19-20) introns interrupting the protein-coding exons (Cho *et al.* 1997; Choi *et al.* 2003). Second, many of the intronless ADAMs are present as multicopy genes in the mouse genome and these gene copies are clustered in the same chromosome (Table 3) (Bolcun *et al.* 2003; Choi *et al.* 2004). Southern blot analysis and genome database searches demonstrated that more than one gene for ADAM 4 (2 copies), ADAM 6 (2 copies), ADAM 25 (3 copies) and ADAM 26 (2 copies) exist in the mouse genome. Two clustered genes (ADAM 4a and ADAM 4b) for mouse ADAM 4 with 90% nucleotide sequence identity are present in the middle region of mouse chromosome 12. Two ADAM 6 genes (ADAM 6a and ADAM 6b) with 95% nucleotide sequence identity are adjacent to each other in the proximal region of mouse chromosome 12. For ADAM 25, three gene copies (ADAM 25a, ADAM 25b and ADAM 25c) with 80-94% nucleotide sequence homology among the copies are present as a cluster in the mid-proximal region of mouse chromosome 8. At a close distance from the ADAM 25 genes on chromosome 8, there exist two genes (ADAM 26a and ADAM 26b) for ADAM 26, having 93% nucleotide sequence homology. In contrast with these ADAM genes, none of the ADAMs belonging to the phylogenetic group II was found to have more than one gene in the mouse genome. Third, the pattern of correspondence between the mouse and human ADAM genes is unusual in more than half of the ADAMs and some human orthologues are pseudogenes (Table 3) (Poindexter *et al.* 1999; Choi *et al.* 2004). Orthology with 1:1 relationship in gene numbers was found in ADAMs 4, 24, 29 and 30 but not in ADAMs 6, 20, 21, 25, 26 and 34. Whereas there are two genes for ADAM 6 in mouse, only a single homologous gene for ADAM 6 is present in the human genome. For ADAM 21, the gene exists as a single copy in the mouse genome, whereas it is present as two copies in the human genome. The most unusual orthology relationship was found for ADAMs 20, 25, 26 and 34. There is no mouse gene for human ADAM 20. The human genome does not contain genes corresponding to the genes for mouse ADAM25 (3 copies), ADAM26 (2 copies) and ADAM34 (1 copy). Like the mouse ADAM genes, all the human ADAM genes do not have introns. The intriguing feature of the human intronless ADAM genes is that many of them (ADAMs 4a, 4b, 6, 21b and 24) are pseudogenes with frameshift or stop codons. As described above (section 2.1), the ADAM 1 gene is a single-copy pseudogene in the human genome but two functional copies for the gene are present in the mouse genome. It was suggested that the ADAM 1 gene in human is derived by genetic recombination between the two ADAM 1 genes present in some primates closely related to human. Although such type of evolutionary selection might explain the relationship of the intronless ADAM genes (group I-2) between mouse and human, a comprehensive, comparative analysis of these ADAM genes from a wide range of rodents and primates would be needed to validate the hypothesis.

*Table 3.* Genomic characteristics of ADAMs belonging to the phylogenetic group I-2

| ADAM | Mouse | | | | Human | | | |
|---|---|---|---|---|---|---|---|---|
| | Gene | Chr* | AA** | F*** | Gene | Chr | AA** | F*** |
| ADAM 4 | 4a | 12 (D2) | 763 | + | 4a | 14q24 | 300 | - |
| | 4b | 12 (D2) | 751 | + | 4b | 14q24 | 375 | - |
| ADAM 6 | 6a | 12 (F2) | 754 | + | 6 | 14q32 | 610 | - |
| | 6b | 12 (F2) | 726 | + | - | | | |
| ADAM 20 | - | | | | 20 | 14q24 | 726 | + |
| ADAM 21 | 21 | 12 (D3) | 729 | + | 21a | 14q24 | 722 | + |
| | - | | | | 21b | 14q24 | 132 | - |
| ADAM 24 | 24 | 8 (B1) | 761 | + | 24 | 8p22 | 3 | - |
| ADAM 25 | 25a | 8 (B1) | 760 | + | - | | | |
| | 25b | 8 (B1) | 756 | + | - | | | |
| | 25c | 8 (B1) | 760 | + | - | | | |
| ADAM 26 | 26a | 8 (B1) | 697 | + | - | | | |
| | 26b | 8 (B1) | 699 | + | - | | | |
| ADAM 29 | 29 | 8 (B3) | 763 | + | 29 | 4q34 | 820 | + |
| ADAM 30 | 30 | 3 (F3) | 732 | + | 30 | 1p12 | 790 | + |
| ADAM 34 | 34 | 8 (B1) | 714 | + | - | | | |

*Chromosomal location of the genes. Letters and numbers in parenthesis indicate subregions in chromosomes.

**Number of amino acids (AA).

***Functionality (F) of the genes based on the integrity of open reading frames. Functional genes and pseudogenes are indicated by + and -, respectively.

The phylogenetic ADAM group I-2 enriched with the active proteases is also unique in gene expression pattern. Northern blot analysis showed that the mouse ADAM 24, 25 and 26 genes are expressed predominantly in postmeiotic germ cells (Zhu *et al.* 1999). RT-PCR analysis revealed that all the ADAM genes belonging to the group I-2 are expressed in mouse testis from postnatal day 6 to adult (Choi *et al.* 2004). Since the majority of cells in the testis at day 6 are somatic cells and the proportion of somatic cells decreases as spermatogenesis proceeds, this indicates that the mouse ADAM genes are expressed by both somatic and germ cells. Furthermore, all of the mouse ADAM genes were found to be transcribed in a Sertoli cell line (Choi *et al.* 2004). Thus it is likely that the group-I-2 ADAM genes are transcribed by both somatic and germ cells with higher expression level in postmeiotic germ cells in the testis. The gene expression pattern of these ADAMs is similar to that of ADAM 1 but not other ADAMs belonging to the separate phylogenetic group (group II) showing a germ-cell specific expression pattern (Wolfsberg *et al.* 1995b; Zhu *et al.* 1999; Choi *et al.* 2003). These suggest a potential relationship between the function of the ADAMs with protease activity and the process of somatic cell -germ cell interaction during spermatogenesis.

## 3.2 Properties of the ADAM 24 protein

To date, among the testicular ADAMs belonging to the phylogenetic group I-2 and having the metalloprotease active site, only ADAM 24 (testase 1) was investigated at the protein level (Zhu *et al.* 2001). Immunoblot analysis revealed that ADAM 24 is synthesized as a precursor (108 kDa) in testicular germ cells and sperm. The protein is partially converted to a processed form (88 kDa) in caput sperm and the processing is complete in corpus sperm, suggesting that the prodomain of ADAM 24 is removed during sperm passage through the caput epididymis. The ADAM 24 precursor and processed form reside on the plasma membrane of testicular and epididymal sperm, respectively. The processing of ADAM 24 is unique in several aspects. First, unlike ADAM 1 whose proteolytic processing removes the pro- and metalloprotease domains, only the prodomain is cleaved in ADAM 24. Second, ADAM 24 does not contain the consensus cleavage site recognized by the furin-type pro-protein convertase (see sections 2.2 and 3.1) known to be responsible for removal of the prodomain in a number of ADAM proteases. Third, the processing of ADAM 24 occurs on the plasma membrane, presenting an apparent contrast to other ADAM proteases of which processing takes place in the secretory pathway before emerging on the cell surface. Thus these indicate the existence of the differential, molecular mechanism underlying the processing of ADAM 24 and insure further investigation of cellular source (sperm or epididymis) and identity of a processing enzyme.

Considering ADAM 24 is a plasma membrane-anchored sperm protease, it is tempting to hypothesize that the protein plays a direct role in the fertilization process requiring protease activity, such as sperm passage through the cumulus cell layer, sperm penetration into the egg ZP and sperm-egg plasma membrane fusion. There exist a number of testicular ADAMs predicted as active proteases (ADAMs 20, 21, 25, 26, 30 and 34) in the phylogenetic group I-2 containing ADAM 24. These ADAMs are highly similar to ADAM 24 in gene structure, gene expression pattern and metalloprotease characteristics based on amino acid sequence analysis (Table 2). While these ADAMs need extensive investigation that concerns processing and function, one can speculate on the dynamic participation of these ADAM proteases in the process of sperm maturation and fertilization.

# 4. CONCLUSION

A number of testicular ADAMs predicted as active proteases have been discovered in various mammalian species. The unique features of these ADAM genes are that all of them belong to the same phylogenetic group and lack introns. Furthermore, many of these ADAM genes are present as multicopy genes in the mouse genome and a number of these mouse genes do not have human orthologues. Even if human orthologues exist, some of them are pseudogenes in human. Noteworthy in relation to this is a global difference between the mouse and human genomes. Comparative genomic analysis revealed that over 90% of the two genomes are syntenic. Nevertheless, each genome is evolving in a non-uniform manner, showing substantial variation across the genome. The proportion of mouse genes with a single identifiable orthologue in the human genome is about 80%. The remainder are genes lacking a strict 1:1 relationship due to differential expansion in at least one of the two genomes. It was found that most genes expanded in the mouse lineage have common features. They are present as a family and clustered in the mouse genome, suggesting that they have been generated by local gene duplication. The significant proportion of these genes seems to be involved in reproduction (Waterson *et al.* 2002). Similarly, the global view of human and mouse proteases revealed that the mouse degradome is more complex and many of genes expanded in the mouse genome correspond to proteases that are involved in reproductive functions (Puente *et al.* 2003). There exist 11 testicular ADAM proteases in mouse (ADAMs 1a, 1b, 21, 24, 25a, 25b, 25c, 26a, 26b, 30 and 34), whereas only 3 ADAM genes (ADAMs 20, 21 and 30) with testicular expression are predicted to encode functional proteases in human. The studies of mouse ADAMs indicate that ADAM 1a is central to the fertilization process by regulating sperm proteins responsible for fertility and ADAM 24 is potentially involved in the various steps of fertilization. These ADAMs may be related to rodent-specific aspects of fertilization. It is possible that the human testicular ADAM proteases have similar functions in fertilization. In this regard, function studies on the ADAM proteases present in both species (ADAMs 21 and 30) should provide more information about the molecular mechanisms underlying mammalian fertilization.

## ACKNOWLEDGEMENTS

I thank Drs. Paul Primakoff and Diana Myles for comments and discussions during the course of ADAM study cited in this chapter. This work was supported by Korea Research Foundation Grant (KRF-2002-070-C0007).

## REFERENCES

Almeida, E.A., Huovila, A.P., Sutherland, A.E., Stephens, L.E., Calarco, P.G., Shaw, L.M., Mercurio, A.M., Sonnenberg, A., Primakoff, P., Myles, D.G., White, J.M., 1995,. Mouse egg integrin α6β1 functions as a sperm receptor. *Cell* **81:** 1095-1104.

Blobel, C.P., Myles, D.G., Primakoff, P., White, J.M., 1990, Proteolytic processing of a protein involved in sperm-egg fusion correlates with acquisition of fertilization competence. *J. Cell Biol.* **111:** 69-78.

Blobel, C.P., Wolfsberg, T.G., Turck, C.W., Myles, D.G., Primakoff, P., White, J.M., 1992, A potential fusion peptide and an integrin ligand domain in a protein active in sperm-egg fusion. *Nature* **356:** 248-252.

Blobel, C.P., 1997, Metalloprotease-disintegrins: links to cell adhesion and cleavage of TNF alpha and Notch. *Cell* **90:** 589-592.

Blobel, C.P., 2000, Functional processing of fertilin: evidence for a critical role of proteolysis in sperm maturation and activation. *Rev. Reprod.* **5:** 75-83.

Black, R.A., White. J.M., 1998, ADAMs: focus on the protease domain. *Curr. Opin. Cell Biol.* **10:** 654-659.

Bolcun, E., Rzymski, T., Nayernia, K., Engel, W., 2003, ADAM family genes testase 2 α and 2 β are chromosomally linked and simultaneously expressed in male germ cells. *Mol. Reprod. Dev.* **65:** 19-22.

Brachvogel, B., Reichenberg, D., Beyer, S., Jehn, B., von der Mark, K., Bielke, W., 2002, Molecular cloning and expression analysis of a novel member of the Disintegrin and Metalloprotease-Domain (ADAM) family. *Gene* **288:** 203-210.

Cerretti, D.P., DuBose, R.F., Black, R.A., Nelson, N., 1999, Isolation of two novel metalloproteinase-disintegrin (ADAM) cDNAs that show testis-specific gene expression. *Biochem. Biophys. Res. Commun.* **263:** 810-815.

Cho, C., Primakoff, P., White, J.M., Myles, D.G., 1996, Chromosomal assignment of four testis-expressed mouse genes from a new family of transmembrane proteins (ADAMs) involved in cell-cell adhesion and fusion. *Genomics* **34:** 413-417.

Cho, C., Turner, L., Primakoff, P., Myles, D.G., 1997, Genomic organization of the mouse fertilin β gene that encodes an ADAM family protein active in sperm-egg fusion. *Dev. Genet.* **20:** 320-328.

Cho, C., Bunch D.O., Faure, J.-E., Goulding E.H., Eddy, E.M., Primakoff, P., Myles, D.G., 1998, Fertilization defects in sperm from mice lacking fertilin β. *Science* **281:** 1857-1859.

Cho, C., Ge, H., Branciforte, D., Primakoff, P., Myles, D.G., 2000, Analysis of mouse fertilin in wild-type and fertilin $\beta^{-/-}$ sperm: evidence for C-terminal modification, α/β dimerization, and lack of essential role of fertilin α in sperm-egg fusion. *Dev. Biol.* **222:** 289-295.

Choi, I., Woo, J.M., Hong, S., Jung, Y.K., Kim D.H., Cho, C., 2003, Identification and characterization of ADAM32 with testis-predominant gene expression. *Gene* **304:** 151-162.

Choi, I., Oh, J., Cho, B.N., Ahnn, J., Jung, Y.K., Kim H.D., Cho, C., 2004, Characterization and comparative genomic analysis of intronless Adams with testicular gene expression. *Genomics* **83:** 636-646.

Eddy, E.M., 1998, Regulation of gene expression during spermatogenesis, *Semin. Cell Dev. Biol.* **9:** 451-457.

Evans, J.P., Schultz, R.M., Kopf, G.S., 1995, Mouse sperm-egg plasma membrane interactions: analysis of roles of egg integrins and the mouse sperm homologue of PH-30 (fertilin) β. *J. Cell Sci.* **108:** 3267-3278.

Evans J.P., Schultz, R.M., Kopf, G.S., 1998, Roles of the disintegrin domains of mouse fertilins α and β in fertilization. *Biol. Reprod.* **59:** 145-152.

Evans, J.P., 2001, Fertilin β and other ADAMs as integrin ligands: insights into cell adhesion and fertilization. *Bioessays* **23:** 628-639.

Frayne, J., Jury, J.A., Barker, H.L., Hall, L., 1997, Rat MDC family of proteins: sequence analysis, tissue distribution, and expression in prepubertal and adult rat testis. *Mol. Reprod. Dev.* **48:** 159-167.

Frayne, J., Jury, J.A., Barker, H.L., Perry, A.C., Jones, R., Hall, L., 1998, Macaque MDC family of proteins: sequence analysis, tissue distribution and processing in the male reproductive tract. *Mol. Hum. Reprod.* **4:** 429-437.

Hardy, C.M., Holland, M.K.,1996, Cloning and expression of recombinant rabbit fertilin. *Mol Reprod. Dev.* **45:** 107-116.

Hooft van Huijsduijnen, R., 1998, ADAM 20 and 21; two novel human testis-specific membrane metalloproteases with similarity to fertilin-α. *Gene* **206:** 273-282.

Hunnicutt, G.R., Koppel, D.E., Myles, D.G., 1997, Analysis of the process of localization of fertilin to the sperm posterior head plasma membrane domain during sperm maturation in the epididymis. *Dev. Biol.* **191:** 146-159.

Ikawa, M., Wada, I., Kominami, K., Watanabe, D., Toshimori, K., Nishimune, Y., Okabe, M., 1997, The putative chaperone calmegin is required for sperm fertility. *Nature* **387:** 607-611.

Ikawa, M., Nakanishi, T., Yamada, S., Wada, I., Kominami, K., Tanaka, H., Nozaki, M., Nishimune, Y., Okabe, M., 2001, Calmegin is required for fertilin α/β heterodimerization and sperm fertility. Dev *Biol.* **240:** 254-261.

Jury, J.A., Frayne, J., Hall, L., 1997, The human fertilin α gene is non-functional: implications for its proposed role in fertilization. *Biochem. J.* **321:** 577-581.

Jury, J.A., Frayne, J., Hall, L., 1998, Sequence analysis of a variety of primate fertilin α genes: evidence for non-functional genes in the gorilla and man. *Mol. Reprod. Dev.* **51:** 92-97.

Kim, E., Nishimura, H., Baba, T., 2003, Differential localization of ADAM1a and ADAM1b in the endoplasmic reticulum of testicular germ cells and on the surface of epididymal sperm. *Biochem. Biophys. Res. Commun.* **304:** 313-319.

Liu, L., Smith, J.W., 2000, Identification of ADAM 31: a protein expressed in Leydig cells and specialized epithelia. *Endocrinology* **141:** 2033-2042.

Lum, L., Blobel, C.P., 1997, Evidence for distinct serine protease activities with a potential role in processing the sperm protein fertilin. *Dev Biol.* **191:** 131-145.

Martin, I., Epand, R.M., Ruysschaert, J.M., 1998, Structural properties of the putative fusion peptide of fertilin, a protein active in sperm–egg fusion, upon interaction with the lipid bilayer. *Biochemistry* **37**: 17030–17039.

McLaughlin, E.A., Frayne, J., Barker, H.L., Jury, J.A., Jones, R., Ford, W.C., Hall, L., 1997, Cloning and sequence analysis of rat fertilin α and β-developmental expression, processing and immunolocalization. *Mol. Hum. Reprod.* **3:** 801-809.

Muga, A., Neugebauer, W., Hirama, T., Surewicz, W.K., 1994, Membrane interactive and conformational properties of the putative fusion peptide of PH-30, a protein active in sperm–egg fusion. *Biochemistry* **33:** 4444–4448.

Myles, D.G., 1993, Molecular mechanisms of sperm-egg membrane binding and fusion in mammals. *Dev. Biol.* **158:** 35-45.

Myles, D.G., Kimmel, L.H., Blobel, C.P., White, J.M., Primakoff, P., 1994, Identification of a binding site in the disintegrin domain of fertilin required for sperm-egg fusion. *Proc. Natl. Acad. Sci. U.S.A.* **91:** 4195-4198.

Myles, D.G., Primakoff, P., 1997, Why did the sperm cross the cumulus? To get to the oocyte. Functions of the sperm surface proteins PH-20 and fertilin in arriving at, and fusing with, the egg. *Biol. Reprod.* **56:** 320-327.

Niidome, T., Kimura, M., Chiba, T., Ohmori, N., Mihara, H., Aoyagi, H., 1997, Membrane interaction of synthetic peptides related to the putative fusogenic region of PH-30α, a protein in sperm–egg fusion. *J. Pept. Res.* **49:** 563–569.

Nishimura, H., Cho, C., Branciforte, D.R., Myles, D.G., Primakoff, P., 2001, Analysis of loss of adhesive function in sperm lacking cyritestin or fertilin β. *Dev. Biol.* **233:** 204-213.

Nishimura, H., Kim, E., Fujimori, T., Kashiwabara, S., Kuroiwa, A., Matsuda, Y., Baba, T., 2002, The ADAM1a and ADAM1b genes, instead of the ADAM1 (fertilin α) gene, are localized on mouse chromosome 5. *Gene* **291:** 67-76.

Nishimura, H., Kim, E., Nakanishi, T., Baba, T., 2004, Possible function of the ADAM1a/ADAM2 fertilin complex in the appearance of ADAM3 on the sperm surface. *J. Biol. Chem.* in press.

Perry, A.C., Jones, R., Hall, L., 1995, Analysis of transcripts encoding novel members of the mammalian metalloprotease-like, disintegrin-like, cysteine-rich (MDC) protein family and their expression in reproductive and non-reproductive monkey tissues. *Biochem. J.* **312:** 239-244.

Phelps, B.M., Koppel, D.E., Primakoff, P., Myles, D.G., 1990, Evidence that proteolysis of the surface is an initial step in the mechanism of formation of sperm cell surface domains. *J. Cell Biol.* **111:** 1839-1847.

Poindexter, K., Nelson, N, DuBose, R.F., Black, R.A., Cerretti, D.P, 1999, The identification of seven metalloproteinase-disintegrin (ADAM) genes from genomic libraries. *Gene* **237:** 61-70.

Primakoff, P., Hyatt, H., Tredick-Kline, J., 1989, Identification and purification of a sperm surface protein with a potential role in sperm-egg membrane fusion. *J. Cell Biol.* **104:** 141-149.

Primakoff, P., Myles, D.G., 2000, The ADAM gene family. *Trends Genet.* **16:** 83-87.

Primakoff, P., Myles, D.G., 2002, Penetration, adhesion, and fusion in mammalian sperm-egg interaction. *Science* **296:** 2183-2185.

Puente, X.S., Sanchez, L.M., Overall, C.M., Lopez-Otin, C., 2003, Human and mouse proteases: a comparative genomic approach. *Nat. Rev. Genet.* **4:** 544-558.

Shamsadin., R., Adham, I.M., Nayernia, K., Heinlein, U.A., Oberwinkler, H., Engel, W., 1999, Male mice deficient for germ-cell cyritestin are infertile. *Biol. Reprod.* **61:** 1445-1451.

Snell, W.J., White, J.M., 1996, The molecules of mammalian fertilization. *Cell* **85:** 629-637.

Talbot, P., Shur, B.D., Myle,s D.G., 2003, Cell adhesion and fertilization: steps in oocyte transport, sperm-zona pellucida interactions, and sperm-egg fusion. *Biol. Reprod.* **68:** 1-9.

Wassarman, P.M., 1999, Mammalian fertilization: molecular aspects of gamete adhesion, exocytosis, and fusion. *Cell* **96:** 175-183.

Waters, S.I., White, J.M., 1997, Biochemical and molecular characterization of bovine fertilin α and β (ADAM 1 and ADAM 2): a candidate sperm-egg binding/fusion complex. *Biol. Reprod.* **56:** 1245-1254.

Waterston, R.H. et al., 2002, Initial sequencing and comparative analysis of the mouse genome. *Nature* **420:** 520-562.

White, J.M., 2003, ADAMs: modulators of cell-cell and cell-matrix interactions. *Curr. Opin. Cell Biol.* **15:** 598-606.

Wolfe, C.A., Cladera, J., Ladha, S., Senior, S., Jones, R., O'Shea, P., 1999, Membrane interactions of the putative fusion peptide (MF αP) from fertilin-α, the mouse sperm protein complex involved in fertilization. *Mol. Membr. Biol.* **16:** 257–263.

Wolfsberg, T.G., Bazan, J.F., Blobel, C.P., Myles, D.G., Primakoff, P., White, J.M., 1993, The precursor region of a protein active in sperm-egg fusion contains a metalloprotease and a disintegrin domain: structural, functional, and evolutionary implications. *Proc. Natl. Acad. Sci. U.S.A.* **90:** 10783-10787.

Wolfsberg, T.G., Primakoff, P., Myles, D.G., White, J.M., 1995a, ADAM, a novel family of membrane proteins containing A Disintegrin And Metalloprotease domain: multipotential functions in cell-cell and cell-matrix interactions. *J. Cell Biol.* **131:** 275-278.

Wolfsberg, T.G., Straight, P.D., Gerena, R.L., Huovila, A.-P.J., Primakoff, P., Myles, D.G., White, J.M., 1995b, ADAM, a widely distributed and developmentally regulated gene family encoding membrane proteins with A Disintegrin And Metalloprotease domain. *Dev. Biol.* **169:** 378-383.

Wolfsberg, T.G., White, J.M., 1996, ADAMs in fertilization and development. *Dev. Biol.* **180:** 389-401.

Wong, G.E., Zhu, X., Prater, C.E., Oh, E., Evans, J.P., 2001, Analysis of fertilin α (ADAM1)-mediated sperm-egg cell adhesion during fertilization and identification of an adhesion-mediating sequence in the disintegrin-like domain. *J. Biol. Chem.* **276:** 24937-24945.

Yanagimachi, R., 1994, Mammalian fertilization. In: The Physiology of Reproduction. Knobil E and Neill JD Eds. Raven, New York, pp189-317.

Yuan, R., Primakoff, P., Myles, D.G., 1997, A role for the disintegrin domain of cyritestin, a sperm surface protein belonging to the ADAM family, in mouse sperm-egg plasma membrane adhesion and fusion. *J. Cell Biol.* **137:** 105-112.

Zhu, G.-Z., Lin, Y., Myles, D.G., Primakoff, O., 1999, Identification of four novel ADAMs with potential roles in spermatogenesis and fertilization. *Gene* **234:** 227-237.

Zhu, G.-Z., Myles, D.G., Primakoff, P., 2001, Testase 1 (ADAM24) a plasma membrane-anchored sperm protease implicated in sperm function during epididymal maturation or fertilization. *J. Cell Sci.* **114:** 1787-1794.

skin cDNA library. The latter screen yielded clones comprising the entire coding sequence for bovine pNPI (Colige *et al.* 1997). Subsequently, the bovine pNPI cDNA clones were used as probes to screen a human dermal fibroblast cDNA library, resulting in isolation of cDNA sequences corresponding to the human full-length pNPI coding sequence (Colige *et al.* 1999). The coding sequence predicted the human protein to be 1,211 amino acids in length, including the signal peptide. This protein appears to be encoded by two different size classes of mRNA of ~7- and ~4.5-kb that differ only in AATAAA polyadenylation sites used and in the consequent lengths of 3'-untranslated regions (Colige *et al.* 1999; Wang *et al.* 2003). A third size class of pNPI RNA of ~ 2.3 kb, also detected on Northern blots of human poly(A+) RNA (Colige *et al.* 1999; Wang *et al.* 2003), encodes a truncated form of pNPI protein that ends at Lys 543 and then continues for an additional 23 amino acids encoded by retained intron sequences (Colige *et al.* 1999). Possible functional significance of this smaller mRNA form is unknown and, unlike the larger RNA species that encode full-length pNPI, there is no comparable bovine mRNA for the ~2.3-kb human form. A mutation in the cognate gene in one EDS VIIC patient has been characterized as a G→A transition that changes the codon for Trp 795 to a premature TAG stop codon (W795X). This mutation results not only in truncation of the coding sequence for full-length pNPI, but also in very low levels of the 7- and 4.5-kb mRNAs, due to nonsense-mediated decay (Colige *et al.* 1999). Nevertheless, this mutation occurs downstream of sequences in the 2.3-kb mRNA, and levels of the latter mRNA species are unaffected in the patient (Colige *et al.* 1999). The other five EDS VIIC patients in which the causal mutation has been characterized have a shared C→T transition, resulting in conversion of the codon for Gln 225 to a premature TAG stop codon (Q225X) and to nonsense-mediated decay of all size classes of mRNA from the corresponding gene (Colige *et al.* 1999). The fact that the phenotype of these latter five patients is not more severe than that of the patient with the W795X mutation suggests that the 2.3-kb mRNA does not encode a protein with N-proteinase activity (Colige *et al.* 1999) or, for that matter, with some other readily apparent physiological role. Dermatosparaxis in inbred cattle in Belgium is due to deletion of genomic sequences, resulting in a frame-shift downstream of the codon for Val 153 (Colige *et al.* 1999). The possibility of nonsense-mediated decay has not been explored for RNA transcripts bearing the latter mutation.

## 4. ADAMTS 2 AND THE ADAMTS FAMILY OF ZINC-DEPENDENT METALLOPROTEINASES

Sequence comparisons with the databases showed pNPI to be a member of the ADAMTS (A Disintegrin And Metalloproteinase with ThromboSpondin motifs) family of metalloproteinases (Kuno *et al.* 1997), of which there are presently 19 known members in vertebrates (Figure 2) (Cal *et al.* 2002; Somerville *et al.* 2003).

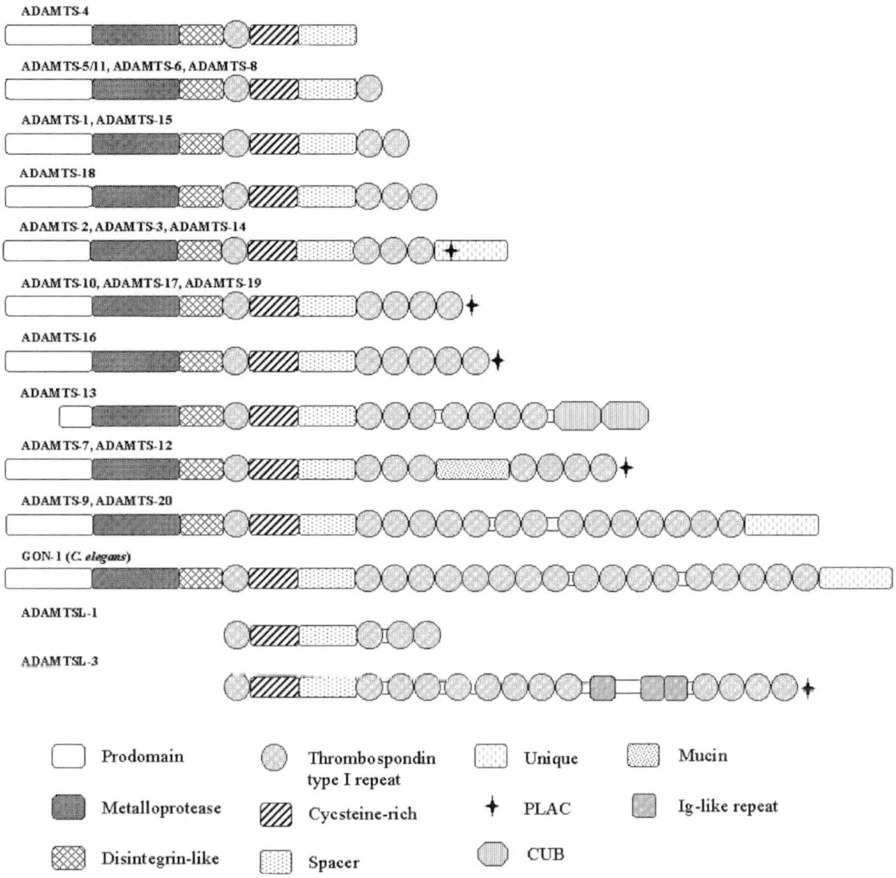


*Figure 2.* Domain structures of the ADAMTS proteinases and ADAMTS-like (ADAMTSL) proteins

The characteristic zinc-binding site, followed by a "methionine-turn" in the metalloprotease domains of these proteins, shows the ADAMTS proteinases to belong to the metzincin superfamily of metallopeptidases

(Stöcker *et al.* 1995). Because pNPI was only the second member of the family to be identified, it has been designated ADAMTS-2. Members of the ADAMTS family of proteinases have a shared domain structure (Figure 2) that resembles the domain structure of the ADAM family of proteinases, in that it includes pro-, adamalysin/reprolysin-like metalloprotease, disintegrin-like and cysteine-rich domains (Kuno *et al.* 1997; Hurskainen *et al.* 1999). However, ADAMTS proteinases differ from ADAM proteinases in lacking EGF-like motifs and, unlike the majority of ADAM proteinases, they lack transmembrane domains. Instead of the latter, ADAMTS proteinases have varying numbers of COOH-terminal thrombospondin type I-like repeats (Kuno *et al.* 1997; Hurskainen *et al.* 1999) which may, at least in some ADAMTS proteinases, be involved in binding ECM components, such as sulfated glycosaminoglycans (Kuno and Matsushima 1998). ADAMTS proteinases also have a "spacer" region comprising sequences that are less conserved between family members than are sequences of other domains, and lacking structural elements or motifs easily identifiable in other proteins. Spacer domains may also be involved in binding elements of the ECM (Kuno and Matsushima 1998).

A motif found in ADAMTS-2 and in several other members of the family is the PLAC (Protease and LACunin) domain, a six cysteine-containing motif first described in the insect protein lacunin (Nardi *et al.* 1999), which is homologous to the *Drosophila* protein papilin (Kramerova *et al.* 2000). ADAMTS-13 is unique among ADAMTS family members in having two CUB (Complement-Uegf-BMP-1) domains, first described in the C1r/C1s components of complement, later found in other proteins such as the procollagen C-proteinase BMP-1 (Kessler *et al.* 1996; Li *et al.* 1996; Tosi *et al.* 1987), and thought to be involved in protein-protein interactions (Bork and Beckmann 1993). ADAMTS-7 and -12 are unique in having carbohydrate-rich mucin domains (Somerville *et al.* 2004). A number of the ADAMTS proteinases have unique protein domains, dissimilar to domains found in other ADAMTS family members or in other proteins in general (Figure 2). Similarities and differences in overall domain structure can be used to divide the ADAMTS family into subfamilies, such as that comprising ADAMTS-2, -3 and -14; and that comprising ADAMTS-9 and -20 (Figure 2). Similarities and differences in sequence homologies have also been used to form the ADAMTS proteinases into subgroups (Figure 3).

A small number of extracellular ADAMTS-like (ADAMTSL) proteins, also designated punctins, have been reported (Hall *et al.* 2003; Hirohata *et al.* 2002), which lack proteinase and disintegrin domains, but which share much domain structure and sequence homology with the ADAMTS family (Figures 2 and 3). Although functions of the ADAMTSL proteins are unknown, it has been speculated that they may modulate the functions of

ADAMTS proteinases via protein-protein interactions (Hall *et al.* 2003; Hirohata *et al.* 2002).

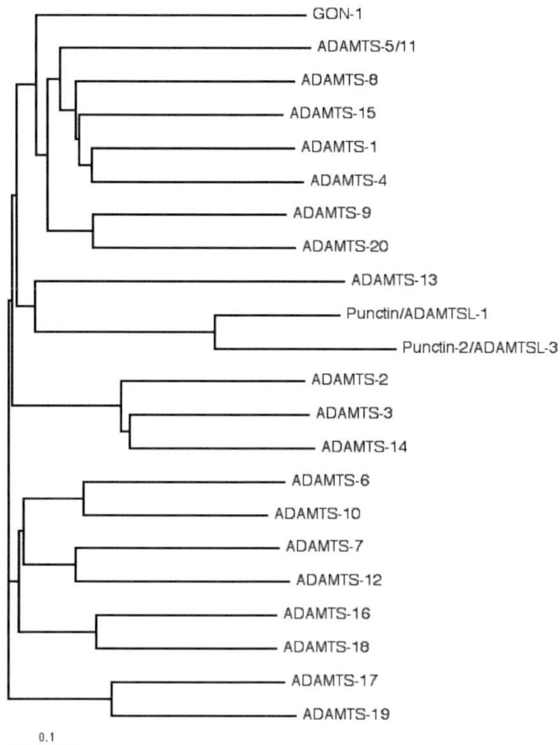


*Figure 3.* Dendrogram of the relatedness of full-length sequences of ADAMTS and ADAMSL proteins, created using Clustalw software (www.ebi.ac.uk/clustalw). All sequences were the most complete versions available in the databases except for that of ADAMTS-7, which was from Somerville *et al.* (2004).

The ADAMTS proteinases appear to play a broad range of roles in development, homeostasis, disease and reproduction. In the case of ADAMTS-2, the phenotype of mice with null alleles for the ADAMTS-2 gene (*Adamts2*) indicates a role for ADAMTS-2 in male fertility, since homozygous null males not only have extreme fragility of the skin, but also have decreased spermatogenesis and are sterile (Li *et al.* 2001). Possible deficiencies in male fertility in EDS VIIC patients and dermatosparactic animals are unreported, presumably because dermatosparactic animals do not generally survive to a reproductive age, and because follow-ups in later life have not been performed on childhood patients who present with the rare

EDS VIIC phenotype. Interestingly, ADAMTS-2 mRNA has been detected at much higher levels at 7 days of gestation, during murine gastrulation, than at later gestational times (Fernandes *et al*. 2001), despite the observation that mRNAs encoding the major fibrillar collagens are at relatively low levels at 7 days of gestation (Schnieke *et al*. 1983). This difference in temporal patterns of expression suggests that ADAMTS-2 may be involved in proteolytic processing of substrates other than procollagens. It is noteworthy, in regard to this possibility, that the procollagen C-proteinases, in addition to their roles in collagen biosynthesis, process a broad range of protein substrates involved in formation of the ECM and involved in other morphogenetic processes, such as signaling by BMPs and other TGF-β-like molecules (Greenspan 2005). Thus, ADAMTS-2, like the procollagen C-proteinases, may serve to orchestrate a number of processes in the course of morphogenetic events.

## 4.1 TGF-β Induction and Post-Translational Modifications of ADAMTS-2

Despite the importance of ADAMTS proteinases in various biological processes and disease states, there has been relatively limited study of the regulation of their expression and activities. However, in a study of the regulation of expression of the procollagen C-proteinases BMP-1 and mTLD by TGF-β1, it was noted that pNPI activity, as well as procollagen C-proteinase activity was up regulated by this growth factor (Lee *et al*. 1997). Using Northern blots, RNase protection assays and ADAMTS-2-specific probes, it was shown that TGF-β1 induces an ~8-fold increase in the levels of ADAMTS-2-specific RNA bands produced by MG-63 human osteosarcoma cells (Wang *et al*. 2003). This induction has kinetics that overlap those for induction of procollagen C-proteinases and induction of expression of various ECM protein components by TGF-β. Induction of ADAMTS-2 mRNA was also cycloheximide-inhibitable, showing that, as with BMP-1 and mTLD mRNAs, induction of ADAMTS-2 mRNA by TGF-β is indirect, requiring prior protein synthesis. Moreover, stability of ADAMTS-2 mRNA is similar in cells treated or untreated with TGF-β1 in the presence of the transcriptional inhibitor actinomycin D, thus indicating the increased steady state levels of ADAMTS-2 mRNA in TGF-β-treated cells to be due to transcriptional effects, rather than to post transcriptional effects on RNA stability (Wang *et al*. 2003).

Western blot analysis with a set of antibodies directed against peptides corresponding to different regions of ADAMTS-2 showed induction of ADAMTS-2 mRNA to be accompanied by parallel induction of an antibody-detectable band of ~132-kDa (Wang *et al*. 2003). This is different from the molecular weight of 107 estimated by SDS-PAGE analysis of active ADAMTS-2 isolated from fetal calf skin (Colige *et al.* 1995), but is similar to the 131,697 molecular weight predicted by the human cDNA sequence (Colige *et al.* 1999) for full-length ADAMTS-2 lacking only the signal peptide. This result thus raised the issue of whether the major form of ADAMTS-2 secreted by cells, at least in response to TGF-β, is an inactive precursor that retains the prodomain. It should be noted that the 132-kDa induced form was not likely to be ADAMTS-3 or –14, both of which have N-proteinase activity and are structurally similar to ADAMTS-2 (Figure 2 and see below), as these proteinases are considerably diverged in sequence from ADAMTS-2 in regions corresponding to the peptides against which antibodies were raised. When recombinant ADAMTS-2 was produced in transfected human embryonic kidney 293-EBNA cells, it was secreted primarily as a 132-kDa form (Wang *et al.* 2003). Edman degradation amino acid sequencing demonstrated the N-terminus of the 132-kDa form to be HAADDDYNIE, showing the 132-kDa form to correspond to a mature, processed form of ADAMTS-2, in which the prodomain has been removed by cleavage at the $Arg^{259}$-$His^{260}$ peptide bond, just downstream of the furin recognition sequence $^{256}RARR^{259}$, which is the more C-terminal of two furin recognition sequences found in the pro-ADAMTS-2 sequence (Figure 4). Thus, the major form of ADAMTS-2 secreted by cells is likely to be the mature, active form of the proteinase.

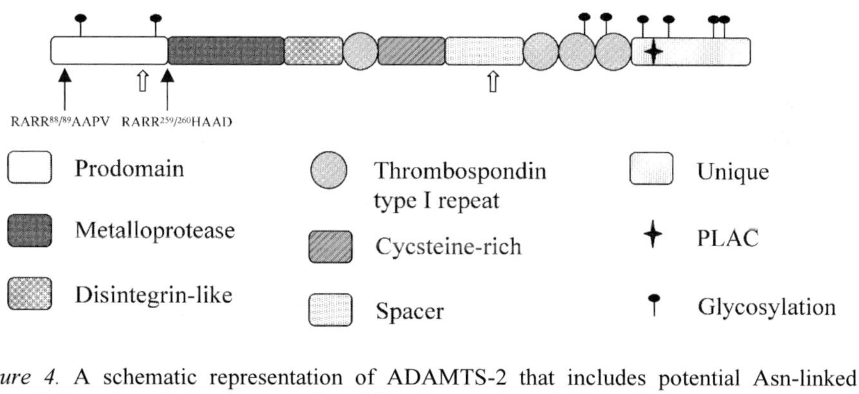


*Figure 4.* A schematic representation of ADAMTS-2 that includes potential Asn-linked glycosylation sites, furin cleavage sites (closed arrows), and sites of premature stop codon mutations in EDS VIIC patients (open arrows)

The cDNA sequence shows human ADAMTS-2 to have eight potential sites for Asn-linked glycosylation (Colige *et al.* 1999) (Figure 4). Treatment of recombinant ADAMTS-2 with peptide *N*-glycosidase F (PNGase F) (which removes virtually all Asn-linked oligosaccharides from glycoproteins) produces a marked mobility shift on SDS-PAGE, indicating glycosylation of at least some of the 8 potential sites (Wang *et al.* 2003). Treatment of ADAMTS-2 with both PNGase F and sialidase A (the latter of which removes mono-, di-, and trisialyl residues, the most commonly occurring modifications, from Galβ(1-3)GalNAc *O*-linked cores) produces a further increase in electrophoretic mobility, indicating the presence of sialylated *O*-linked oligosaccharides. Endo-*O*-glycosidase removes Galβ(1-3)GalNAc cores from Ser and Thr residues, but only after modifying monosaccharides have first been removed by exoglycosidases. Treatment of ADAMTS-2 with endo-*O*-glycosidase, in addition to PNGase F and sialidase A, does not produce a detectable additional increase in electrophoretic mobility, suggesting an absence of simple sialylated *O*-linked Galβ(1-3)GalNAc *O*-linked cores. In contrast, β-1-4-galactosidase and glucosaminidase, which remove less common modifying β-1,4-linked galactose and β-1-6-linked *N*-acetylglucosamine residues, respectively, from Galβ(1-3)GalNAc *O*-linked cores, enable endo-O-glycosidase to induce a detectable mobility increase in ADAMTS-2 (Wang *et al.* 2003). Thus, in addition to Asn-linked carbohydrates, ADAMTS-2 appears to be furnished with *O*-linked Galβ(1-3)GalNAc cores, modified with β-1,4-linked galactose and β-1-6-linked *N*-acetylglucosamine residues. Deglycosylation of ADAMTS-2 with a mixture of all five of the enzymes described above increases mobility of the ~132-kDa form to ~110-kDa (Wang *et al.* 2003). This is similar in size to the 107,450 molecular weight predicted by the cDNA coding sequence for human ADAMTS-2 processed at the $Arg^{259}$-$His^{260}$ bond, further supporting the likelihood that mature, active ADAMTS-2 processed at this site, is the major form of ADAMTS-2 secreted by cells. A 141-kDa form of ADAMTS-2 is produced by cleavage of 150-kDa full-length pro-ADAMTS-2 at the more N-terminal of the two furin recognition sites (Figure 4). However, this 141-kDa species remains associated with cell layers, and appears to be solely an intracellular form (Wang *et al.* 2003). Possible significance for this processing intermediate remains to be determined.

## 4.2 ADAMTS-2 is a Procollagen III N-Proteinase

Surprisingly, purified recombinant ADAMTS-2 was found to have N-proteinase activity against not only procollagen types I and II (pNPI activity), but against procollagen type III as well (Wang *et al.* 2003). Moreover, levels of activity against procollagen III were similar to the levels of activity against procollagens I and II. If ADAMTS-2 truly acts as an *in vivo* pNPIII, then processing of procollagen III N-propeptides should be deficient in EDS VIIC patient tissues. To test for this possibility, processing of procollagen III was compared in cultures of dermal fibroblasts from the skin of a patient with EDS VIIC and from the skin of a normal control. Such comparisons are best made in the cell layers of fibroblasts cultured in the presence of neutral polymers. These conditions potentiate normal processing of procollagen, and association of processed forms with cell layers, apparently via a volume exclusion mechanism, such that defects in processing are accentuated when mutant and control fibroblast cell layers are compared (Bateman *et al.* 1986; Smith *et al.* 1992). Under these conditions, pNα1(III) chains (processing intermediates in which the C-propeptide has been removed, but the N-propeptide has been retained) were found to accumulate in the cell layers of EDS VIIC, but not normal dermal fibroblasts (Wang *et al.* 2003). Thus, ADAMTS-2 appears to have a physiologically relevant role in the provision of pNPIII activity, suggesting that the phenotypes of EDS VIIC and dermatosparaxis result from impairment of procollagen III processing, as well from impairment in the processing of procollagen I.

## 5. KNOWN ROLES OF OTHER ADAMTS PROTEINASES

On the bases of domain structure and sequence homologies, ADAMTS-2 can be seen to form a subfamily of ADAMTS proteinases with ADAMTS-3 and -14 (Figures 2 and 3). Both ADAMTS-3 and -14 have been shown to have pNPI activity (Colige *et al.*, 2002; Fernandes *et al.*, 2001) and may thus provide residual pNPI activity noted in dermatosparactic and EDS VIIC tissues. ADAMTS-3 and -14 are described in detail in a separate chapter in this book.

ADAMTS-5 and -11 designate the same proteinase, cloned and differently named by two separate groups (Abbaszade *et al.* 1999; Hurskainen *et al.* 1999). ADAMTS-5/11, -4, -8, and -15, which are collected together as a subgroup in dendrograms that relate ADAMTS proteinases by sequence homology (Figure 3), have been shown to provide the sought after "aggrecanase" activity that cleaves aggrecan, the major proteoglcycan of cartilage, between residues $Glu^{373}$-$Ala^{374}$ (Abbaszade *et al.* 1999; Collins-Racie *et al.* 2004; Tortorella *et al.* 1999), which is thought to represent a pivotal early step in the degeneration of cartilage associated with osteoarthritis and inflammatory joint disease. This cleavage of aggrecan is probably important not only in pathological conditions, but also in the turnover of agrrecan important to tissue remodeling and homeostasis. ADAMTS-5/11 and -4 are dealt with in detail in a separate chapter of this book.

ADAMTS-1, which is grouped together with ADAMTS-5/11, -4, -8, and -15 by sequence homology (Figure 3) has been shown capable of cleaving aggrecan at the same $Glu^{373}$-$Ala^{374}$ site and at additional sites as well (Kuno *et al.* 2000; Rodríguez-Manzaneque *et al.* 2002). ADAMTS-1 and -4 are also capable of cleaving versican (Sandy *et al.* 2001), and ADAMTS-4 can cleave brevican (Matthews *et al.* 2000), both of which are proteoglycans closely related to aggrecan. As a point of interest regarding the specificity of ADAMTS proteinases, it has been reported that ADAMTS-1 lacks pNPI activity (Rodriguez-Manzaneque *et al.* 2002). Although originally identified as an inflammation-induced protein associated with cachexia (Kuno *et al.* 1997), the phenotype of mice null for the cognate *Adamts1* gene also suggests roles for ADAMTS-1 in growth, organogenesis and female fertility (Shindo *et al.* 2000). ADAMTS-1 has also been shown to have anti-angiogenic activity, as has ADAMTS-8 (Vázquez *et al.* 1999), and in the case of ADAMTS-1 this may involve binding and sequestration of the angiogenic factor $VEGF_{165}$ by the COOH-terminal thrombospondin-like domains (Luque *et al.* 2003).

The probability that ADAMTS proteinases play roles in fertility in a broad range of species is underscored by the finding that the *Caenorhabditis elegans* gene *gon-1*, which is essential for correct gonadal morphogenesis, encodes an ADAMTS proteinase (Blelloch and Kimble 1999). The *gon-1* gene is expressed by leader cells, which normally direct migration of the tips of the gonad to properly shape the organ. It has been speculated that GON-1, the protein product of the *gon-1* gene, may be necessary for remodeling basement membranes and/or overall penetration of the ECM by migrating cells (Blelloch and Kimble 1999). Interestingly, vertebrates possess two ADAMTS proteinases, ADAMTS-9 and -20, which are quite similar in domain structure to GON-1 (Somerville *et al.* 2003) (Figure 2). It has been found that ADAMTS-9 is retained at or near the cell surface, and within ECM associated with transfected cells expressing the recombinant protein. Thus, ADAMTS-9 and -20 have been suggested to adhere to the surfaces of migrating cells, permitting them to transverse the ECM in a manner analogous to that of GON-1 (Somerville *et al.* 2003). ADAMTS-9 has also been shown to have the capability of cleaving aggrecan and versican (Somerville *et al.* 2003). However, a substrate has yet to be identified for GON-1, and it has yet to be demonstrated that either ADAMTS-9 or -20 plays a role in cell migration.

Further adding to the scope of roles that ADAMTS proteinases appear to play in normal biological processes and in disease has been the finding that mutations in the gene for the plasma protein ADAMTS-13 are causal for thrombotic thrombocytopenic purpura, perhaps due to a demonstrated ability to process von Willebrand factor (Levy *et al.* 2001). The latter result indicates that at least one ADAMTS proteinase, ADAMTS-13, plays an important role in human vascular homeostasis. A separate chapter in this book deals with ADAMTS-13 at length.

## 6. CONCLUDING REMARKS

The ADAMTS family of metalloproteinases was identified relatively recently and characterization of its members has only recently begun. Nevertheless, it is already apparent that ADAMTS proteinases play key and widely varied roles in development and in the homeostasis of various adult systems, in a broad range of species. Notably however, the majority of ADAMTS proteinases have yet to be assigned *in vivo* roles, and substrates for these members of the ADAMTS family have yet to be identified. It also seems likely that additional substrates and roles will be discovered for those ADAMTS proteinases for which roles and/or substrates have already been identified. As noted above, for example, it is likely that ADAMTS-2 cleaves some substrate other than procollagens in early vertebrate embryogenesis. It must also be stressed that substrates shown to be processed by certain ADAMTS proteinases *in vitro*, must be demonstrated to be substrates for these same ADAMTS proteinases *in vivo*. Such demonstrations await characterization of protein processing in mouse "knockout" lines engineered to have null alleles for genes encoding the various ADAMTS proteinases. In fact, knockout lines with null alleles for multiple genes encoding different ADAMTS proteinases may be necessary in such genetic approaches, due to possible functional overlap between different ADAMTS proteinases *in vivo*.

Another area to be addressed comprises the mechanisms whereby ADAMTS proteinases are targeted or partitioned to certain extracellular sites, and details regarding their interactions with other extracellular macromolecules. It has been reported that ADAMTS-1 and -2 interact with sulfated glycosaminoglycans associated with cell layers in culture (Kuno and Matsushima 1998; Wang *et al.* 2003), that ADAMTS-9 is located at or near the surface of transfected cells (Somerville *et al.* 2003), that ADAMTS-2 binds collagen type XIV (Colige *et al.* 1995), that ADAMTS-4 may bind glycosaminoglycans (Tortorella *et al.* 1999) and that various ADAMTS proteinases may bind the basement membrane protein papilin/lacunin which, like the ADAMTSL proteins, bears some similarities of domain structure to ADAMTS proteinases (Kramerova *et al.* 2000). ADAMTS-7 has been found to localize to cell surfaces, whereas a form of ADAMTS-7 modified by attachment of chondroitin sulfate to its mucin domain does not (Somerville *et al.* 2004). Yet, the nature and specificity of interactions between the various ADAMTS proteinases, ECM, and cell surface components remain to be determined, as does the significance of these interactions in regard to the proper *in vivo* functioning of these proteinases. The possible significance of RGD sequences reported within a number of ADAMTS proteinases remains to be determined, since this sequence in other proteins has been shown to mediate their interactions with cell surface integrins. ADAMTS-13 is unique among ADAMTS proteinases in having CUB domains, which are found on complement components, and in being a plasma protein. It will be of interest to determine whether there is a connection between these two attributes or, if no connection exists, what determines the distribution of ADAMTS-13 to plasma, rather than to tissue ECM or cell surfaces.

It has variously been suggested that ADAMTS-2 operates primarily as a monomer (Colige *et al.* 1995), or as a homomeric or heteromeric oligomer (Hojima *et al.* 1989; Hojima *et al.* 1994). The question of whether ADAMTS-2 operates primarily as monomer or oligomer remains to be answered, and the same question applies to other ADAMTS proteinases. Another question to be addressed is the identities, specificities and numbers of endogenous inhibitors that may exist for ADAMTS proteinases. Such a role has been suggested for the basement membrane protein papilin/lacunin (Kramerova *et al.* 2000), and the existence of at least some endogenous inhibitors is supported by the demonstration that TIMP-3 (Hashimoto *et al.* 2001; Kashiwagi *et al.* 2001) and $\alpha_2$-macroglubulin (Tortorella *et al.* 2004) are capable of inhibiting the aggrecanase activity of ADAMTS-4 and –5/11. It has also been speculated that the ADAMTSL proteins might either inhibit or enhance the activities of ADAMTS proteinases, via protein-protein interactions (Hall *et al.* 2003; Hirohata *et al.* 2002).

Finally, it remains somewhat of a mystery as to why various studies have reported a pNPI activity that is unable to cleave procollagen III and a pNPIII activity that is unable to cleave procollagen I (Halila and Peltonen 1984; Halila and Peltonen 1986; Hojima *et al.* 1989; Hojima *et al.* 1994; Nusgens *et al.* 1980; Tuderman and Prockop 1982), and shown procollagens I and III to be processed with different kinetics in tissues and tissue culture (Fessler *et al.* 1981), when a single protein, ADAMTS-2, appears to have similar levels of pNPI and pNPIII activities (Wang *et al.* 2003).

## REFERENCES

Abbaszade, I., Liu, R.-Q., Yang, F., Rosenfeld, S.A., Ross, O.H., Link, J.R., Ellis, D.M., Tortorella, M.D., Pratta, M.A., Hollis, J.M., Wynn, R., Duke, J.L., George, H.J., Hillman, M.C., Jr., Murphy, K., Wiswall, B.H., Copeland, R.A., Decicco, C.P., Bruckner, R., Nagase, H., Itoh, Y., Newton, R.C., Magolda, R.L., Trzaskos, J.M., Hollis, G.F., Arner, E.C., Burn, T.C., 1999, Cloning and characterization of *ADAMTS11*, an aggrecanase from the ADAMTS family. *J. Biol. Chem.* **274:** 23443-23450.

Bateman, J.F., Cole, W.G., Pillow, J.J., Ramshaw, J.A.M., 1986, Induction of procollagen processing in fibroblast cultures by neutral polymers. *J. Biol. Chem.* **261:** 4198-4203.

Blelloch, R., Kimble, J., 1999, Control of organ shape by a secreted metalloprotease in the nematode Caenorhabditis elegans. *Nature* **399:** 586-590.

Bork, P., Beckmann, G., 1993, The CUB domain, A widespread module in developmentally regulated proteins. *J. Mol. Biol.* **231:** 539-545.

Byers, P.H., 1995, Disorders of collagen biosynthesis and structure. In: The Metabolic And Molecular Bases of Inherited Disease, 7th Edition. Scriver CR, Beaudet AL, Sly WS, and Valle D. Eds. McGraw-Hill, pp. 4029-4078.

Cal, S., Obaya, A.J., Llamazares, M., Garabaya, C., Quesada, V., Lopez-Otin, C., 2002, Cloning, expression analysis, and structural characterization of seven novel human ADAMTSs, a family of metalloproteinases with disintegrin and thrombospondin-1 domains. *Gene* **283:** 49-62.

Clark, T.G., Conway, S.J., Scott, I.C., Labosky, P.A., Winnier, G., Bundy, J., Hogan, B.L.M., Greenspan, D.S., 1999, The mammalian Tolloid-like gene, *Tll1*, is necessary for normal septation and positioning of the heart. *Development* **126:** 2631-2642.

Colige, A., Beschin, A., Samyn, B., Goebels, Y., Van Beeumen, J., Nusgens, B.V., Lapière, C.M., 1995, Characterization and partial amino acid sequencing of a 107-kDa procollagen I N-proteinase purified by affinity chromatography on immobilized type XIV collagen. *J. Biol. Chem.* **270:** 16724-16730.

Colige, A., Li, S.-W., Sieron, A.L., Nusgens, B.V., Prockop, D.J., Lapière, C.M., 1997, cDNA cloning and expression of bovine procollagen I N-proteinase: A new member of the superfamily of zinc-metalloproteinases with binding sites for cells and other matrix components. *Proc. Natl. Acad. Sci. USA* **94:** 2374-2379.

Colige, A., Sieron, A.L., Li, S.-W., Schwarze, U., Petty, E., Wertelecki, W., Wilcox, W., Krakow, D., Cohn, D.H., Reardon, W., Byers, P.H., Lapière, C.M., Prockop, D.J., Nusgens, B.V., 1999, Human Ehlers-Danlos syndrome type VII C and bovine dermatosparaxis are caused by mutations in the procollagen I N-proteinase gene. *Am. J. Hum. Genet.* **65:** 308-317.

Colige, A., Vandenberghe, I., Thiry, M., Lambert, C.A., Beeumen, J.V., Li, S.-W., Prockop, D.J., Lapière, C.M., Nusgens, B.V., 2002, Cloning and characterization of ADAMTS-14, a novel ADAMTS displaying high homology with ADAMTS-2 and ADAMTS-3. *J. Biol. Chem.* **277:** 5756-5766.

Collins-Racie, L.A., Flannery, C.R., Zeng, W., Corcoran, C., Annis-Freeman, B., Agostino, M.J., Arai, M., DiBlasio-Smith, E., Dorner, A.J., Georgiadis, K.E., Jin, M., Tan, X.-Y., Morris, E.A., LaVallie, E.R., 2004, ADAMTS-8 exhibits aggrecanase activity and is expressed in human articular cartilage. *Matrix Biol.* **23:** 219-230.

D'Alessio, M., Ramirez, F., Blumberg, B.D., Wirtz, M.K., Rao, V.H., Godfrey, M.D., Hollister, D.W., 1991, Characterization of a COL1A1 splicing defect in a case of Ehlers-Danlos syndrome type VII: Further evidence of molecular homogeneity. *Am. J. Hum. Genet.* **49:** 400-406.

Fernandes, R.J., Hirohata, S., Engle, J.M., Collige, A., Cohn, D.H., Eyre, D.R., and Apte, S.S., 2001, Procollagen II amino propeptide processing by ADAMTS-3. Insights on dermatosparaxis. *J. Biol. Chem.* **276:** 31502-31509.

Fessler, L.I., Timpl, R., Fessler, J.H., 1981, Assembly and processing of procollagen type III in chick embryo blood vessels. *J. Biol. Chem.* **256:** 2531-2537.

Fjölstad, M., Helle, O., 1973, A hereditary dysplasia of collagen tissues in sheep. *J. Pathol.* **112:** 183-188.

Greenspan, D.S., 2005, Biosynthetic processing of collagen molecules. *Top. Curr. Chem.* **247:** In press.

Halila, R., Peltonen, L., 1984, Neutral protease cleaving the N-terminal propeptide of type III procollagen: Partial purification and characterization of the enzyme from smooth muscle cells of bovine aorta. *Biochemistry* **23:** 1251-1256.

Halila, R., Peltonen, L., 1986, Purification of human procollagen type III N-proteinase from placenta and preparation of antiserum. *Biochem. J.* **239:** 47-52.

Hall, N.G., Klenotic, P., Anand-Apte, B., Apte, S.S., 2003, ADAMTSL-3/punctin-2, a novel glycoprotein in extracellular matrix related to the ADAMTS family of metalloproteases. *Matrix Biol.* **22:** 501-510.

Hanset, R., Ansay, M., 1967, Dermatosparaxie (peau déchirée) chez le veau: un défaut general du tissu conjonctif, de nature héréditaire. *Ann. Med. Vet.* **7:** 451-470.

Hanset, R., 1971, Dermatosparaxis of the calf, a genetic defect of the connective tissue. 1. Genetic aspects. *Hoppe-Seyler's Z. Physiol. Chem.* **352:**13.

Hashimoto, G., Aoki, T., Nakamura, H., Tanzawa, K., Okada, Y., 2001, Inhibition of ADAMTS4 (aggrecanaase-1) by tissue inhibitors of metalloproteinases (TIMP-1, 2, 3 and 4). *FEBS Lett.* **494:** 192-195.

Helseth, D.L., Jr., Veis, A., 1981, Collagen self-assembly in vitro. Differentiating specific telopeptide-dependent interactions using selective enzyme modification and the addition of free amino telopeptide. *J. Biol. Chem.* **256:** 7118-7128.

Hirohata, S., Wang, L.W., Miyagi, M., Yan, L., Seldin, M.G., Keene, D.R., Crabb, J.W., Apte, S.S., 2002, Punctin, a novel ADAMTS-like molecule, ADAMTSL-1, in extracellular matrix. *J. Biol. Chem.* **277:** 12182-12189.

Hojima, Y., McKenzie, J., van der Rest, M., Prockop, D..J., 1989, Type I procollagen N-proteinase from chick embryo tendons. Purification of a new 500-kDa form of the enzyme and identification of the catalytically active polypeptides. *J. Biol. Chem.* **264:** 11336-11345.

Hojima, Y., Mörgelin, M.M., Engel, J., Boutillon, M.M., van der Rest, M., McKenzie, J., Chen, G.-C., Rafi, N., Romani, A.M., Prockop, D.J., 1994, Characterization of type I procollagen N-proteinase from fetal bovine tendon and skin. *J. Biol. Chem.* **269:** 11381-11390.

Hojima, Y., van der Rest, M., Prokop, D.J., 1985, Type I procollagen carboxyl-terminal proteinase from chick embryo tendons. *J. Biol. Chem.* **260:** 15996-16003.

Holbrook, K.A. , Byers, P.H., Counts, D.F., Hegreberg, G.A., 1980, Dermatosparaxis in a Himalayan cat. II. Ultrastructural studies of dermal collagen. *J. Invest. Dermatol.* **74:** 100-104.

Holmes, D.H., Watson, R.B., Steinmann, B., Kadler, K.E., 1993, Morphology of type I collagen fibrils formed in vivo and in vitro is determined by the conformation of the retained N-Propeptide. *J. Biol. Chem.* **268:** 15758-15765.

Hurskainen, T.L., Hirohata, S., Seldin, M.F., Apte, S.S., 1999, ADAM-TS5, ADAM-TS6, and ADAM-TS7, novel members of a new family of zinc metalloproteases: General features and genomic distribution of the ADAM-TS family. *J. Biol. Chem.* **274:** 25555-25563.

Kashiwagi, M., Tortorella, M., Nagase, H., Brew, K., 2001, TIMP-3 is a potent inhibitor of aggrecanase 1 (ADAM-TS4) and aggrecanasae 2 (ADAM-TS5). *J. Biol. Chem.* **276:** 12501-12504.

Kessler, E., Adar, R., Goldberg, B., Niece, R., 1986, Partial purification and characterization of a procollagen C-proteinase from the culture medium of mouse fibroblasts. *Collagen Relat Res* **6:** 249-266.

Kessler, E., Takahara, K., Biniaminov, L., Brusel, M., Greenspan, D.S., 1996, Bone morphogenetic protein-1: The type I procollagen C-proteinase. *Science* **271:** 360-362.

Kramerova, I.A., Kawaguchi, N., Fessler, L.I., Nelson, R.E., Chen, Y., Kramerov, A.A., Kusche-Gullberg, M., Kramer, J.M., Ackley, B.D., Sieron, A.L., Prockop, D.J., Fessler, J.H., 2000, Papilin in development; a pericellular proein with a homology to the ADAMTS metalloproteinases. *Development* **127:** 5475-5485.

Kuhn, K., 1987, The classical collagens: types I, II and III. In: Structure and function of collagen types. Mayne R and Burgeson RE. Eds. Academic Press, Orlando, pp.1-42.

Kuno, K., Matsushima, K., 1998, ADAMTS-1 protein anchors at the extracellular matrix through the thrombospondin type I motifs and its spacing region. *J. Biol. Chem.* **273:** 13912-13917.

Kuno, K., Kanada, N., Nakashima, E., Fujiki, F., Ichimura, F., Matsushima, K., 1997, Molecular cloning of a gene encoding a new type of metalloproteinase-disintegrin family protein with thrombospondin motifs as an inflammation associated gene. *J. Biol. Chem.* **272:** 556-562.

Kuna, K., Okada, Y., Kawashima, H., Nakamura, H., Miyasaka, M., Ohno, H., Matsushima, K., 2000, ADAMTS-1 cleaves a cartilage proteoglycan, aggrecan. *FEBS Lett.* **478:** 241-245.

Lapière, C.M., Lenaers, A., Kohn, L.D., 1971, Procollagen peptidease: An enzyme excising the coordination peptides of procollagen. *Proc. Natl. Acad. Sci. USA* **68:** 3054-3058.

Lee, S., Solow-Cordero, D.E., Kessler, E., Takahara, K., Greenspan, D.S., 1997, Transforming growth factor-β regulation of bone morphogenetic protein-1/procollagen C-proteinase and related proteins in fibrogenic cells and keratinocytes. *J. Biol. Chem.* **272:** 19059-19066.

Lenaers, A., Ansay, M., Nusgens, B.V., Lapiere, C.M., 1971, Collagen made of extended α-chains, procollagen, in genetically-defective dermatosparaxic calves. *Eur. J. Biochem.* **23:** 533-543.

Levy, G.G., Nichols, W.C., Lian, E.C., Foroud, T., McClintick, J.N., McGeen B.M., Yang, A.Y., Siemleniak, D.R., Stark, K.R., Gruppo, R., Sarode, R., Shurin, S.B., Chandrasekaran, V., Stabler, S.P., Sabio, H., Bouhassira, E.E., Upshaw, J.D., Jr., Ginsburg, D., Tsai, H.-M., 2001, Mutattions in a member of the *ADAMTS* gene family cause thrombotic thrombocytopenic purpura. *Nature* **413:** 488-494.

Li, S.-W., Arita, M., Fertala, A., Bao, Y., Kopen, G.C., Långsjö, T.K., Hyttinen, M.M., Helminen, H.J., Prockop, D.J., 2001, Transgenic mice with inactive alleles for procollagen N-proteinase (ADAMTS-2) develop fragile skin and male sterility. *Biochem. J.* **355:** 271-278.

Li, S.-W., Sieron, A.L., Fertala, A., Hojima, Y., Arnold, W.V., Prockop, D.J., 1996, The C-proteinase that processes procollagens to fibrillar collagens is identical to the protein previously identified as bone morphogenic protein-1. *Proc. Natl. Acad. Sci. USA* **93:** 5127-5130.

Luque, A., Carpizo, D.R., Iruela-Arispe, M.L., 2003, ADAMTS1/METH1 inhibits endothelial cell proliferation by direct binding and sequestration of $VEGF_{165}$. *J. Biol. Chem.* **278:** 23656-23665.

Matthews, R.T., Gary, S.C., Zerillo, C., Pratta, M., Solomon, K., Arner, E.C., Hockfield, S., 2000, Brain-enriched hyaluronan binding (BEHAB)/brevican cleavage in a glioma cell line is mediated by a disintegrin and metalloproteinase with thrombospondin motifs (ADAMTS) family member. *J. Biol. Chem.* **275:** 22695-22703.

Mould, A.P., Hulmes, D.J.S., 1987, Surface-induced aggregation of type I procollagen. *J. Mol. Biol.* **195:** 543-553.

Myllyharju, J., Kivirikko, K.I., 2004, Collagens, modifying enzymes and their mutations in humans, flies and worms. *Trends Genet.* **20:** 33-43.

Nardi, J. B., Martos, R., Walden, K.K., Lampe, D.J., Robertson, H.M., 1999, Expression of lacunin, a large multidomain extracellular matrix protein, accompanies morphogenesis of epithelial monolayers in Manduca sexta. *Insect Biochem. Mol. Biol.* **29:** 883-897.

Nusgens, B.V., Goebels, Y., Shinkai, H., Lapière, C.M., 1980, Procollagen type III N-terminal endopeptidase in fibroblast culture *Biochem. J.* **191:** 699-706.

Nusgens, B.V., Verellen-Dumoulin, G., Hermans-Le, T., De Paepe, A., Nuytinck, L., Piérard, G.E., Lapière, C.M., 1992, Evidence for a relationship between Ehlers-Danlos type VIIC in humans and bovine dermatosparaxis. *Nat. Genet.* **1:** 214-221.

Pappano, W.N., Steiglitz, B.M., Scott, I.C., Keene, D.R., Greenspan, D.S., 2003, Use of *Bmp1/Tll1* doubly homozygous null mice and proteomics to identify and validate in vivo substrates of bone morphogenetic protein 1/Tolloid-like metalloproteinases. *Mol. Cell. Biol.* **23:** 4428-4438.

Piérard, G.E. and Lapière, C.M., 1976, Skin in dermatosparaxis: dermal microarchitecture and biomechanical properties. *J. Invest. Dermatol.* **66:** 2-7.

Prockop, D.J., Hulmes, D.J.S., 1994, Assembly of collagen fibrils de novo from soluble precursors: polymerization and copolymerization of procollagen, pN-collagen, and mutated collagens. In: Extracellular matrix assembly and structure. Yurchenceo PD, Birk DE, and Mechan RP. Eds. Academic Press, New York, pp. 47-90.

Prockop, D.J., Kivirikko, K.I., 1995, Collagens: Molecular biology, diseases, and potentials for therapy. *Annu. Rev. Biochem.* **64:** 403-434.

Prockop, D.J., Sieron, A.L., Li, S.-W., 1998, Procollagen N-proteinase and procollagen C-proteinase. Two unusual metalloproteinases that are essential for procollagen processing probably have important roles in development and cell signaling. *Matrix Biol.* **16:** 399-408.

Rodríguez-Manzaneque, J.C., Westling, J., Thai, S.N.-M., Luque, A., Knauper, V., Murphy, G., Sandy, J.D., Iruela-Arispe, M.L., 2002, ADAMTS1 cleaves aggrecan at multiple sites and is differentially inhibited by metalloproteinase inhibitors. *Biochem. Biophys. Res. Commun.* **293:** 501-508.

Sandy, J.D., westling, J., Kenagy, R.D., Iruela-Arispe, M.L., Verscharen C., Rodriguez-Mazaneque, J.C., Zimmermann, D.R., Lemire, J.M., Fischer, J.W., Wight, T.N., and Clowes, A.W., 2001, Versican V1 proeolysis in human aorta in vivo occurs at the Glu441-Ala442 bond, a site that is cleaved by recombinant ADAMTS-1 and ADAMTS-4. *J. Biol. Chem.* **276:** 13372-13378.

Schnieke, A., Harbers, K., Jaenisch, R., 1983, Embryonic lethal mutation in mice induced by retrovirus insertion into the alpha 1(I) collagen gene. *Nature* **304:** 315-320.

Scott, I.C., Blitz, I.L., Pappano, W.N., Imamura, Y., Clark, T.G., Steiglitz, B.M., Thomas, C.L., Maas, S.A., Takahara, K., Cho, K.W.Y., Greenspan, D.S., 1999, Mammalian BMP-1/Tolloid-related metalloprroteinases, including novel family member mammalian Tolloid-like 2, have differential enzymatic activities and distributions of expression relevant to patterning and skeletogenesis. *Dev. Biol.* **213:** 283-300.

Shindo, T., Kurihara, H., Kuno, K., Yokoyama, H., Wada, T., Kurihara, Y., Imai, T., Wang, Y., Ogata, M., Nishimatsu, H., Moriyama, N., Oh-hashi, Y., Morita, H., Ishikawa, T., Nagai, R., Yazaki, Y., Matsushima, K., 2000, *J. Clin. Invest.* **105:** 1345-1352.

Smith, L.T., Wertelecki, W., Milstone, L.M., Petty, E.M., Seashore, M.R., Braverman, I.M., Jenkins, T.G., Byers, P.H., 1992, Human dermatosparaxis: A form of Ehlers-Danlos syndrome that results from failure to remove the amino-terminal propeptide of type I procollagen. *Am. J. Hum. Genet.* **51:** 235-244.

Somerville, R.P.T., Longpré, J.-M., Apel, E.D., Lewis, R.M., Wang, L.W., Sanes, J.R., Leduc, R., Apte, S.S., 2004, ADAMTS7B, the full-length product of the *ADAMTS7* gene, is a chodroitin sulfate proteoglycan containing a mucin domain. *J. Biol. Chem.* **279:** 35159-35175.

Somerville, R.P.T., Longpre, J.-M., Jungers, K.A., Engle, J.M., Ross, M., Evanko, S., Wight, T.N., Leduc, R., Apte, S.S., 2003, Characterization of ADAMTS-9 and ADAMTS-20 as a distinct ADAMTS subfamily related to *Caenorhabditis elegans* GON-1. *J. Biol. Chem.* **278:** 9503-9513.

Stöcker, W., Grams, F., Baumann, U., Reinemer, P., Gomis-Rüth, F.-X., McKay, D.B., Bode, W., 1995, The metzincins- topological and sequential relations between the astacins, adamalysins, serralysins, and matrixins (collagenases) define a superfamily of zinc-peptidases. *Protein Sci.* **4:** 823-840.

Suzuki, S., Labosky, P.A., Furuta, Y., Hargett, L., Dunn, R., Fogo, A.B., Takahara, K., Peters, D.M.P., Greenspan, D.S., Hogan, B.L.M., 1996, Failure of ventral body wall closure in mouse embryos lacking a procollagen C-proteinase encoded by *Bmp1*, a mammalian gene related to *Drosophila tolloid. Development* **122:** 3587-3595.

Tortorella, M.D., Arner, E.C., Hills, R., Easton, A., Korte-Sarfaty, J., Fok, K., Wittwer, AJ., Liu, R.-Q., Malfait, A.-M., 2004, $\alpha_2$-Macroglobulin is a novel substrate for ADAMTS-4 and ADAMTS-5 and represents an endogenous inhibitor of these enzymes. *J. Biol. Chem.* **279:** 17554-17561.

Tortorella, M.D., Burn, T.C., Pratta, M.A., Abbaszade, I., Hollis, J.M., Liu, R., Rosenfeld, S.A., Copeland, R.A., Decicco, C.P., Wynn, R., Rockwell, A., Yang, F., Duke, J.L., Solomon, K., George, H., Bruckner, R., Nagase, H., Itok, Y., Ellis, D.M., Ross, H., Wiswall, B.H., Murphy, K., Hillman, M.C., Jr, Hollis, G.G., Newton, R.C., Magolda, R.L., Trzaskos, J.M., Arner, E.C., 1999, Purifiction and cloning of aggrecanase-1: A member of the ADAMTS family of proteins. *Science* **284:** 1664-1666.

Tosi, M., Duponchel, C., Meo, T., Julier, C., 1987, Complete cDNA sequence of human complement C1s and close physical linkage of the homologous genes C1s and C1r. *Biochemistry* **26:** 8516-8524.

Tuderman, L., Kivirikko, K.I., Prockop, D.J., 1978, Partial purification and characterization of a neutral protease which cleaves the N-terminal propeptides from procollagen. *Biochemistry* **17:** 2948-2954.

Tuderman, L. and Prockop, D.J., 1982, Procollagen N-proteinase: properties of the enzyme purified from chick embryo tendons. *Eur. J. Biochem.* **125:** 545-549.

Vázquez, F., Hastings, G., Ortega, M.-A., Lane, T.F., Oikemus, S., Lombardo, M., Iruela-Arispe, M.L., 1999, METH-1, a human ortholog of ADAMTS-1 and METH-2 are members of a new family of proteins with angio-inhibitory activity. *J. Biol. Chem.* **274:** 23349-23357.

Vitagliano, L., Nemethy, G., Zagari, A., Scheraga, H.A., 1995, Structure of the type I collagen molecule based on conformational endrgy computations: the triple-stranded helix and the N-terminal telopeptide. *J. Mol. Biol.* **247:** 69-80.

Wang, W.-M., Lee, S., Steiglitz, B.M., Scott, I.C., Lebares, C.C., Allen, M.L., Brenner, M.C., Takahara, K., Greenspan, D.S., 2003, Transforming growth factor-β induces secretion of activated ADAMTS-2. *J. Biol. Chem.* **278:** 19549-19557,

Weil, D., Bernard, M., Combata, N., Wirtz, M.K., Hollister, D.W., Steinmann, B., Ramirez, F., 1988, Identification of a mutation that causes exon-skipping during collagen pre-mRNA splicing in an Ehlers-Danlos syndrome variant. *J. Biol. Chem.* **263:** 8561-8564.

Yamaji, N., Nishimura, K., Abe, K., Ohara, O., Nagase, T., Nomura, N., inventors: Yamanouchi Pharmaceutical Co. Ltd., assignee. Novel Metalloprotease Having Aggrecanase Activity. European Patent 00974894.8. 2000 Oct 11.

Chapter 13

# ADAMTS3 AND ADAMTS14

## *Procollagen amino-propeptidases that provide insight into the clinical presentation of dermatosparaxis*

Carine Le Goff and Suneel S. Apte

*Department of Biomedical Engineering and Orthopaedic Research Center, Cleveland Clinic Foundation (ND20), 9500 Euclid Avenue, Cleveland, OH 44195, USA*

Abstract: ADAMTS3 and ADAMTS14 belong to the procollagen aminopropeptidase subfamily of ADAMTS proteases that also includes ADAMTS2. These enzymes appear to have co-evolved with their substrates, the major fibrillar collagen types I, II and III by gene duplication from a primitive precursor. Mutations in ADAMTS2 cause an inherited connective tissue disorder, named dermatosparaxis or the Ehlers-Danlos syndrome type VIIC. Compensatory procollagen processing by ADAMTS3 and ADAMTS14 may explain why this disorder has its most severe manifestation in skin, whereas other collagen-containing tissues are relatively unaffected.

Key words: ADAMTS; Collagen; Metalloprotease; Procollagen; Aminopropeptidase; Dermatosparaxis.

## 1. INTRODUCTION

The quantitatively major fibrillar collagens I, II and III are among the most abundant proteins in the body and have critical tissue-specific mechanical roles: collagen I in the dermis, bone, tendons, ligaments and a variety of visceral organs, collagen II in cartilage and vitreous humor and collagen III in blood vessels. Like other fibrillar collagens, their crucial functional domain consists of a long, uninterrupted triple-helical region that forms a rod-like structure. These collagens are synthesized as precursors (procollagens) with sizeable amino (N) and carboxy (C) propeptides. Each propeptide undergoes proteolytic excision extracellularly. The mature collagen, retaining only very short non-triple helical terminal peptides



*N. M. Hooper and U. Lendeckel (eds.), The ADAM Family of Proteases,* 283-298.

(telopeptides), then assembles into fibrils. The removal of the propeptides is essential for proper assembly. Identification of the recessively inherited connective tissue disorder, dermatosparaxis (DS), first in cattle and then in other animals, disclosed the existence of the procollagen N-propeptidase (PNP), and demonstrated that excision of the N-propeptide is essential for proper collagen fibril assembly (Lapiere *et al.* 1971; Lenaers *et al.* 1971; Hanset and Lapiere, 1974; Holbrook *et al.* 1980; Lapiere and Nusgens, 1993). In this disorder, and its human counterpart, the Ehlers-Danlos syndrome (type VIIC or dermatosparactic type), a significant proportion of procollagen I in skin retains the N-propeptide (Lapiere *et al.* 1971; Lenaers *et al.* 1971; Hanset and Lapiere, 1974; Nusgens *et al.* 1992; Smith *et al.* 1992; Wertelecki *et al.* 1992; Lapiere and Nusgens, 1993; Petty *et al.* 1993). However, complete excision of the C-propeptide occurs, since that is executed by BMP-1, an astacin-like proteinase (Kessler *et al.* 1996). The presence of the intact N-propeptide is believed to sterically hinder fibril assembly, and abnormally thin or ribbon-shaped fibrils that are "hieroglyphic" on cross section are formed instead (Lenaers *et al.* 1971; Watson *et al.* 1998). Thus, the resulting collagen network is mechanically incompetent, and severe skin fragility ensues.

In 1997, Colige *et al.* cloned the cDNA encoding the PNP (designated ADAMTS2) from bovine fibroblasts, and shortly thereafter obtained the corresponding human sequence (Colige *et al.* 1997; Colige *et al.* 1999). They have subsequently identified a number of causative ADAMTS2 mutations in human and bovine dermatosparaxis (Colige *et al.* 1999; Colige *et al.* 2004). Since the first description of dermatosparaxis over 3 decades ago (Hanset and Ansay, 1967), several aspects of its presentation have puzzled clinicians and biochemists. Collagen I is a major component of numerous tissues other than dermis, such as bone, tendon, ligament, the outer wall of blood vessels (adventitia) and the eye (sclera), yet these tissues were not functionally compromised by *ADAMTS2* mutations to the same extent as dermis (Lenaers *et al.* 1971; Lapiere and Nusgens, 1993). The presence of some completely processed procollagen I in skin was at first felt to be reflective of a gene defect that did not completely inactivate the PNP. However, the observation that a number of other collagen I-rich tissues and organs were not as severely affected as the dermis, supported the existence of one or more compensating PNPs in those tissues (Lenaers *et al.* 1971; Lapiere and Nusgens, 1993). In fact, the degree of functional impairment in a tissue correlated quantitatively with the defect in procollagen I processing; Thus, processing was maximally impaired in dermis, and less so in bone, ligament, sclera and aorta (Lenaers *et al.* 1971; Lapiere and Nusgens, 1993). Where processed procollagen could be found in dermatosparactic mice (Li *et al.* 2001), it was cleaved at the same peptide bond attacked by ADAMTS2, indicating processing by closely related enzymes (Colige *et al.* 2002). In addition, although ADAMTS2 could process procollagen II (Tuderman *et al.* 1978), dermatosparactic nasal cartilage collagen II was completely processed (Fernandes *et al.* 2001). Thus, a search for functionally related proteases was undertaken, that resulted in the discovery of ADAMTS3 and ADAMTS14. In this chapter, we provide a detailed discussion of the structural features of ADAMTS3 and ADAMTS14, and discuss their likely biological roles in the context of ADAMTS2 and dermatosparaxis.

## 2. AN ADAMTS SUBFAMILY OF PROCOLLAGEN N-PROPEPTIDASES

There are 19 ADAMTS proteases in the human and mouse genomes (Apte, 2004). ADAMTS2, ADAMTS3 and ADAMTS14, have an identical domain organization and gene structure, and highly similar primary sequences (Apte, 2004). In many respects, these three enzymes are clearly different from other ADAMTS proteases, suggesting that they share a specialized function, constituting a distinct subfamily of PNPs. Indeed, like ADAMTS2, ADAMTS3 and ADAMTS14 have now been shown to process the N-propeptide of one or more fibrillar procollagens. Of course, this does not exclude the possibility that each of these enzymes may have additional non-collagenous substrates. The procollagen substrates are however, highly specialized molecules whose processing requires a specific tertiary structure, such that the N-propeptide of denatured procollagen cannot be efficiently processed (Tuderman *et al.* 1978). That and the specialized cysteine signature of the PNP catalytic domain (discussed in 2.3) render it unlikely that non-PNP ADAMTS proteases will process procollagen.

### 2.1 Discovery and functional characterization of ADAMTS3

The KIAA0366 cDNA was obtained from a large-scale transcript sequencing project, whose goal was to identify full-length human brain cDNA clones (Nagase *et al.* 1997). The sequence of this cDNA was deposited in GenBank by the Kazusa DNA Institute, but its protein product had not been characterized in any way. The sequence encoded a close homolog of ADAMTS2 that we named ADAMTS3. The missing 5′ end of this sequence was completed by RACE and the complete ADAMTS3 ORF was obtained (Fernandes *et al.* 2001). To test a role for ADAMTS3 in procollagen processing, a cell system that was defective in procollagen II processing was used. The RCS-LTC chondrosarcoma cell line produces all the typical matrix components of cartilage, but unlike the parental RCS cell line, the collagen II is deposited as thin filaments and not as normal fibrils (Fernandes *et al.* 1997). This phenomenon is similar to that seen in dermatosparaxis, since the N-propeptide of RCS-LTC collagen is not excised (Fernandes *et al.* 1997). The defect appears to be a lack of the processing enzyme, since the processing site in procollagen II is intact, and processing can be rescued by addition of conditioned medium from normal chondrocytes (Fernandes *et al.* 1997). The RCS-LTC cells were stably transfected with either ADAMTS3, ADAMTS2, or *lacZ* plasmid (as a control). Both ADAMTS3 and ADAMTS2 transfected cells showed a partial, but quantitatively equivalent rescue of the processing defect, suggesting that ADAMTS3 was a procollagen II N-propeptidase (Fernandes *et al.* 2001). Since dermatosparactic cartilage failed to show a procollagen II processing defect, we next asked whether ADAMTS3 was expressed in cartilage. Quantitative RT-PCR showed that ADAMTS3 was expressed at higher levels than ADAMTS2 in cartilage, whereas ADAMTS2 was preferentially expressed in skin (Fernandes *et al.* 2001). Together with its pN-collagen II N-propeptidase activity, this suggested that ADAMTS3 was a major procollagen II processing enzyme. Recent unpublished data from our laboratory using *in situ* hybridization of mouse embryos confirms that the *Adamts3* mRNA is highly expressed in developing cartilage throughout skeletal development, but not in dermis. The data also show *Adamts3* expression in collagen I rich tissues such as bone, tendon and ligament (Carine Le Goff and Suneel Apte, unpublished data). This highly selective expression of *Adamts3* suggests that it is a major procollagen I and II processing enzyme during mouse embryogenesis.

## 2.2 Discovery and functional characterization of ADAMTS14

ADAMTS14 was independently discovered by three groups. Bolz *et al.* reported its identification during a positional cloning effort to identify the causative gene for Usher syndrome type 1D, which was linked to human chromosome 10q in close proximity to *ADAMTS14* (Bolz *et al.* 2001). Cal *et al.* cloned ADAMTS14 by a genome-wide database search (Cal *et al.* 2002). Colige *et al.* presented the most detailed characterization of ADAMTS14 hitherto reported, starting with cDNA cloning directed specifically at finding ADAMTS2 related proteins, and followed this by functional characterization of the protease (Colige *et al.* 2002). Northern blot analysis of multiple tissue mRNA has not shown significant *ADAMTS14* mRNA levels in any tissues, although it is detectable in mRNA from cultured skin fibroblasts (Colige *et *al.* 2002). However, RT-PCR analysis indicates that *ADAMTS14* mRNA is present in heart, brain, prostate, placenta, lung and retina (Bolz *et al.* 2001; Cal *et al.* 2002; Colige *et al.* 2002). ADAMTS14 is detectable in human skin fibroblasts, by northern analysis, indicating it may account for the partial processing of procollagen I seen in DS skin (Colige *et al.* 2002). Thus, overall expression of ADAMTS14 is apparently quite low, and it appears to be expressed at the highest levels in skin fibroblasts.

Bolz *et al.* identified four non-synonymous single nucleotide repeat polymorphisms (SNPs) in *ADAMTS14* (Bolz *et al.* 2001). Since these SNPs encode non-conservative amino acid changes (S270L, P590L, S1017N, E1049G), it is possible that they lead to subtle changes in function. It would be interesting to ask whether individuals who are homozygous for any of these *ADAMTS14* SNPs or whose *ADAMTS14* alleles each have a different SNP, could have mild skin fragility. Additional genetic variation in *ADAMTS14* is introduced by alternative splicing. Colige *et al.* described alternative splicing of exon 6 that led to deletion of 9 bp at the 3′ end of this exon and an in-frame deletion of three amino acids (MQG) within the catalytic domain (Colige *et al.* 2002). The corresponding mouse region lacks these amino acids, suggesting that the MGQ-deleted form is likely to be the major functional variant. Colige also reported complex splicing of the 5′ end of the mRNA (Colige *et al.* 2002). The major transcript variant, ADAMTS14A, encodes a secreted protein whereas ADAMTS14B and ADAMTS14C encode proteases lacking a signal peptide (Colige *et al.* 2002). While the A variant thus seems to be physiologically relevant, the significance of the B and C variants is not known.

As previously done for ADAMTS3, the N-propeptidase activity of ADAMTS14 was investigated in a co-culture assay. Control 293 cells or 293 cells expressing ADAMTS2 or ADAMTS14 were co-cultured with fibroblasts derived from dermatosparactic calf skin. In these experiments, the ADAMTS14 transfected cells, but not the control cells, were able to partly rescue the defect in procollagen processing (Colige *et al.* 2002). Interestingly, 293 cells expressing ADAMTS2 completely rescued the processing defect in co-cultures with dermatosparactic skin fibroblasts suggesting that ADAMTS2 might be a more efficient PNP than ADAMTS14 (Colige *et al.* 2002). Recombinant procollagen I processing activity recovered from the medium of 293 cells expressing ADAMTS14 was low, suggesting perhaps, that the purified enzyme was not fully activated or that activity may have been lost during purification.

```
1    MDPPAGAARRLLCPALLLLLLLLPPPLLPPPPPPANARLAAAADPPGGPLGHGAERILAV ADAMTS2
1    M---------------VLLSLWLIAAALVEVRTSADGQAGNEEMVQIDLPIKRYREYELVT ADAMTS3
1    M---------------A--PLRALLSYLLPLHCALCAAAGSRT-PELHLSGK-LSDYGVTV ADAMTS14

61   PVRTDAQGRLVSHVVSAATSRAGVRARRAAP------------VRTPSFPGGN--EEEPG ADAMTS2
47   PVSTNLEGRYLSHTLSASHKKRSARDVSSNP----------------------------- ADAMTS3
43   PCSTDFRGRFLSHVVSGPAAASAGSMVVDTPPTLPRHSSHLRVARSPLHPGGTLWPGRVG ADAMTS14

107  SH-LFYNVTVFGRDLHLRLRPNARLVAPGATMEWQGEK------------------GTTRV ADAMTS2
78   -EQLFFNITAFGKDFHLRLKPNTQLVAPGAVVEWHETSLVPGNITDPINNHQPGSATYRI ADAMTS3
103  RHSLYFNVTVFGKELHLRLRPNRRLVVPGSSVEWQED--------------------FRE ADAMTS14

149  ---EPLLGSCLYVGDVAGLAEASSVALSNCDGLAGLIRMEEEEFFIEPLEKGLAAQEAEQ ADAMTS2
137  RKTEPLQTNCAYVGDIVDIP-GTSVAISNCDGLAGMIKSDNEEYFIEPLERGKQMEE-EK ADAMTS3
143  LFRQPLRQECVYTGGVTGMP-GAAVAISNCDGLAGLIRTDSTDFFIEPLERGQQEKE-AS ADAMTS14

206  GRVHVVYRRPPTSPPLGGPQA-LD-TGASLDSLDSLSRALGVLEEHANSSRRRARRHAAD ADAMTS2
195  GRIHVVYKRSAVEQAPIDMSKDFHYRESDLEGLDDLGTVYGNIHQQLNETMRR-RRHAGE ADAMTS3
201  GRTHVVYRREAVQQEWAEPDGDLH---NEAFGLGDLPNLLGLVGDQLGDTERK-RRHAKP ADAMTS14

264  DDYNIEVLLGVDDSVVQFHGKEHVQKYLLTLMNIVNEIYHDESLGAHINVVLVRIILLSY ADAMTS2
254  NDYNIEVLLGVDDSVVRFHGKEHVQNYLLTLMNIVNEIYHDESLGVHINVVLVRMIMLGY ADAMTS3
257  GSYSIEVLLVVDDSVVRFHGKEHVQNYVLTLMNIVDEIYHDESLGVHINIALVRLIMVGY ADAMTS14

324  GKSMSLIEIGNPSQSLENVCRWAYLQQKPDTGHDEYHDHAIFLTRQDFGPSGMQGYAPVT ADAMTS2
314  AKSISLIERGNPSRSLENVCRWASQQQRSDINHSEHHDHAIFLTRQDFGPAGMQGYAPVT ADAMTS3
317  RQSLSLIERGNPSRSLEQVCRWAHSQQRQDPSHAEHHDHVVFLTRQDFGPSGMQGYAPVT ADAMTS14

384  GMCHPVRSCTLNHEDGFSSAFVVAHETGHVLGMEHDGQGNRCGDEVRLGSIMAPLVQAAF ADAMTS2
374  GMCHPVRSCTLNHEDGFSSAFVVAHETGHVLGMEHDGQGNRCGDETAMGSVMAPLVQAAF ADAMTS3
377  GMCHPLRSCALNHEDGFSSAFVIAHETGHVLGMEHDGQGNGCADETSLGSVMAPLVQAAF ADAMTS14
                                                      Dis
444  HRFHWSRCSQQELSRYLHSYDCLLDDPFAHDWPALPQLPGLHYSMNEQCRFDFGLGYMMC ADAMTS2
434  HRYHWSRCSGQELKRYIHSYDCLLDDPFDHDWPKLPELPGINYSMDEQCRFDFGVGYKMC ADAMTS3
437  HRFHWSRCSKLELSRYLPSYDCLLDDPFDPAWPQPPELPGINYSMDEQCRFDFGSGYQTC ADAMTS14

504  TAFRTFDPCKQLWCSHPDNPYFCKTKKGPPLDGTMCAPGKHCFKGHCIWLTPD-ILKRDG ADAMTS2
494  TAFRTFDPCKQLWCSHPDNPYFCKTKKGPPLDGTECAAGKWCYKGHCMWKNANQQ-KQDG ADAMTS3
497  LAFRTFEPCKQLWCSHPDNPYFCKTKKGPPLDGTECAPGKWCFKGHCIWKSPEQTYGQDG ADAMTS14

563  SWGAWSPFGSCSRTCGTGVKFRTRQCDNPHPANGGRTCSGLAYDFQLCSRQDCPDSLADF ADAMTS2
553  NWGSWTKFGSCSRTCGTGVRFRTRQCNNPMPINGGQDCPGVNFEYQLCNTEECQKHFEDF ADAMTS3
557  GWSSWTKFGSCSRSCGGGVRSRSRSCNNPSPAYGGRPCLGPMFEYQVCNSEECPGTYEDF ADAMTS14
                                  1
623  REEQCRQWDLYFEHGDAQHHWLPHEHRDAKERCHLYCESRETGEVVSMKRMVHDGTRCSY ADAMTS2
613  RAQQCQQRNSHFEYQNTKHHWLPYEHPDPKKRCHLYCQSKETGDVAYMKQLVHDGTHCSY ADAMTS3
617  RAQQCAKRNSYYVHQNAKHSWVPYEPDDDAQKCELICQSADTGDVVFMNQVVHDGTRCSY ADAMTS14

683  KDAFSLCVRGDCRKVGCDGVIGSSKQEDKCGVCGGDNSHCKVVKGTFTRSPKKHGYIKMF ADAMTS2
673  KDPYSICVRGECVKVGCDKEIGSNKVEDKCGVCGGDNSHCRTVKGTFTRTPRKLGYLKMF ADAMTS3
677  RDPYSVCARGECVPVGCDKEVGSMKADDKCGVCGGDNSHCRTVKGTLGKASKQAGALKLV ADAMTS14
                                             Spacer

743  EIPAGARHLLIQEVDATSHHLAVKNLETGKFILNEENDVDASSKTFIAMGVEWEYRDEDG ADAMTS2
733  DIPPGARHVLIQEDEASPHILAIKNQATGHYILNGKGE-EAKSRTFIDLGVEWDYNIEDD ADAMTS3
737  QIPAGARHIQIEALEKSPHRIVVKNQVTGSFILNPKGK-EATSRTFTAMGLEWEDAVEDA ADAMTS14

803  RETLQTMGPLHGTITVLVIPV--GDTRVSLTYKYMIHEDSLN-VDDNNVLEEDSVVYEWA ADAMTS2
792  IESLHTDGPLHDPVIVLIIPQ-ENDTRSSLTYKYIIHEDSVPTINSNNVIQEELDTFEWA ADAMTS3
796  KESLKTSGPLPEAIAILALPPTEGGPRSSLAYKYVIHEDLLPLIGSNNVLLEEMDTYEWA ADAMTS14

860  LKKWSPCSKPCGGGSQFTKYGCRRRLDHKMVHRGFCAALSKPKAIRRACNPQECSQPVWV ADAMTS2
851  LKSWSQCSKPCGGGFQYTKYGCRRKSDNKMVHRSFCEANKKPKPIRRMCNIQECTHPLWV ADAMTS3
856  LKSWAPCSKACGGGIQFTKYGCRRRRDHHMVQRHLCDHKKRPKPIRRRCNQHPCSQPVWV ADAMTS14
                                          2
920  TGEWEPCSQTCGRTGMQVRSVRCIQPLHDNTTRSVHAKHCNDARPESRRACSRELCPGRW ADAMTS2
911  AEEWEHCTKTCGSSGYQLRTVRCLQPLLDGTNRSVHSKYCMGDRPESRRPCNRVPCPAQW ADAMTS3
916  TEEWGACSRSCGKLGVQTRGIQCLLPLSNGTHKVMPAKACAGDRPEARRPCLRVPCPAQW ADAMTS14
                                          3
980  RAGPWSQCSVTCGNGTQERPVPCRTADDSFGICQEERPETARTCRLGPCPRNISDPSKKS ADAMTS2
971  KTGPWSECSVTCGEGTEVRQVLCRAG----DHCDGEKPESVRACQLPPCND--------- ADAMTS3
976  RLGAWSQCSATCGEGIQQRQVVCRTNANSLGHCEGDRPDTVQVCSLPACGGNHQNSTVRA ADAMTS14
                                          4
1040 YV-------VQWLSRPDPDSPIRKISSKGHCQGDKSIFCRMEVLSRYCSIPGYNKLSCKS ADAMTS2
1018 ----------------------------EPCLGDKSIFCQMEVLARYCSIPGYNKLCCES ADAMTS3
1036 DVWELGTPEGQWVPQSEPLHPINKISSTEPCTGDRSVFCQMEVLDRYCSIPGYHRLCCVS ADAMTS14

1093 CN----------LYNNLTNVEGRIEPPPG------KHNDIDVFMPTLPVPTVAMEVRPS- ADAMTS2
1050 CSKRSSTLPPPYLLEAAETHDDVISNPSDLPRSLVMPTSLVPYHSETPAKKMSLSSISSV ADAMTS3
1096 CIKKASGPNPG-------------PDPG--------PTSLPPF--STPGSPLP------- ADAMTS14

1136 -----------PSTPLEVPLNASSTNA---TEDHPETNAVDEPYKIHGLEDEVQ------ ADAMTS2
1110 GGPNAYAAFRPNSKPDGANLRQRSAQQAGSKTVRLVTVPSSPPTKRVHLSSASQMAAASF ADAMTS3
1126 GPQDPADAAEPPGKPTGSE-----DHQHGRATQLPGALDTSSPGTQHPFAPETPIPGASW ADAMTS14
```

```
1176 --PPNLIPRRPSPYEKTRNQRIQELI---DEMRKKEM-----LGKF             ADAMTS2
1170 FAASDSIGA----SSQARTSKKDGKII---DNRRPTRS---STLER             ADAMTS3
1181 SISPTTPGGLPWGWTQTPTPVPEDKGQPGEDLRHPGTSLPAASPVT            ADAMTS14
```

*Figure 1.* Alignment of the primary structures of human ADAMTS2 (GenBank accession no. AJ003125), ADAMTS3 (GenBank Accession no. AAK28400)and ADAMTS14 (GenBank Accession no. AAL79814). Alignment was performed using the MegAlign software (DNAStar, Madison, WI). Single-letter amino acid code is used. Significant sequence features are highlighted as follows: The active-site sequence HETGHVLGMEHD is boxed. Arrows indicate the start of the disintegrin-like (Dis) and spacer module. The cysteine-rich module is shown with a gray background. The four thrombospondin type 1 repeats (TSRs) are underlined and numbered sequentially. The PLAC motif is shown by the thick underline and lies within the C-terminal extension from TSR4 through to the C-terminus. Consensus N-linked oligosaccharide attachment sites (NXS/T, where X is any amino acid other than proline) are boxed. The cysteine residues in the pro/catalytic domain are indicated by the solid ovals and those within the cysteine-rich module are in white boldface on gray background. Furin processing sites (RXRR, where X is any amino acid) are shown by the bold underline.

## 2.3 Distinctive features of the PNP subfamily

The PNPs have a domain structure consisting of a pro-domain, a catalytic domain and an ancillary domain, and they retain the general architecture of the ADAMTS family. However, the sequences of each of these three domains and the modular composition of the ancillary domain are highly conserved within the PNPs and one can thus identify several characteristics that are the hallmarks of this subfamily (Fernandes *et al.* 2001; Colige *et al.* 2002). The PNPs are the only ADAMTS proteases to possess three C-terminal TSRs and a C-terminal extension containing an embedded PLAC motif (Fernandes *et al.* 2001; Cal *et al.* 2002; Colige *et al.* 2002). Their signal peptides are unremarkable and are followed by a pro-domain that constitutes the most variable region of the PNPs. This is not surprising, since these pro-domains are fully excised by pro-protein convertases such as furin during secretion, and are therefore not present in the mature enzymes, or essential for catalytic activity. Like many ADAMTS proteases, each PNP has two consensus furin recognition sites (one near the amino terminus of the pro-domain and the other at its junction with the catalytic domain) (Fig. 1) (Fernandes *et al.* 2001; Colige *et al.* 2002; Wang *et al.* 2003). Wang *et al.* showed that the excision of the ADAMTS2 pro-domain occurred in a stepwise process that utilized both furin sites (Wang *et al.* 2003). In all PNPs the final activation step likely occurs at the C-terminal furin recognition sequence (Fig. 1) following which active enzyme is released from the cell. However, unlike other ADAMTS proteases, whose prodomains have three cysteines, the PNP pro-domains contain only two cysteine residues (Fernandes *et al.* 2001; Colige *et al.* 2002). The missing cysteine residue is in a position that corresponds to a putative 'cysteine-switch" residue in matrix metalloproteases. The "cysteine-switch" mechanism of matrix metalloprotease activation involves the maintenance of latency of the zymogen by interaction of this cysteine with the active-site zinc (Van Wart and Birkedal-Hansen, 1990). Proteolytic cleavage of the pro-domain destabilizes the interaction and permits access of water into the active-site, rendering it catalytically active. This mechanism is not known to be operative in the ADAMTS proteases and is irrelevant to the PNPs because they lack the unpaired cysteine.

Like other reprolysin metalloproteases, the PNPs have the typical active-site sequence of HEXXH + HD (Rawlings and Barrett, 1995). The active site sequence of PNPs is completely conserved, but it differs from that of the other ADAMTS proteases (Fig. 1). Within the highly conserved catalytic domain active site, the PNPs each have a Thr that is not present in other ADAMTS (Fig. 1). The active site is present within an ~70 amino acid sequence context that demonstrates exceptional similarity in the PNP subfamily, but is different from other ADAMTS proteases. Another specific feature of the PNPs is that their catalytic domain has a distinctive layout of cysteines. The PNPs contain only six cysteine residues while all other ADAMTS (except, ADAMTS13, the von Willebrand factor protease) have eight (Fernandes *et al.* 2001). The two deleted cysteines are both upstream of the active site. Thus the 3-dimensional structure of the PNP catalytic domain may be quite different from other ADAMTS proteases, perhaps to accommodate the unique requirements of the fibrillar procollagen extended recognition site. The cysteine-rich module of ADAMTS2, but not ADAMTS3 or ADAMTS14, contains an RGDC motif (ADAMTS3 and ADAMTS14 have the sequence RGEC at this location) (Fig. 1) (Colige *et al.* 1997; Colige *et al.* 1999) which could be an integrin-binding motif. The C-terminal extension of each PNP is unique. These extensions lack detectable homology with the C-terminal unique domain of the *gon-1* related proteases, ADAMTS9 and ADAMTS20 (Somerville *et al.* 2003), or to the CUB-domain containing region of ADAMTS13 (Zheng *et al.* 2001). However, a PLAC (protease and lacunin) motif that is highly conserved in the PNPs, to almost the same extent as the catalytic domain, is embedded within this C-terminal PNP extension (Fig. 1). In contrast, the remainder of the C-terminal extension is very poorly conserved among the PNPs.

Sites for N-linked glycosylation are present in each PNP (Fig. 1), suggesting that like ADAMTS2, ADAMTS3 and ADAMTS14 are glycoproteins and therefore amenable, like it, to affinity purification by lectins such as concanavalin A. Each PNP also has numerous regions rich in basic amino acids and TSR1 of ADAMTS14 has aWSSW motif that has the potential to bind heparin. Thus heparin-sepharose chromatography could be used for purification of these enzymes and indeed we find that ADAMTS3 binds avidly to heparin-sepharose (Carine Le Goff and Suneel Apte, unpublished data). Heparin-binding suggests a high degree of likelihood of interaction with cell surface proteoglycans.

Comparison of the PNPs suggests that the catalytic domain, disintegrin-like module and PLAC motif, which are the most highly conserved regions of the PNPs, are likely to underlie their specialized function. Observations made in ADAMTS9 and ADAMTS4 demonstrate the critical role of the ancillary domain in proteolysis, probably through its requirement for cell surface localization, substrate binding, or protease activation (Somerville *et al.* 2003; Kashiwagi *et al.* 2004). Thus we propose that while the ancillary domain is indispensable for procollagen binding, the specialized ability to subsequently cleave the procollagen N-propeptide is an intrinsic property of the PNP catalytic domain. These hypotheses can be tested by generation of truncated and chimeric proteins and module exchange within the PNP subfamily and between it and the ADAMTS family at-large.

# 3. BIOLOGICAL CONSIDERATIONS

Specific biological functions have not been identified *in vivo* for ADAMTS3 and ADAMTS14. The following discussion addresses a number of issues raised by the existence of ADAMTS3 and ADAMTS14 in the context of ADAMTS2 and DS.

## 3.1 Which of ADAMTS3 and ADAMTS14 is most similar to ADAMTS2 and therefore, most likely to compensate for its loss in dermatosparaxis?

Based on the sequence alignment in Figure 1, we undertook a systematic quantitative comparison of the different modules of these proteases (Table 1). Next, we compared each module of human ADAMTS3 and ADAMTS14 with the corresponding module of ADAMTS2 (Fig. 2). Except for the pre-prodomain and the disintegrin-like module, ADAMTS3 is fractionally more closely related to ADAMTS2 than is ADAMTS14. Nevertheless, the calculated differences are quite small and if one includes conserved amino acid substitutions, they are negligible. Thus, we conclude that for all practical purposes, ADAMTS3 and ADAMTS14 are structurally and functionally equivalent to each other and to ADAMTS2.

| | | % IDENTITY | | |
|---|---|---|---|---|
| | | ADAMTS2 | ADAMTS3 | ADAMTS14 |
| PREPRODOMAIN | ADAMTS2 | | 37 | 39 |
| | ADAMTS3 | 37 | | 36 |
| | ADAMTS14 | 39 | 36 | |
| METALLOPROTEASE DOMAIN | ADAMTS2 | | 85 | 79 |
| | ADAMTS3 | 85 | | 78 |
| | ADAMTS14 | 79 | 78 | |
| DISINTEGRIN-LIKE MODULE | ADAMTS2 | | 77 | 82 |
| | ADAMTS3 | 77 | | 83 |
| | ADAMTS14 | 82 | 83 | |
| CENTRAL TSR | ADAMTS2 | | 63 | 51 |
| | ADAMTS3 | 63 | | 69 |
| | ADAMTS14 | 51 | 69 | |
| CYSTEINE-RICH MODULE | ADAMTS2 | | 67 | 55 |
| | ADAMTS3 | 67 | | 68 |
| | ADAMTS14 | 55 | 68 | |
| SPACER MODULE | ADAMTS2 | | 56 | 46 |
| | ADAMTS3 | 56 | | 53 |
| | ADAMTS14 | 46 | 53 | |
| TSR 2-4 | ADAMTS2 | | 64 | 56 |
| | ADAMTS3 | 64 | | 56 |
| | ADAMTS14 | 56 | 56 | |
| PLAC MODULE | ADAMTS2 | | 83 | 70 |
| | ADAMTS3 | 83 | | 77 |
| | ADAMTS14 | 70 | 77 | |

*Table 1.* A numerical comparison of the percent amino acid sequence identity for individual modules of ADAMTS2, ADAMTS3 and ADAMTS14

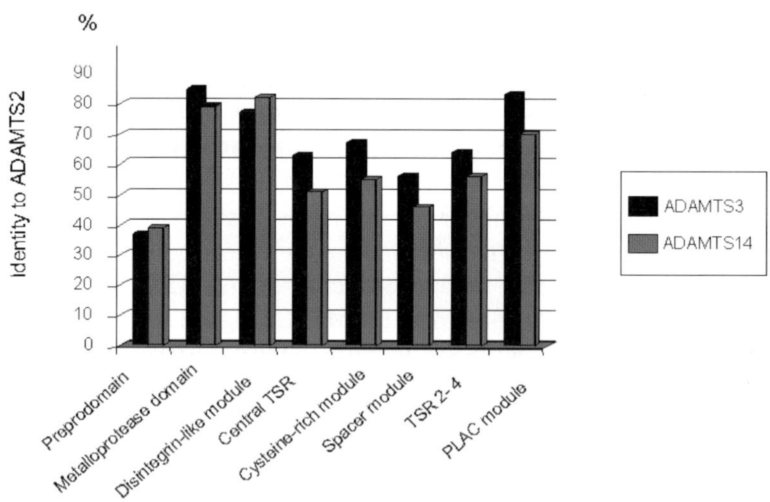


*Figure 2.* Amino acid sequence identity of individual modules of ADAMTS3 and ADAMTS14 to the respective modules of ADAMTS2. The histograms were constructed using the data in Table 1.

## 3.2 What accounts for the restricted phenotype of DS?

We believe that the answer to this question is provided by differential tissue-specific gene regulation. Thus the paradigm we propose (and which may be broadly applicable to other ADAMTS subfamilies) is that the PNPs are structurally and functionally identical, but it is differential gene regulation that defines distinct biological roles for each of them. Emerging data suggests that this may well be the case. We previously showed that *ADAMTS2* was highly expressed in skin but that *ADAMTS3* was expressed at higher levels in human fetal cartilage (Fernandes *et al.* 2001). More recently, we undertook a systematic developmental profiling of these ADAMTS proteases with their known procollagen substrates. The results are intriguing because they demonstrate exquisite co-regulation of these enzymes with their substrates in a tissue-specific fashion, such as co-expression of *Adamts3* (but not *Adamts2* and *Adamts14),* with *Col2a1* in cartilage (Carine Le Goff and Suneel S. Apte, unpublished data). Expression of ADAMTS3 and ADAMTS14 in procollagen I expressing tissues other than dermis appears to explain why the effect of ADAMTS2 mutations is most evident in the skin (Fernandes *et al.* 2001; Colige *et al.* 2002).

## 3.3 Why are there three PNPs?

The limited tissue-specific expression of each PNP may underlie the need to have three distinct PNPs, each expressed at different sites of procollagen synthesis. Another possibility is that each of these enzymes may prefer one procollagen substrate to others. Thus ADAMTS3 may process procollagen II more efficiently than it processes procollagen I or III. Such preferences appear unlikely because of the highly restricted expression profiles of each of the PNPs and its coexpression with just one or two of the procollagens (Carine Le Goff and Suneel Apte, unpublished data). Another explanation may be that each PNP processes additional non-collagenous substrates and so they exist for other specialized functions. Some support for this possibility comes from expression data that indicate a discrepancy between the levels of procollagen I and ADAMTS2/3/14 mRNA in tissues such as testis, heart, lung and brain (Li *et al.* 2001; Colige *et al.* 2002) (Carine Le Goff and Suneel Apte, unpublished data).

## 3.4 How did the PNPs evolve?

The specialized function of the PNPs predicts that their evolution would be intimately linked to that of their substrates, the A-clade procollagens I, II and III. The A-clade collagens are also found, naturally, in other mammalian species, in avian (*Gallus domesticus*), amphibian (*Xenopus laevis*) and piscean (*Fugu rubripes*) species, as well as in the invertebrate chordate *Ciona intestinalis* (Boot-Handford *et al.* 2003). However, the cuticular collagens of non-chordate invertebrate organisms such as *Caenorhabditis elegans* and *Drosophila melanogaster* are very different from A-clade collagens (Boot-Handford *et al.* 2003). To investigate if PNPs are present in genomes that also contain the A-clade collagens, we first analyzed the primary sequence alignment of the PNPs to determine which sequence features were not only highly conserved, but also most distinct from other ADAMTS subfamilies (Fig. 1). Although each PNP contains a C-terminal extension that is not found in non-PNP ADAMTS proteases, it is quite divergent among the PNPs and apart from the PLAC motif, it does not have sequence landmarks that could be used for such an analysis. On the other hand, the PNP pro-domain and catalytic domain is clearly unique within the ADAMTS family; in particular, the characteristic arrangement of Cys residues in these regions and the sequence and context of the active site is highly conserved (Fig. 1). Accordingly, the prodomain and catalytic domain of ADAMTS3 was used for BLAST searches of a variety of genomes. The searches demonstrated highly similar sequences (with complete conservation of the active site sequence and >90% conservation of the sequence context) in *Fugu rubripes* (four predicted PNPs), *Xenopus laevis* (three predicted PNPs), *Gallus domesticus* (three predicted PNPs) and *Ciona intestinalis* (one predicted PNP), but not in *Caenorhabditis elegans* or *Drosophila melanogaster*. This is similar to the evolutionary distribution of the classic fibrillar collagens, where both arthropods and nematodes have lost their fibrillar collagens during the evolution away from the other protostomes, such as annelids and molluscs (Boot-Handford *et al.* 2003). Thus it appears that PNP function has evolved coordinately with the A-clade collagens.

The observations further suggest that the primitive chordate PNP (represented by the single *Ciona* gene) probably underwent successive rounds of gene duplication to result in the three PNPs found in most vertebrates. The presence of four PNPs in *Fugu rubripes* can be explained by the extra round of genome duplication that occurred early in the divergence of teleosts from other fish (Boot-Handford *et al.* 2003; Christoffels *et al.* 2004). Despite being so closely related, however, the three PNPs are dispersed in the human genome. ADAMTS2, ADAMTS3 and ADAMTS14 are located on human chromosomes 5, 4 and 10, respectively (Hurskainen *et al.* 1999; Fernandes *et al.* 2001; Colige *et al.* 2002). Nevertheless, their gene structures are identical (comprising 22 coding exons with similar intron-exon junctions) (Colige *et al.* 2004), supporting an origin by gene duplication. Gene duplication may have occurred quite recently, since the enzymes have diverged very little from each other; alternatively there may be strong negative selection against divergence.

## 3.5 Could ADAMTS3 or ADAMTS14 be used to treat individuals with DS?

Following further characterization of their procollagen I processing ability, it is conceivable that ADAMTS3 and ADAMTS14 may be useful for rescuing the cutaneous phenotype of DS. The logic behind using these two proteases instead of ADAMTS2 is that there is a lesser likelihood of immune rejection of enzymes that are already present in individuals with DS. An intriguing therapeutic possibility is that of enhanced production of ADAMTS3 and ADAMTS14 in skin through transcriptional activation of their genes. For instance, Wang *et al.* demonstrated that TGFβ up-regulated *ADAMTS2*, consistent with its role in enhancing collagen biosynthesis (Wang *et al.* 2003). However, results from our laboratory and from Colige *et al.,* have failed to disclose a similar response of the *ADAMTS3* and *ADAMTS14* genes (Colige *et al.* 2002). Nevertheless, if *ADAMTS3* and *ADAMTS14* inductive factors could be identified, it might be possible to use them topically on skin and produce an increase in dermal collagen strength with relatively few side effects in other locations. Alternatively, one could achieve targeted expression of ADAMTS3 or ADAMTS14 in collagen I-rich tissues using a collagen I promoter-*ADAMTS3/ADAMTS14* construct for gene therapy. DS is an extremely rare disorder, such that investment in intensive therapeutic efforts is probably a low priority for the pharmaceutical or biotechnology industries. Nevertheless, the existence of ADAMTS3 and ADAMTS14 does offer the potential for a cure, but a much better understanding of their overall biology and substrate preferences must first be achieved.

# 4. CONCLUSIONS

In summary, the existing literature strongly supports a role for ADAMTS3 and ADAMTS14 in procollagen processing and suggests that *ADAMTS3* is more widely expressed and has a broader developmental role in mice than *ADAMTS14*. In particular, it is likely to be the dominant, if not only, procollagen N-propeptidase in cartilage. Inherited disease-causing mutations in *ADAMTS3* and *ADAMTS14* are not known. We predict that *ADAMTS3* mutations will cause a severe, perhaps lethal chondrodysplasia, whereas *ADAMTS14* mutations are likely to cause minor skin fragility with other syndromic features reflecting its tissue-specific expression. Such functions can be disclosed by analysis of transgenic mice with targeted inactivation of *Adamts3* and *Adamts14* alleles. In addition, population genetic analysis of SNPs in ADAMTS2, ADAMTS3 and ADAMTS14 from individuals with mild skin fragility, or altered rheological properties of skin, may provide interesting insight into the effects of genetic variation in PNPs in humans.

## ACKNOWLEDGEMENTS

The authors are supported by a grant from the National Institutes of Health (AR49590). We thank Dr. Ray Boot-Handford for information regarding the evolution of fibrillar collagens.

## REFERENCES

Apte, S.S. (2004). A disintegrin-like and metalloprotease (reprolysin type) with thrombospondin type 1 motifs: the ADAMTS family. *Int J Biochem Cell Biol* **36,** 981-985.

Bolz, H., Ramirez, A., von Brederlow, B. and Kubisch, C. (2001). Characterization of ADAMTS14, a novel member of the ADAMTS metalloproteinase family. *Biochim Biophys Acta* **1522**, 221-225.

Boot-Handford, R.P., Tuckwell, D.S., Plumb, D.A., Rock, C.F. and Poulsom, R. (2003). A novel and highly conserved collagen (pro(alpha)1(XXVII)) with a unique expression pattern and unusual molecular characteristics establishes a new clade within the vertebrate fibrillar collagen family. *J Biol Chem* **278**, 31067-31077.

Cal, S., Obaya, A.J., Llamazares, M., Garabaya, C., Quesada, V. and Lopez-Otin, C. (2002). Cloning, expression analysis, and structural characterization of seven novel human ADAMTSs, a family of metalloproteinases with disintegrin and thrombospondin-1 domains. *Gene* **283**, 49-62.

Christoffels, A., Koh, E.G., Chia, J.M., Brenner, S., Aparicio, S. and Venkatesh, B. (2004). Fugu genome analysis provides evidence for a whole-genome duplication early during the evolution of ray-finned fishes. *Mol Biol Evol* **21**, 1146-1151.

Colige, A., Li, S.W., Sieron, A.L., Nusgens, B.V., Prockop, D.J. and Lapiere, C.M. (1997). cDNA cloning and expression of bovine procollagen I N-proteinase: a new member of the superfamily of zinc-metalloproteinases with binding sites for cells and other matrix components. *Proc Natl Acad Sci U S A* **94**, 2374-2379.

Colige, A., Nuytinck, L., Hausser, I., Van Essen, A.J., Thiry, M., Herens, C., Ades, L.C., Malfait, F., Paepe, A.D., Franck, P., Wolff, G., Oosterwijk, J.C., Smitt, J.H., Lapiere, C.M. and Nusgens, B.V. (2004). Novel Types of Mutation Responsible for the Dermatosparactic Type of Ehlers-Danlos Syndrome (Type VIIC) and Common Polymorphisms in the ADAMTS2 Gene. *J Invest Dermatol* **123**, 656-663.

Colige, A., Sieron, A.L., Li, S.W., Schwarze, U., Petty, E., Wertelecki, W., Wilcox, W., Krakow, D., Cohn, D.H., Reardon, W., Byers, P.H., Lapiere, C.M., Prockop, D.J. and Nusgens, B.V. (1999). Human Ehlers-Danlos syndrome type VII C and bovine dermatosparaxis are caused by mutations in the procollagen I N-proteinase gene. *Am J Hum Genet* **65**, 308-317.

Colige, A., Vandenberghe, I., Thiry, M., Lambert, C.A., Van Beeumen, J., Li, S.W., Prockop, D.J., Lapiere, C.M. and Nusgens, B.V. (2002). Cloning and characterization of ADAMTS-14, a novel ADAMTS displaying high homology with ADAMTS-2 and ADAMTS-3. J Biol Chem **277**, 5756-5766.

Fernandes, R.J., Hirohata, S., Engle, J.M., Colige, A., Cohn, D.H., Eyre, D.R. and Apte, S.S. (2001). Procollagen II amino propeptide processing by ADAMTS-3. Insights on dermatosparaxis. *J Biol Chem* **276**, 31502-31509.

Fernandes, R.J., Schmid, T.M., Harkey, M.A. and Eyre, D.R. (1997). Incomplete processing of type II procollagen by a rat chondrosarcoma cell line. *Eur J Biochem* **247**, 620-624.

Hanset, R. and Ansay, M. (1967). Dermatospaxie (peau dechiree) chez le veau: un defaut general du tissu conjonctif. *Ann Med vet* **7**, 451-470.

Hanset, R. and Lapiere, C.M. (1974). Inheritance of dermatosparaxis in the calf. A genetic defect of connective tissues. *J Hered* **65**, 356-358.

Holbrook, K.A., Byers, P.H., Counts, D.F. and Hegreberg, G.A. (1980). Dermatosparaxis in a Himalayan cat: II. Ultrastructural studies of dermal collagen. *J Invest Dermatol* **74**, 100-104.

Hurskainen, T.L., Hirohata, S., Seldin, M.F. and Apte, S.S. (1999). ADAM-TS5, ADAM-TS6, and ADAM-TS7, novel members of a new family of zinc metalloproteases. General features and genomic distribution of the ADAM-TS family. *J Biol Chem* **274**, 25555-25563.

Kashiwagi, M., Enghild, J.J., Gendron, C., Hughes, C., Caterson, B., Itoh, Y. and Nagase, H. (2004). Altered proteolytic activities of ADAMTS-4 expressed by C-terminal processing. *J Biol Chem* **279**, 10109-10119.

Kessler, E., Takahara, K., Biniaminov, L., Brusel, M. and Greenspan, D.S. (1996). Bone morphogenetic protein-1: the type I procollagen C-proteinase. *Science* **271**, 360-362.

Lapiere, C.M., Lenaers, A. and Kohn, L.D. (1971). Procollagen peptidase: an enzyme excising the coordination peptides of procollagen. *Proc Natl Acad Sci U S A* **68**, 3054-3058.

Lapiere, C.M. and Nusgens, B.V. (1993). Ehlers-Danlos type VII-C, or human dermatosparaxis. The offspring of a union between basic and clinical research [editorial; comment]. *Arch Dermatol* **129**, 1316-1319.

Lenaers, A., Ansay, M., Nusgens, B.V. and Lapiere, C.M. (1971). Collagen made of extended -chains, procollagen, in genetically-defective dermatosparaxic calves. *Eur J Biochem* **23**, 533-543.

Li, S.W., Arita, M., Fertala, A., Bao, Y., Kopen, G.C., Langsjo, T.K., Hyttinen, M.M., Helminen, H.J. and Prockop, D.J. (2001). Transgenic mice with inactive alleles for procollagen N-proteinase (ADAMTS-2) develop fragile skin and male sterility. *Biochem J* **355**, 271-278.

Nagase, T., Ishikawa, K., Nakajima, D., Ohira, M., Seki, N., Miyajima, N., Tanaka, A., Kotani, H., Nomura, N. and Ohara, O. (1997). Prediction of the coding sequences of unidentified human genes. VII. The complete sequences of 100 new cDNA clones from brain which can code for large proteins in vitro. *DNA Res* **4**, 141-150.

Nusgens, B.V., Verellen-Dumoulin, C., Hermanns-Le, T., De Paepe, A., Nuytinck, L., Pierard, G.E. and Lapiere, C.M. (1992). Evidence for a relationship between Ehlers-Danlos type VII C in humans and bovine dermatosparaxis. *Nat Genet* **1**, 214-217.

Petty, E.M., Seashore, M.R., Braverman, I.M., Spiesel, S.Z., Smith, L.T. and Milstone, L.M. (1993). Dermatosparaxis in children. A case report and review of the newly recognized phenotype [see comments]. *Arch Dermatol* **129**, 1310-1315.

Rawlings, N.D. and Barrett, A.J. (1995). Evolutionary families of metallopeptidases. *Methods Enzymol* **248**, 183-228.

Smith, L.T., Wertelecki, W., Milstone, L.M., Petty, E.M., Seashore, M.R., Braverman, I.M., Jenkins, T.G. and Byers, P.H. (1992). Human dermatosparaxis: a form of Ehlers-Danlos syndrome that results from failure to remove the amino-terminal propeptide of type I procollagen. *Am J Hum Genet* **51**, 235-244.

Somerville, R.P., Longpre, J.M., Jungers, K.A., Engle, J.M., Ross, M., Evanko, S., Wight, T.N., Leduc, R. and Apte, S.S. (2003). Characterization of ADAMTS-9 and ADAMTS-20 as a distinct ADAMTS subfamily related to Caenorhabditis elegans GON-1. *J Biol Chem* **278**, 9503-9513.

Tuderman, L., Kivirikko, K.I. and Prockop, D.J. (1978). Partial purification and characterization of a neutral protease which cleaves the N-terminal propeptides from procollagen. *Biochemistry* **17**, 2948-2954.

Van Wart, H.E. and Birkedal-Hansen, H. (1990). The cysteine switch: a principle of regulation of metalloproteinase activity with potential applicability to the entire matrix metalloproteinase gene family. *Proc Natl Acad Sci U S A* **87**, 5578-5582.

Wang, W.M., Lee, S., Steiglitz, B.M., Scott, I.C., Lebares, C.C., Allen, M.L., Brenner, M.C., Takahara, K. and Greenspan, D.S. (2003). Transforming growth factor-beta induces secretion of activated ADAMTS-2. A procollagen III N-proteinase. *J Biol Chem* **278**, 19549-19557.

Watson, R.B., Holmes, D.F., Graham, H.K., Nusgens, B.V. and Kadler, K.E. (1998). Surface located procollagen N-propeptides on dermatosparactic collagen fibrils are not cleaved by procollagen N-proteinase and do not inhibit binding of decorin to the fibril surface. *J Mol Biol* **278**, 195-204.

Wertelecki, W., Smith, L.T. and Byers, P. (1992). Initial observations of human dermatosparaxis: Ehlers-Danlos syndrome type VIIC. *J Pediatr* **121**, 558-564.

Zheng, X., Chung, D., Takayama, T.K., Majerus, E.M., Sadler, J.E. and Fujikawa, K. (2001) Structure of von Willebrand factor cleaving protease (ADAMTS13), a metalloprotease involved in thrombotic thrombocytopenic purpura. *J Biol Chem* **276**, 41059-41063

# Chapter 14

# ADAMTS-4 AND ADAMTS-5

## *Aggrecanases*

Anne-Marie Malfait, Micky Tortorella and Elizabeth Arner
*Joint Biology, Pfizer Global Research and Development, St Louis Research Laboratories, 700 Chesterfield Parkway West, St Louis, MO, USA*

Abstract: Osteoarthritis (OA) is characterized by articular cartilage erosion as a consequence of proteolytic cleavage of its two major functional macromolecules, type II collagen and aggrecan. Aggrecan degradation in OA and rheumatoid arthritis is attributed to cleavage at the $Glu^{373}$-$Ala^{374}$ bond by the aggrecanases. Two aggrecanases, purified from IL-1-stimulated cartilage explants, were identified as members of the a disintegrin and metalloproteinase with thrombospondin motifs (ADAMTS) family, ADAMTS-4 and ADAMTS-5, and work from a number of groups has begun to provide insight into the molecular basis for the role of these proteases in aggrecan catabolism. The expression of the aggrecanases can be up-regulated by a number of factors including cytokines, retinoic acid, and fragments of the extracellular matrix molecule, fibronectin. To date two endogenous inhibitors of aggrecanase activity have been identified, TIMP-3 and α2-macroglobulin. However, recent studies suggest that activity may also be controlled by the ability of aggrecanases to access the core protein of the heavily glycosylated aggrecan substrate. In addition, post-translational processing is another means of controlling activity of these proteases. Removal of the propeptide domain is required for activity as well as potentially C-terminal truncation. Knowledge continues to accumulate on the expression pattern of these proteases in different tissues and their potential role in normal physiological mechanisms and in disease.

Key words: aggrecanase, ADAMTS-4, ADAMTS-5, cartilage, osteoarthritis.

# 1. INTRODUCTION

Osteoarthritis (OA), a degenerative joint disease affecting more than 234 million people worldwide, is the most common form of arthritis. In OA, an



*N. M. Hooper and U. Lendeckel (eds.), The ADAM Family of Proteases,* 299-322.

imbalance in cartilage matrix synthesis and breakdown leads to the destruction and eventual loss of articular cartilage, which results in restricted joint movement, joint instability, pain and chronic disability. Current therapies alleviate the mild to moderate pain and inflammation associated with OA. However, they do not protect the articular cartilage and have not demonstrated utility in modifying disease progression. Therefore, there is a great interest in identifying and characterizing the proteases responsible for cartilage degradation as potential targets in developing therapeutics to prevent joint destruction in arthritis.

The major components of the cartilage extracellular matrix are type II collagen and aggrecan. Because aggrecan provides cartilage with its properties of compressibility and elasticity, by swelling and hydrating the collagen framework, the loss of aggrecan from the matrix is a key event in cartilage destruction. One of the earliest changes observed in osteoarthritis is indeed the loss of aggrecan from the cartilage matrix (Mankin and Lipiello, 1970). Aggrecan exists within cartilage as huge aggregates with hyaluronan. Each aggrecan monomer is approximately 2.5 million Daltons and consists of a 250-kD protein core with chondroitin sulfate and keratan sulfate glycosaminoglycan (GAG) side chains attached to it. The amino-terminal region of the aggrecan core protein contains two globular domains, G1 and G2, which are separated by an interglobular domain (IGD) that spans about 150 amino residues. The G2 region is followed by a long GAG attachment region and by the carboxyl terminal globular domain G3.

Aggrecan depletion in arthritic cartilage has been ascribed to increased proteolytic cleavage of the core protein. Two major cleavage sites have been identified in the IGD, one at $Asn^{341}$-$Phe^{342}$ and one at the $Glu^{373}$-$Ala^{374}$ bond. Matrix metalloproteinases (MMP) act primarily on the former bond and all MMPs present in cartilage including MMP-1, MMP-2, MMP-3, MMP-7, MMP-8, MMP-9, MMP-13, and MT1-MMP, are capable of cleaving aggrecan at this site *in vitro* (Flannery *et al.* 1992; Fosang *et al.* 1992; Fosang *et al.* 1993; Fosang *et al.*. 1996; Buttner *et al.* 1998). Cleavage at the second site, $Glu^{373}$-$Ala^{374}$, has been attributed to an enzymatic activity referred to as aggrecanase (Sandy *et al.* 1991; Ilic *et al.* 1992; Loulakis *et al.* 1992; Lark *et al.* 1995).

## 2. IDENTIFICATION OF AGGRECANASES

Since the discovery of aggrecanase activity, two cartilage aggrecanases, aggrecanase-1 (Tortorella *et al.* 1999) and aggrecanase-2 (Abbaszade *et al.* 1999) were purified from IL-1-stimulated bovine nasal cartilage and were identified by following their activity with an assay using the neoepitope antibody, BC3, which detects the new N-terminus, ARGS, formed by specific cleavage at the Glu$^{373}$-Ala$^{374}$ bond (Hughes *et al.* 1995). Aggrecanase-1 was isolated as a 62-kDa protein doublet with the N-terminus $^{213}$FASL, whereas aggrecanase-2 was detected as two species of 64 and 50 kDa with the N-terminus $^{262}$SISR (Abbaszade *et al.* 1999; Tortorella *et al.* 1999).

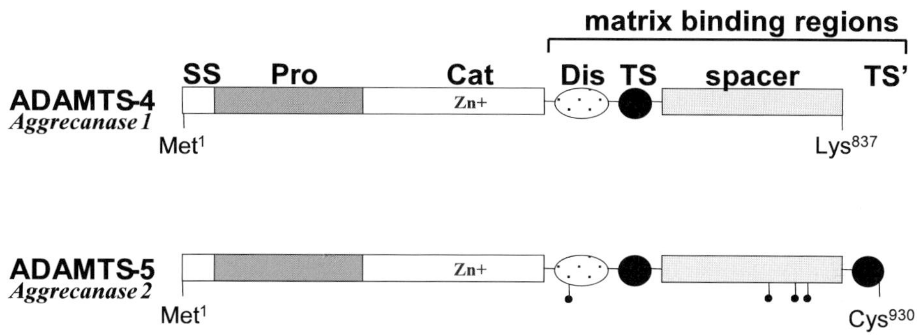


*Figure 1.* Domain structure of ADAMTS-4 and ADAMTS-5. ADAMTS-4 and ADAMTS-5 are comprised of a signal sequence (SS), pro-domain (Pro), catalytic domain (Cat), disintegrin-like domain (Dis), spacer region, and thrombospondin motifs (TS) and submotifs (TS'). The small filled circles show the relative positions of glycosylation sites.

Both enzymes are members of the a disintegrin and metalloprotease (ADAM) family of zinc metalloproteases that consist of an amino-terminal signal sequence, a propeptide domain, a catalytic domain with a zinc binding motif, HEXXHXXGXXH, similar to the matrix metalloproteinases and ADAMs, and a disintegrin-like domain, resembling the structure of the reprolysin family of metalloproteases (reviewed in Kaushal and Shah, 2000). However, while most ADAM proteins are membrane-anchored and have a transmembrane and a cytoplasmic domain in the carboxyl terminal region, the aggrecanases lack these domains. Instead, the C-terminal portion of aggrecanase-1 contains one thrombospondin (TSP) type-1 motif, the sequence of which is the conserved motif in thrombospondin 1 and 2, while aggrecanase-2 contains a similar TSP type-1 motif, and in addition a TSP-like motif (Figure 1). ADAM proteins containing TSP motifs are referred to as ADAMTS proteins, and aggrecanase-1 and aggrecanase-2 have been designated ADAMTS-4 and ADAMTS-5, respectively.

# 3. ENZYMATIC PROFILE OF ADAMTS-4 AND ADAMTS-5

## 3.1 Enzyme kinetics

Cleavage of aggrecan at $Glu^{373}$-$Ala^{374}$ by ADAMTS-4 and –5 is dependent on pH, NaCl and $CaCl_2$ concentration (Arner *et al.* 1999b; Tortorella *et al.* unpublished data). Optimal activity for both enzymes is achieved at approximately 100 mM NaCl, but at 250 mM NaCl, activity decreases by ~50% and is completely lost at 500 mM NaCl or higher. The pH optimum for aggrecanase is 7.5. However, a broad pH range, between 6.5 and 9.5, supports greater than 75% of the activity seen at the pH optimum, whereas at pH below 6.0 complete loss of activity is observed. Full catalytic activity requires 1-5 mM $CaCl_2$, although in the absence of $CaCl_2$ approximately 10% of the catalytic activity can still be detected, suggesting that Ca is a cofactor required for full catalytic activity, as has been described for many metalloproteinases. Sequence and structural alignment with Acutolysin A and Adamalysin II reveal that both ADAMTS-4 and ADAMTS-5 contain at least one putative Ca-binding motif in the catalytic domain, $Glu^{9}$, $Asp^{99}$, $Cys^{211}$, $Asp^{214}$ and $Glu^{9}$, $Asp^{99}$, $Cys^{210}$, $Asp^{213}$, respectively (Stallings, personal communication). Both enzymes are inhibited by the metalloprotease inhibitor, EDTA, and by the peptidic hydroxamate, BB16. A panel of inhibitors of serine, cysteine, and aspartic proteinases such as aprotinin, bestatin, chymostatin, E64, leupeptin, pefabloc, and pepstatin has little effect on aggrecanase activity. The $K_m$ of ADAMTS-4 and ADAMTS-5 for cleaving aggrecan at $Glu^{373}$-$Ala^{374}$ is 11 and 50 nM, respectively (unpublished observations).

## 3.2 Additional Aggrecan Cleavage Sites

Although aggrecanase activity is characterized by its specific cleavage of cartilage aggrecan at $TEGE^{373}$ – $^{374}ARGS$, human recombinant ADAMTS-4 and ADAMTS-5 cleave aggrecan more readily at four additional sites located in the chondroitin sulfate-rich region between the G2 and G3 globular domains (Tortorella *et al.* 2000b; Tortorella *et al.* 2002). These sites have a glutamic acid in the P1 position and a non-polar residue (Ala, Leu, Gly) in the P1' position, which include $Glu^{1480}$-$Gly^{1481}$, $Glu^{1666}$-$Gly^{1667}$, $Glu^{1771}$-$Ala^{1772}$ and $Glu^{1871}$-$Leu^{1872}$ bonds (bovine sequence) and they are highly conserved in aggrecan from various species (Figure 2). The $K_m$ of ADAMTS-4 and ADAMTS-5 for cleaving aggrecan at these four C-terminal sites was calculated to be at least 20-fold lower than at $Glu^{373}$-$Ala^{374}$

(Tortorella *et al.* 2000b; Tortorella *et al.* 2002). These kinetics suggest that the $Glu^{373}$-$Ala^{374}$ scissile bond is cryptic and C-terminal truncation is required before cleavage at this site can occur. This is physiologically relevant, because cleavage at $Glu^{373}$-$Ala^{374}$ is more detrimental than cleavage at the other sites since it results in loss of the bulk of the GAG-bearing portion of the aggrecan molecule, compromising cartilage function. The catalytic domain of ADAMTS-4 is only 49% identical in amino acid sequence with ADAMTS-5, suggesting that these proteins may cleave cartilage aggrecan at different rates. In fact, recombinant human ADAMTS-4 cleaves aggrecan at the $Glu^{373}$-$Ala^{374}$ and $Glu^{1667}$-$Gly^{1668}$ bonds at a two-fold greater rate than ADAMTS-5, suggesting that the expression of ADAMTS-4 may be more detrimental to the health of cartilage. However, determining the levels of both active ADAMTS-4 and ADAMTS-5 in arthritic versus normal cartilage will be important for elucidating the relative importance of these enzymes in accelerated turnover of aggrecan in arthritis.

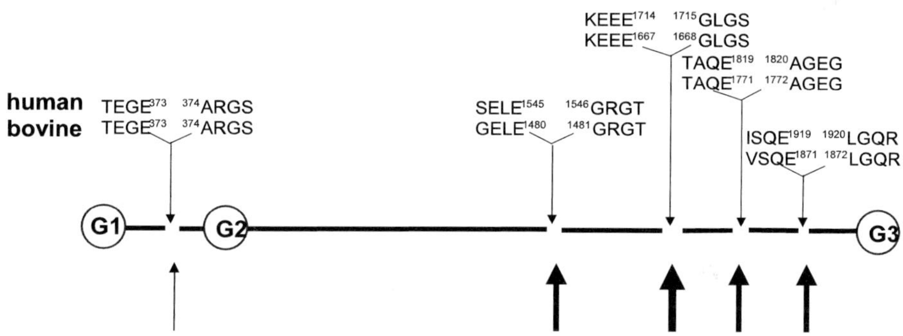


*Figure 2.* Aggrecanase-sensitive cleavage sites in the aggrecan core protein. Sites of human (top line) and bovine (bottom line) aggrecan cleavage by ADAMTS-4 by and ADAMTS-5. Thicker arrows indicate a greater rate of cleavage.

## 3.3 Enzyme:substrate binding

The TSP-1 motif located within the C-terminus of ADAMTS-4 is important for aggrecan binding and cleavage. ADAMTS-4 binds its substrate through interaction of its TSP-1 motif with the keratan sulfate and chondroitin sulfate side chains of aggrecan with a $K_d$ of ~10.6 nM (Tortorella *et al.* 2000a). Removal of the GAG from aggrecan results in a significant decrease of ADAMTS-4 binding and cleavage (Pratta *et al.* 2000). Peptides corresponding to regions of the TSP-1 motif block both binding of ADAMTS-4 and subsequent cleavage of native aggrecan. For example, the peptide $^{521}$GGWGPWGPWGD$^{531}$, corresponding to the sequence implicated in the interaction of thrombospondin with sulfated GAG chains, was effective in blocking cleavage of aggrecan with an $IC_{50}$ of 17 μM. Extension of the peptide to include the sequence $^{532}$CSRTCG$^{537}$ representing a putative CD36-binding region in thrombospondin, resulted in a peptide that was more potent in inhibiting aggrecan cleavage ($IC_{50}$ = 3 μM) (Tortorella *et al.* 2000a).

In addition to thrombospondin peptides, it has been shown that free chondroitin sulfate and fragments of heparin sulfate block ADAMTS-4 cleavage of aggrecan (Sugimoto *et al.* 1999; Mousa *et al.* 2003; Tortorella *et al.* unpublished observations), most likely by binding to the TSP-1 motif of ADAMTS-4 and interfering with substrate binding. Additional GAG binding motifs have been described outside the TSP-1 motif, in the spacer domain, including $^{668}$SKKKFDKC$^{675}$, $^{692}$FRKFRY$^{697}$, and $^{829}$RRRPWAGRK$^{837}$, and these are also important in mediating aggrecan binding (Flannery *et al.* 2002).

Finally, a recombinant truncated form of ADAMTS-4 lacking the C-terminal domains$^{213\text{-}431}$ does not bind or cleave aggrecan, consistent with the fact that the C-terminal region of the enzyme is important for substrate recognition and cleavage (Tortorella *et al.* 2000a).

## 3.4 Substrate specificity

Full length ADAMTS-4 and ADAMTS-5 are ineffective at cleaving a broad range of matrix proteins including fibronectin, thrombospondin, type I and type II collagen, the zymogen of stromelysin-1, and the more general protease substrates, casein and gelatin, indicating a high degree of substrate specificity. However, ADAMTS-4 cleaves other aggregating proteoglycans such as versican V1 (Sandy *et al.* 2001b), versican V2 (Westling *et al.* 2004), and brevican (Matthews *et al.* 2000) at Glu$^{441}$, Glu$^{405}$and Glu$^{393}$, respectively (Table 1). More recently, it has been demonstrated that ADAMTS-4 and ADAMTS-5 are not exclusively glutamyl endopeptidases and not limited to cleavage of proteoglycans. ADAMTS-4 has been reported to cleave COMP (Dickinson *et al.* 2003), TIMP-4 (Zeng *et al.* 2004), and matrilin-2 (Wang *et al.* 2004b). In addition, both ADAMTS-4 and ADAMTS-5 cleave α2-macroglobulin (α2M) within the bait region following Met$^{690}$ (Tortorella *et al.* 2004). Unlike cleavage of aggrecan, cleavage of the α2M molecule is not dependent on the TSP-1 motif, as a truncated form of ADAMTS-4 lacking the C-terminal domains$^{213\text{-}431}$ is equally active to the full-length active form.

It has also been shown that ADAMTS-4 undergoes autocatalytic C-terminal truncation through cleavage at Lys$^{694}$ and Thr$^{581}$ (Flannery *et al.* 2002), and ADAMTS-5 through cleavage at Glu$^{753}$ (Georgiadis *et al.* 2002).

Finally, it was recently reported that recombinant ADAMTS-4 lacking the spacer domain[688-837] is more promiscuous and cleaves transferrin, decorin and fibromodulin, (Kashiwagi *et al.* 2004), suggesting that, in addition to mediating aggrecan catabolism, ADAMTS-4 may cleave other matrix molecules that are critical for cartilage integrity.

*Table 1.* Alignment of the known aggrecanase cleavage sites for human aggrecan (Agg), brevican (Brev), versican (Ver), α2-macroglobulin (α2M), and autocatalytic sites of cleavage.

| Substrate | Aggrecanase Cleavage Site (Indicated by Space Following Bolded Letter) |
|---|---|
| α2M (M690) | F Y E S D V **M** G R G H A R |
| Agg (E373) | R N I T E G **E** A R G S V I |
| Agg (E1545) | S T A S E L **E** G R G T I G |
| Agg (E1714) | T T F K E E **E** G L G S V E |
| Agg (E1819) | Q A P T A Q **E** A G E G P S |
| Agg (E1919) | E P T I S Q **E** L G Q R P P |
| Agg (N341) | F V D I P E **N** F F G V G G |
| Brev (E393) | Q E A V E S **E** S R G A I S |
| Ver (E441) | K D P E A A **E** A R R G Q Y |
| Ver2 (E405) | G N I V S F **E** N Q K A T V |
| **Autocatalytic sites** | |
| ADAMTS5 (E753) | D V V R I P **E** G A T H I K |
| ADAMTS4 (T581) | P H G S A L **T** F R E E Q C |
| ADAMTS4 (K694) | G S G S F R **K** F R Y G Y N |

# 4. REGULATION OF ADAMTS-4/ADAMTS-5 ACTIVITY

ADAMTS-4/ADAMTS-5 activity is regulated at multiple levels including alteration of transcription, translation, endogenous inhibitors, substrate accessibility and post-translational processing.

## 4.1 Transcription and translation

Expression of ADAMTS-4 mRNA and protein in bovine cartilage can be induced by interleukin-1 (IL-1) (Tortorella *et al.* 2001), tumor necrosis factor (TNF) (Tortorella *et al.* 2001), retinoic acid (Hughes *et al.* 2003), and fibronectin fragments (Homandberg *et al.* 1992; Stanton *et al.* 2002). Stimulation of bovine cartilage explants with IL-6 and/or sIL-6R potentiated aggrecan catabolism and release above that seen in the presence of IL-1α or TNF alone. This catabolism was associated with aggrecanase (but not MMP) activity, with correlative mRNA expression for ADAMTS-4 (Flannery *et al.* 2000). In addition, oncostatin M in combination with IL-1 (Koshy *et al.*

2002), transforming growth factor β (Moulharat *et al.* 2004), and interleukin-17 (Sylvester *et al.* 2004) induce ADAMTS-4 mRNA (corresponding protein levels were not evaluated). Interestingly, IL-1 and TNF do not trigger aggrecanase-mediated aggrecan catabolism in human cartilage explants (Sztrolovics *et al.* 2002; own unpublished observations), suggesting that human chondrocytes respond differently from bovine chondrocytes to these cytokines. These recent findings corroborate older reports in the literature that normal human cartilage is relatively unresponsive to the degradative effects of IL-1 (Ismaiel *et al.* 1991; Hollander *et al.* 1993). However, recent publications indicate that human osteoarthritic cartilage may be more responsive to IL-1 (Curtis *et al.* 2002; Hardy *et al.* 2002).

In addition, other catabolic factors may be involved in inducing aggrecanase-mediated aggrecan breakdown in human disease. One of the most likely sources for such catabolic factors is within the cartilage extracellular matrix (ECM) itself. In particular, fragments of fibronectin (Fn) have been implicated in cartilage breakdown (Homandberg, 1999). Fn is a glycoprotein present at low levels in the ECM of normal articular cartilage. In OA cartilage, Fn levels increase significantly and, moreover, it is mostly fragmented (Clemmensen *et al.* 1982; Chevalier *et al.* 1992; Xie *et al.* 1992; Chevalier *et al.* 1996). The 45-kDa collagen-binding fragment of fibronectin induces MMP-13 synthesis by porcine chondrocytes and aggrecan degradation by aggrecanases (Stanton *et al.* 2002). In a separate study, a specific Fn fragment containing the COOH-terminal heparin-binding domain induced MMP-13-mediated breakdown of type II collagen involving an IL-1 dependent pathway but aggrecan catabolism induced by the same fragment could not be blocked with IL-1ra (Yasuda *et al.* 2002).

## 4.2 Endogenous inhibitors

To date, two endogenous inhibitors of ADAMTS-4 and –5 have been identified, tissue inhibitor of matrix metalloproteinases-3 (TIMP-3) found in the cartilage extracellular matrix and α2-macroglobulin (α2M) found in the synovial fluid (SF). TIMP-3 is a potent inhibitor with K*i*(app) values of 3.30 nM for ADAMTS-4 and 0.66 nM for ADAMTS-5 (Kashiwagi *et al.* 2001; Hashimoto *et al.* 2001). The binding affinity of TIMP-3 for ADAMTS-4 and ADAMTS-5 is greater than for MMP-1, MMP-2 and MMP-3, suggesting that a primary physiological function is inhibition of the aggrecanases. Moreover, TIMP-3 has been detected in cartilage (Su *et al.* 1999) and shown to have a high binding affinity for chondroitin sulfate polysaccharides (Yu *et al.* 2000), unlike the other 3 members of the family (TIMP-1, -2, -4), which do not inhibit either ADAMTS-4 or ADAMTS-5 at concentrations capable of inhibiting MMPs. This suggests that TIMP-3 co-localizes with the aggrecanases in the cartilage extracellular matrix. TIMP-3 mRNA is elevated in osteoarthritic cartilage (Kevorkian *et al.* 2004), perhaps as a mechanism to counter the increased aggrecanase-mediated aggrecan catabolism.

α2M is another endogenous inhibitor of ADAMTS-4 and –5 and inhibits the activity of both proteinases in a concentration-dependent manner, demonstrating 1:1 stoichiometry with second-order rate constants on the order of $10^6$ and $10^5$ $M^{-1}s^{-1}$, respectively (Tortorella *et al.* 2004). Inhibition of the aggrecanases is mediated by proteolysis of the "bait" region. Both ADAMTS-4 and ADAMTS-5 cleave α2M at $Met^{690}$-$Gly^{691}$ within α2M, resulting in a conformational change of the α2M molecule and subsequent physical entrapment of these proteinases. α2M levels are elevated in the synovial fluid of OA patients (Hadler *et al.* 1981). It is interesting to speculate that, upon cleavage of aggrecan, ADAMTS-4/-5, bound to aggrecan, diffuse into the synovial fluid where they are trapped by α2M. Because α2M is a large molecule (720 kDa), it cannot diffuse into the cartilage matrix and the α2M:aggrecanase complex will thus be cleared from the joint very rapidly, through binding to the lipoprotein receptor-related protein on macrophages and other cells (Aschom *et al.* 1990; Andersen *et al.* 2000).

## 4.3 Substrate accessibility

In order for ADAMTS-4 to cleave aggrecan, access to the substrate is required. Cartilage aggrecan is surrounded by and interacts with many ECM proteins that may interfere with this accessibility and could therefore act as ADAMTS-4/-5 antagonists. For example, relatively high levels of thrombospondin-1 (TSP) are present in the matrix (approximately 0.1-1 μM). TSP has three TSP-1 motifs and binds to the GAG chains of aggrecan by the same mechanism employed by ADAMTS-4/-5. Therefore, TSP bound to aggrecan will prevent cleavage by the aggrecanases through steric hindrance. In fact, intact thrombospondin has been shown to block cleavage of aggrecan by recombinant ADAMTS-4 and ADAMTS-5 *in vitro* in a concentration-dependent fashion with an $IC_{50}$ of approximately 400 nM (Tortorella, unpublished observations).

The spacer domain of ADAMTS-4 ($Arg^{693}$- $Lys^{837}$) binds to the C-terminal domain of fibronectin, which may account for pericellular localization of ADAMTS-4 in cartilage (Hashimoto *et al.* 2004). When bound, ADAMTS-4 is unable to cleave aggrecan due to steric hindrance, consistent with the finding that ADAMTS-4 is inhibited by fibronectin, as well as by a C-terminal 40kDa fibronectin fragment in a concentration-dependent manner, with an $IC_{50}$ of 110 and 170 nM, respectively (Hashimoto *et al.* 2004). Therefore, removal of the spacer domain of ADAMTS-4 or breakdown of fibronectin may be required before ADAMTS-4 can cleave aggrecan.

## 4.4 Post-translational processing

A final means of controlling aggrecanase activity is through post-translational processing. Like all mammalian matrix metalloproteinases, ADAMTS-4 and ADAMTS-5 are synthesized as latent zymogens, with a bulky prodomain and a putative cysteine-switch. When ADAMTS-4 and ADAMTS-5 were first purified from IL-1-stimulated bovine nasal cartilage, one active species containing the N-terminus $^{213}$FASL of ADAMTS-4 and two active species of ADAMTS-5 both with the N-terminus $^{262}$SISR (Tortorella, 1999) were detected. Also, when recombinant full-length ADAMTS-4 or -5 are expressed in insect cells or mammalian cells, the recombinant protein contains the N-terminus $^{213}$FASL and $^{262}$SISR, respectively, suggesting activation/cleavage between the pro- and catalytic domains at RAKR$^{212}$-$^{213}$FASL and RRRR$^{261}$-$^{262}$SISR, respectively. The P1, P2, P3, and P4 amino acids encode a recognition motif, RX(K/R)R, for furin and other members of the proprotein convertase (PC) family, which have been shown to activate many precursor proteins (reviewed in Taylor *et al.* 2003). Furin efficiently removes the prodomain of latent ADAMTS-4 and ADAMTS-5 through cleavage at Arg$^{212}$-Phe$^{213}$ and Arg$^{261}$-Ser$^{262}$, respectively, resulting in an active enzyme capable of cleaving native aggrecan (Wang *et al.* 2004a and unpublished observations). In the case of ADAMTS-4, removal of the prodomain can be mediated at multiple furin recognition sites, including RPRR$^{209}$, RAKR$^{212}$, or KR$^{212}$ (Wang *et al.* 2004a). ADAMTS-4 is co-localized with furin in the *trans*-Golgi network in HEK293, SW1353 and MDCK cells. The processing of proADAMTS-4 is completely blocked by brefeldin A treatment, suggesting that processing occurs in the *trans*-Golgi network. Finally, proADAMTS4 but not mature ADAMTS-4 co-precipitates with furin, suggesting that furin physically interacts with the prodomain of ADAMTS-4 (Wang *et al.* 2004a). Whether furin is the proprotein convertase responsible for activation of pro-aggrecanase in cartilage has yet to be determined.

After furin-mediated prodomain removal, which converts full-length enzyme (aa 52-837), to the ~68-kDa form (aa 213-837), this intermediate has been shown to associate with GPI-anchored MT4-MMP (MMP-17) in ADAMTS-4-transfected human chondrosarcoma cells (Gao *et al.* 2004). In this system, MT4-MMP then truncates the C-terminus of ADAMTS-4 through cleavage at Lys$^{694}$-Phe$^{695}$, resulting in a ~53-kDa species (aa 213-

694) that has a high binding affinity for cell surface syndecan-1. MT4-MMP can further truncate the ~53-kDa species (aa 213-694) through proteolytic cleavage at $Thr^{581}$-$Phe^{582}$ to a 40 kDa species (aa 213-581) which is released from syndecan-1. ADAMTS-4 processing is illustrated in Figure 3. These observations support the notion that the 40-kDa species is mainly responsible for aggrecanolysis in the cartilage extracellular matrix distant from the cells. This may be physiologically important because recombinant ADAMTS-4 lacking the spacer domain (aa 213-687) was found to cleave the $Glu^{373}$-$Ala^{374}$ bond in the aggrecan interglobular domain more efficiently than the recombinant ~68 kDa form (aa 213-837) with the C-terminus intact. However, the recombinant ~68 kDa form had greater overall activity against native aggrecan via cleavage at the C-terminal sites between the G2 and G3 globular domains at $Glu^{1480}$-$Gly^{1481}$, $Glu^{1667}$-$Gly^{1668}$, $Glu^{1771}$-$Ala^{1772}$ and $Glu^{1871}$-$Leu^{1872}$ (Kashiwagi *et al.* 2004). In contrast, a different research group recently demonstrated that recombinant 68-kDa ADAMTS-4 containing the spacer domain has equal activity compared to the 40-kDa species in cleaving aggrecan at $Glu^{373}$-$Ala^{374}$ (Hashimoto *et al.* 2004). Thus, further studies are required to determine the importance of spacer domain removal for catalytic activity. The primary role of removing the spacer may not be to increase catalytic activity, but rather to release/free the enzyme from the cell surface, where it was bound to syndecan-1 or fibronectin, into the cartilage ECM resulting in aggrecan catabolism.

C-terminal truncation may occur by other mechanisms, including autocatalysis and cleavage by proprotein convertases. Both ADAMTS-4 and ADAMTS-5 have a putative proprotein convertase consensus motif at the C-terminus, $RRTR^{565}$ in ADAMTS-4 and $RAIYR^{614}$ in ADAMTS-5, but no data exist to show that these cleavages occur *in vivo* or *in vitro*. Interestingly, ADAMTS-5 protein is constitutively expressed in healthy synovium, and detected as a 70-kDa species. The protein is not upregulated in synovium of rheumatoid arthritis (RA) patients but is detected as a 53-kDa species, suggesting that ADAMTS-5 protein may be predominantly regulated by a C-terminal truncation by an as yet unidentified proteinase in OA and RA (Vankemmelbeke *et al.* 2001; Yamanishi *et al.* 2002).

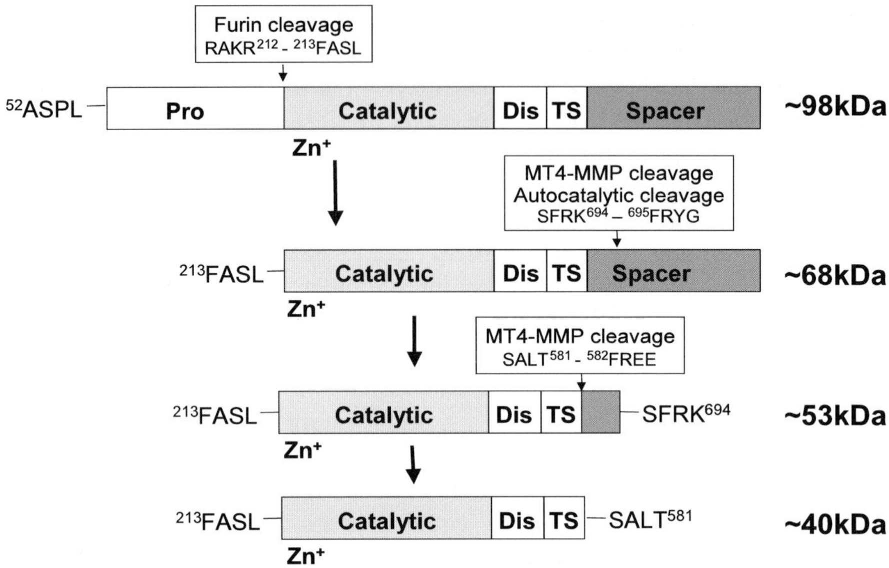


*Figure 3.* ADAMTS-4 processing. Schematic representation of N-terminal and C-terminal cleavage of ADAMTS-4 by furin and MMP-17, respectively.

## 5. OTHER ENDOGENOUS AGGRECANASES

Since aggrecanase activity was initially defined as the ability to cleave aggrecan at the $Glu^{373}$-$Ala^{374}$ bond, any protease capable of this cleavage could be considered an aggrecanase. ADAMTS-1, –8, -9, -15, and –20 share sequence homology with ADAMTS-4 and ADAMTS-5, raising the possibility that they may also represent endogenous aggrecanases. However, to date only five proteinases other than ADAMTS-4 and ADAMTS-5 have been reported to be capable of this cleavage *in vitro*. These include atrolysin C (Tortorella *et al.* 1998), MMP-8 (Fosang *et al.* 1994), MT1-MMP (Buttner *et al.* 1998), ADAMTS-1, and ADAMTS-9 (Kuno *et al.* 2000; Sandy and Verscharen, 2001a; Somerville *et al.* 2003). However, the extent to which they contribute to aggrecan catabolism in disease remains unclear. MMP-8 cleaves aggrecan at the $Glu^{373}$-$Ala^{374}$ site 1000-fold less efficiently than ADAMTS-4/-5 and, moreover, it does not cleave at the 4 other aggrecanase-sensitive sites between G2 and G3 (Figure 2). MMP-8 selective inhibitors that do not block ADAMTS-4/-5 are not capable of blocking IL-1 induced aggrecanase-mediated catabolism (Arner *et al.* 1997). Taken together, these data suggest that MMP-8 does not represent a cartilage aggrecanase. Although MT1-MMP has been shown to cleave a recombinant protein representing the IGD of aggrecan, it does not cleave native aggrecan at any of the aggrecanase-sensitive sites, ruling it out as a cartilage aggrecanase (Malfait *et al.* 2002). ADAMTS-1 cleaves aggrecan at all the aggrecanase-sensitive sites, but its specific activity is approximately 100-fold less than ADAMTS-4 and ADAMTS-5. Moreover, preliminary data indicate that small MW inhibitors of ADAMTS-1 that do not block ADAMTS-4/-5 are not capable of blocking aggrecanase-mediated cartilage catabolism *in vitro* (unpublished observations). ADAMTS-9 has recently been shown to cleave native aggrecan at one of the sites, namely the $Glu^{1480}$-$Gly^{1481}$ (Somerville *et al.* 2003). The role of ADAMTS-15 is undefined at this time, and finally, ADAMTS-20 message is not expressed in knee articular cartilage (unpublished observations), but has been found in hip cartilage (Kevorkian *et al.* 2004). Therefore, further studies are needed to elucidate the role of this ADAMTS in cartilage catabolism.

## 6. SMALL MOLECULE INHIBITORS OF AGGRECANASES

To date, only small molecule active site inhibitors that chelate the nucleophilic Zn, have been described. Similar to many metalloproteinases, it was found that hydroxamate is a preferred chelator for the preparation of aggrecanase inhibitors demonstrating nanomolar affinity. In contrast, other potential chelators, such as thiols, phosphinic acid and carboxylic acid, all result in a decline in potency. Early on, the aggrecanase inhibitors identified were hydroxamates with broad-spectrum activity against many metalloproteinases, including MMPs, ADAMs, and ADAMTSs (Arner *et al.* 1999a). Although they may have increased potential for efficacy through inhibition of multiple proteases, broad-spectrum metalloproteinase inhibitors are not good drug candidates for the treatment of chronic diseases such as osteoarthritis, due to likely toxicity. More recently, inhibitors selective for aggrecanases over MMPs have been designed, such as α-aminohydroxamates described by Bristol-Myers-Squibb (Yao *et al.* 2001; Cherney *et al.* 2003). It is still unknown whether these small molecules are selective against other ADAMTSs, but this is likely to be important because several members of the ADAMTS family have been shown to have critical functions. Transgenic mice with ADAMTS-2 (procollagen N-proteinase) knocked out develop fragile skin and male sterility (Li, 2001), and null mutations cause dermatosparaxis in cattle and in humans, characterized by fragile skin (Colige *et al.* 1999). In addition, mutations in ADAMTS-13, the enzyme responsible for cleavage of von Willebrand factor, result in thrombotic thrombocytopenic purpura (Levy *et al.* 2001; Soejima *et al.* 2003).

Recently, it has been reported that catechin gallate esters, extracted from green tea, inhibit recombinant ADAMTS-4 and ADAMTS-5 in a competitive manner (Vankemmelbeke *et al.* 2003). Other inhibitors that have been reported include mannosamine and glucosamine, which have been shown to block aggrecanase-mediated cleavage in explant or chondrocyte cultures (Sandy *et al.* 1998; Patwari *et al.* 2000), although the mechanism of inhibition has not been elucidated.

Based on ADAMTS-4 and ADAMTS-5 biochemistry, one can envisage the design of three classes of inhibitors: (1) Active site inhibitors that bind to the nucleophilic $Zn^{++}$ or to the amino acids that coordinate the metal in the catalytic domain, as exemplified by the hydroxamic acids discussed above. (2) Exosite inhibitors, which bind to motifs within ADAMTS-4 and -5 that mediate binding to aggrecan, such as that found in the TSP-1 motif of ADAMTS-4; examples include heparin sulfate and chondroitin sulfate (Sugimoto *et al.* 1999). (3) Exosite antagonists that bind to motifs on the glycosaminoglycan chains of aggrecan, recognized by ADAMTS-4 and -5 and required for substrate binding. Molecules capable of binding the glycosaminoglycan chains of aggrecan and inhibiting aggrecanase-mediated cleavage include proteins and peptides containing the amino acid sequence $^{521}$GGWGPWGPWGD$^{531}$ (see section 3.3).

## 7. ADAMTS-4 AND ADAMTS-5 EXPRESSION

Although originally purified from and most widely studied in cartilage, ADAMTS-4 and ADAMTS-5 are expressed in other joint tissues. ADAMTS-4/ADAMTS-5 mRNA is present in RA and OA synovial tissue, while the ADAMTS-5 gene is also detected in non-arthritic synovium (Yamanishi *et al.* 2002). In addition, ADAMTS-4 is expressed in human tendons (Tsuzaki *et al.* 2003) and aggrecanase activity can be induced in bovine joint capsules (Ilic *et al.* 2000). ADAMTS-5 has been detected in growth plate cartilage, where its expression and proteoglycan degradation can be enhanced by thyroid hormone (Makihira *et al.* 2003).

In addition to joint tissues, ADAMTS-4 mRNA is present in ovary, spinal cord, uterus, bladder, brain, heart, and monocytes (Abbaszade *et al.* 1999), and is induced in PMA stimulated THP-1 cells (Worley *et al.* 2003). ADAMTS-5 transcripts are seen in a variety of normal tissues including cervix, uterus, bladder, esophagus, and placenta (Abbaszade *et al.* 1999). The physiological role of ADAMTS-4 and ADAMTS-5 in other tissues is unknown. However, the structural domains, including the disintegrin-like domain, TSP-1 motif, and spacer region (Figure 1), suggest that these proteins are involved in cell signaling and protein-protein interactions. For example, the TSP-1 motif contains a GAG-binding and a putative CD36-binding site, raising the possibility that these enzymes bind to the cell surface or to matrix heparan sulfate proteoglycans and CD36. At this time, it is known that ADAMTS-4 or ADAMTS-5 embryonic knock-out mice are viable, and detailed analysis of these mice is underway (Glasson *et al.* 2004; and own observations).

# 8. THE ROLE OF ADAMTS-4 AND ADAMTS-5 IN DISEASE

## 8.1 ADAMTS-4/-5 in Arthritis

Osteoarthritis is characterized by articular cartilage erosion as a consequence of proteolytic cleavage of its two major functional macromolecules, aggrecan and type II collagen, by aggrecanases and collagenases, respectively.

Early studies, analyzing aggrecan fragments in OA and RA synovial fluids by N-terminal sequencing, found that the major high-buoyant density fragments were generated by cleavage at the TEGE$^{373}$- $^{374}$ARGS site in the aggrecan core protein and thus suggested a crucial role for aggrecanases in aggrecan catabolism in these diseases (Sandy *et al.* 1992; Lohmander *et al.* 1993). More detailed studies have provided strong evidence that aggrecanases are responsible for aggrecan degradation in arthritis and for a major role of ADAMTS-4 and ADAMTS-5 in OA, based on the following findings:

1. Unlike healthy cartilage, human OA cartilage in culture releases aggrecan fragments that are generated by cleavage at all five aggrecanase-sensitive sites but not at the MMP-sensitive site (Malfait *et al.* 2002).
2. Inhibitors selective for ADAMTS-4/-5 over MMPs completely block the release of these fragments from human OA cartilage *in vitro*, but MMP inhibitors have no effect (Malfait *et al.* 2002, and unpublished observations).
3. Immunohistochemistry shows that ADAMTS-4 expression co-localizes with areas of proteoglycan depletion and generation of the NITEGE neoepitope in human OA cartilage, as demonstrated in a recent study with human cartilage with early or severe OA (Figure 4).

4. All the ADAMTS-4- and ADAMTS-5-generated fragments are detected in OA synovial fluid (Sandy *et al.* 2001; Malfait *et al.* 2002).
5. Cartilage from knock-in mice where the EGE$^{373}$- $^{374}$ALG (mouse aggrecan contains Leu instead of Arg in the P2' position) site in the IGD has been mutated to EGE$^{373}$- $^{374}$NVY, is resistant to aggrecan degradation *in vitro* (Fosang *et al.* 2004) and pilot studies suggest that these mice are protected from the development of OA lesions after destabilizing surgery (Fosang and Little, personal communication), suggesting that aggrecanase-mediated cleavage of aggrecan drives the destructive process in OA.

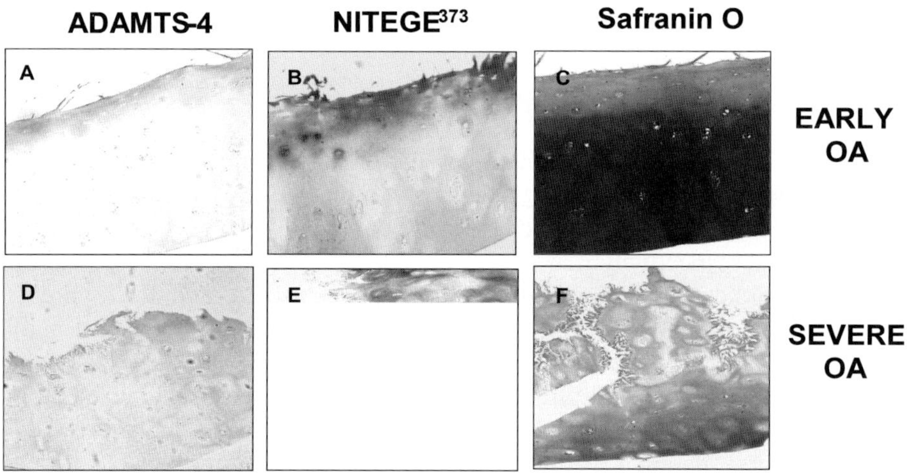


*Figure 4.* Localization of ADAMTS-4 protein in human OA cartilage. (Immuno)histochemistry of human cartilage explants, taken from patients with early OA (top row) or severe OA (bottom row), showing co-localization of ADAMTS-4 (A,D) with the NITEGE neoepitope (B,E), and with areas of aggrecan loss as indicated by loss of Safranin O staining (C,F).

Whether other ADAMTS proteins that share significant sequence homology with ADAMTS-4 and ADAMTS-5 (ADAMTS-1, -8, -9, -15, and - 20) play a role in cartilage degradation in arthritis is currently being investigated. Studies with traditional ADAMTS-4 KO mice showed that these mice are not protected against surgically induced osteoarthritis (Glasson *et al.* 2004), suggesting that ADAMTS-4 alone is not responsible for cartilage destruction in the disease process or that overcompensation by ADAMTS-5 has occurred. Results in ADAMTS-5 knock-out and in ADAMTS-4/TS-5 double knock out mice will clarify these issues.

In addition to preventing cleavage and loss of aggrecan, it was recently demonstrated *in vitro* that an ADAMTS4/ADAMTS-5-selective inhibitor was capable of preventing the breakdown of type II collagen, the other major structural component of cartilage (Pratta *et al.* 2003). In the presence of aggrecan, collagenases exhibit poor proteolytic activity on cartilage explants, due to steric and charge hindrance exhibited by the large aggrecan aggregates that surround the collagen fibrils. Once aggrecan is depleted from the cartilage matrix, the collagen is more readily hydrolyzed. Consequently, selective aggrecanase inhibitors may have a clear advantage over selective collagenase inhibitors. This bodes well for the efficacy of small molecular weight selective aggrecanase-inhibitors as therapeutics in arthritis (Yao *et al.* 2001).

## 8.2 ADAMTS-4/-5 in Other Diseases

ADAMTS-4/-5 may be involved in aggrecan degradation in other tissues, such as the degenerating intervertebral disc, as suggested by the presence of these enzymes and their breakdown products (Roberts *et al.* 2000). Immobilization and/or dynamic compression of intervertebral discs in a rat tail model results in enhanced ADAMTS-4 as well as MMP-3 and MMP-13 mRNA expression, as measured by RT-PCR (MacLean *et al.* 2003).

As discussed earlier, other chondroitin sulfate proteoglycans are substrates of the aggrecanases, and these include brevican and versican. Brevican is a major brain matrix proteoglycan, and its cleavage by ADAMTS-4 may be key to the invasiveness of malignant gliomas (Matthews *et al.* 2000; Nakamura *et al.* 2000). Versican V1, a demonstrated native substrate for ADAMTS-4 (Table I), is expressed in the aorta wall, and its expression and degradation are increased in atherosclerotic lesions. During all stages of atherogenesis, monocytes and macrophages influence plaque composition and stability through secretion of proteolytic enzymes. It was recently shown that PMA-stimulated THP-1 cells, a monocytic cell line, produced ADAMTS-4, suggesting a potential role for this enzyme in atherosclerosis (Worley *et al.* 2003).

There has also been a report that ADAMTS-4 is transcriptionally induced by β-amyloid treatment in rat astrocytes, suggesting that ECM degradation is promoted in the brain during Alzheimer's disease. This may contribute to neuronal damage and development of chronic neurodegenerative disorders (Satoh *et al.* 2000). It was recently shown that simian retroviral neuroinfection in rats induces ADAMTS-4 and ADAMTS-1 in brain macrophages, leading to destruction of ECM proteoglycans (Medina-Flores *et al.* 2004).

# 9. CONCLUSIONS

Increasing evidence is accumulating for the importance of the aggrecanases, ADAMTS-4 and –5, in cartilage degradation in arthritis. Work from a number of laboratories has begun to provide insight into the molecular basis of the role of these proteases in aggrecan catabolism, and recent advances have identified several potential regulatory mechanisms for controlling aggrecan degradation by aggrecanases, including induction of protease expression, post-translational modification, endogenous inhibitors and substrate glycosylation. Knowledge continues to evolve on the expression of these proteases in various tissues and their potential role in normal function and disease. Increasing evidence that the aggrecanases are responsible for aggrecan degradation in arthritis, as well as evidence that the presence of aggrecan protects collagen from cleavage by the matrix metalloproteinases, makes ADAMTS-4 and ADAMTS-5 attractive targets for the development of disease-modifying therapeutics for the treatment of osteoarthritis. However, although several small molecule inhibitors of these proteases have been described, whether this approach will provide an effective treatment for disease awaits demonstration of efficacy in clinical trials.

## ACKNOWLEDGEMENTS

The authors would like to thank Dr. Christopher Little (University of Sydney) and Dr. Amanda Fosang (Melbourne University) for sharing their unpublished data on aggrecan knock-in mice, Dr. William Stallings (Pfizer Global Research and Development) for sharing his unpublished observations on the calcium binding domain of ADAMTS-4/-5, Stella Markosyan for immunohistochemistry of cartilage, and Maureen Ellenberger and Jen Multari for assistance with preparation of the manuscript.

## REFERENCES

Abbaszade, I., Liu, R.Q., Yang, F., Rosenfeld, S.A., Ross, O.H., Link, J.R., Ellis, D.M., Tortorella, M.D., Pratta, M.A., Hollis, J.M., *et al.*. 1999. Cloning and characterization of ADAMTS11, an aggrecanase from the ADAMTS family. *J. Biol. Chem.* **274**:23443-23450.

Andersen, O., Christensen, P., Christensen, L., Jacobsen, C., Moestrup, S., Etzerodt, M., and Thogersen, H, 2000. Specific binding of alpha-macroglobulin to complement-type repeat CR4 of the low-density lipoprotein receptor-related protein. *Biochemistry* **39**:10627-10633.

Arner, E., Decicco, C., Cherney, R., and Tortorella, M, 1997. Cleavage of native cartilage aggrecan by neutrophil collagenase (MMP-8) is distinct from endogenous cleavage by aggrecanase. *J. Biol. Chem.* **272**:9294-9299.

Arner, E., Pratta, M., Decicco, C., Xue, C., Newton, R., Trzaskos, J., Magolda, R., and Tortorella, M. 1999a, Aggrecanase. A target for the design of inhibitors of cartilage degradation. *Ann. N Y Acad. Sci.* **878**:92-107.

Arner, E., Pratta, M., Trzaskos, J., Decicco, C., and Tortorella, M, 1999b. Generation and characterization of aggrecanase. A soluble, cartilage-derived aggrecan-degrading activity. *J. Biol. Chem.* **274**:6594-6601.

Aschom, J.D., Tiller, S.E., Dickerson, K., Cravens, J.L., Argraves, W.S., and Strickland, D.K, 1990. The human 2-macroglobulin receptor: identification of a 420-kD cell surface glycoprotein specific for the activated conformation of 2-macroglobulin. *J. Cell Biol.* **110**:1041.

Buttner, F.H., Hughes, C.E., Margerie, D., Lichte, A., Tschesche, H., Caterson, B., and Bartnik, E, 1998. Membrane type 1 matrix metalloproteinase (MT1-MMP) cleaves the recombinant agrecan substrate rAgg1mut at the 'aggrecanase'and the MMP sites. Characterization of MT1-MMP catabolic activities on the interglobular domain of aggrecan. *Biochem. J.* **333**:159-165.

Cherney, R., Mo, R., Meyer, D., Wang, L., Yao, W., Wasserman, Z., Liu, R., Covington, M., Tortorella, M., Arner, E., *et al.*. 2003. Potent and selective aggrecanase inhibitors containing cyclic P1 substituents. *Bioorg. Med. Chem. Lett.***13**:1297-1300.

Chevalier, X., Claudepierre, P., Groult, N., Zardi, L., and Hornebeck, W, 1996. Presence of ED-A containing fibronectin in human articular cartilage from patients with osteoarthritis and rheumatoid arthritis. *J. Rheumatol.* **23**:1022-1030.

Chevalier, X., Groult, N., and Labat-Robert, J, 1992. Biosynthesis and distribution of fibronectin in normal and osteoarthritic human cartilage. *Clin. Physiol. Biochem.* **9**:1-6.

Clemmensen, I., Holund, B., Johansen, N., and Andersen, R.B, 1982. Demonstration of fibronectin in human articular cartilage by an indirect immunoperoxidase technique. *Histochemistry* **76**:51-56.

Colige, A., Sieron, A., Li, S., Schwarze, U., Petty, E., Wertelecki, W., Wilcox, W., Krakow, D., Cohn, D., Reardon, W., *et al.*. 1999. Human Ehlers-Danlos syndrome type VII C and bovine dermatosparaxis are caused by mutations in the procollagen I N-proteinase gene. *Am. J. Hum. Genet.* **65**:308-317.

Curtis, C., Rees, S., Little, C., Flannery, C., Hughes, C., Wilson, C., Dent, C., Otterness, I., Harwood, H., and Caterson, B. 2002. Pathologic indicators of degradation and inflammation in human osteoarthritic cartilage are abrogated by exposure to n-3 fatty acids. *Arthritis Rheum.* **46**:1544-1553.

Dickinson, S.C., Vankemmelbeke, M.N., Buttle, D.J., Rosenberg, K., Heinegard, D., and Hollander, A.P, 2003. Cleavage of cartilage oligomeric matrix protein (thrombospondin-5) by matrix metalloproteinases and a disintegrin and metalloproteinase with thrombospondin motifs. *Matrix Biology* **22**:267-278.

Flannery, C.R., Lark, M.W., and Sandy, J.D, 1992. Identification of a stromelysin cleavage site within the interglobular domain of human aggrecan. Evidence for proteolysis at this site in vivo in human cartilage. *J. Biol. Chem.* **267**:1008-1014.

Flannery, C., Little, C., Hughes, C., Curtis, C., Caterson, B., and Jones, S, 2000. IL-6 and its soluble receptor augment aggrecanase-mediated proteoglycan catabolism in articular cartilage. *Matrix Biol.* **19**:549-553.

Flannery, C., Zeng, W., Corcoran, C., Collins-Racie, L., Chockalingam, P., Hebert, T., Mackie, S., McDonagh, T., Crawford, T., Tomkinson, K., *et al.*. 2002. Autocatalytic

Makihira, S., Yan, W., Murakami, H., Furukawa, M., Kawai, T., Nikawa, H., Yoshida, E., Hamada, T., Okada, Y., Kato, Y, 2003. Thyroid hormone enhances aggrecanase-2/ADAM-TS5 expression and proteoglycan degradation in growth plate cartilage. *Endocrinology* **144**: 2480-8.

Malfait, A.M., Liu, R.-Q., Ijiri, K., Setsuro, K., and Tortorella, M, 2002. Inhibition of ADAM-TS4 and ADAM-TS5 prevents aggrecan degradation in osteoarthritic cartilage. *J. Biol. Chem.* **277**:22201-22208.

Mankin, H.J., and Lipiello, L, 1970. Biochemical and metabolic abnormalities in articular cartilage from osteoarthritic human hips. *J. Bone. Joint Surg. [Am]* **52**:424-434.

Matthews, R., Gary, S., Zerillo, C., Pratta, M., Solomon, K., Arner, E., and Hockfield, S, 2000. Brain-enriched hyaluronan binding (BEHAB)/brevican cleavage in a glioma cell line is mediated by a disintegrin and metalloproteinase with thrombospondin motifs (ADAMTS) family member. *J. Biol. Chem.* **275**:22695-22703.

Medina-Flores, R., Wang, G., Bissel, SJ., Murphey-Corb, M., Wiley, CA, 2004, Destruction of extracellular matrix proteoglycans is pervasive in simian retroviral neuroinfection. *Neurobiol. Dis.* **16**: 604-1.

Moulharat, N., Lesur, C., Thomas, M., Rolland-Valognes, G., Pastoureau, P., Anract, P., De Ceuninck, F., and Sabatini, M, 2004. Effects of transforming growth factor-beta on aggrecanase production and proteoglycan degradation by human chondrocytes in vitro. *Osteoarthritis Cartilage* **12**:296-305.

Mousa, S., and Liu, R, 2003. Effect of LMWH and Different Heparin Molecular Weight Fractions on Aggrecanase Activity: Structure-Function Relationships (abstract). *IUA Congress, April 7-11 2002. New York, New York.*

Nakamura, H., Fujii, Y., Inoki, I., Sugimoto, K., Tanzawa, K., Matsuki, H., Miura, R., Yamaguchi, Y., and Okada, Y, 2000. Brevican is degraded by matrix metalloproteinases and aggrecanase-1 (ADAMTS4) at different sites. *J. Biol. Chem.* **275**:38885-38890.

Patwari, P., Kurz, B., Sandy, J.D., and Grodzinsky, A.J, 2000. Mannosamine inhibits aggrecanase-mediated changes in the physical properties and biochemical composition of articular cartilage. *Arch. Biochem. Biophys.* **374:**79-85.

Pratta, M.A., Tortorella, M.D., and Arner, E.C. 2000. Age-related changes in aggrecan glycosylation affect cleavage by aggrecanase. *J. Biol. Chem.* **275**:39096-39102,.

Pratta, M., Yao, W., Decicco, C., Tortorella, M., Liu, R., Copeland, R., Magolda, R., Newton, R., Trzaskos, J., and Arner, E, 2003. Aggrecan protects cartilage collagen from proteolytic cleavage. *J. Biol. Chem.* **278**:45539-45545.

Roberts, S., Caterson, B., Menage, J., Evans, E., Jaffray, D., and Eisenstein, S, 2000. Matrix metalloproteinases and aggrecanase: their role in disorders of the human intervertebral disc. *Spine* **25**:3005-3013.

Sandy, J.D., Flannery, C.R., Neame, P.J., and Lohmander, L.S, 1992. The structure of aggrecan fragments in human synovial fluid. Evidence for the involvement in osteoarthritis of a novel protease which cleaves the Glu373-Ala374 bond of the interglobular domain. *J. Clin. Invest.* **89**:1512-1516.

Sandy, J.D., Gamett, D., Thompson, V., and Verscharen, C, 1998. Chondrocyte-mediated catabolism of aggrecan: aggrecanase-dependent cleavage induced by interleukin-1 or retinoic acid can be inhibited by glucosamine. *Biochem. J.* **335**:59-66.

Sandy, J.D., Neame, P.J., Boynton, P.L., and Flannery, C.R, 1991. Catabolism of aggrecan in cartilage explants. Identification of a major cleavage site within the interglobular domain. *J. Biol. Chem.* **266**:8683-8685.

Sandy, J.D., and Verscharen, C, 2001a. Analysis of aggrecan in human knee cartilage and synovial fluid indicates the aggrecanase (ADAMTS) activity is responsible for the turnover and loss of whole aggrecan whereas other protease activity is required for C-terminal processing in vivo. *Biochem. J.* **358**:615-626.

Sandy, J.D., Westling, J., Kenagy, R., Iruela-Arispe, M., Verscharen, C., Rodriguez-Mazaneque, J., Zimmermann, D., Lemire, J., Fischer, J., Wight, T., *et al.*. 2001b, Versican V1 proteolysis in human aorta in vivo occurs at the Glu441-Ala442 bond, a site that is cleaved by recombinant ADAMTS-1 and ADAMTS-4. *J. Biol. Chem.* **276:** 13372-13378.

Satoh, K., Suzuki, N., and Yokota, H, 2000. ADAMTS-4 (a disintegrin and metalloproteinase with thrombospondin motifs) is transcriptionally induced in beta-amyloid treated rat astrocytes. *Neurosci. Lett.* **289**:177-180.

Soejima, K., Matsumoto, M., Kokame, K., Yagi, H., Ishizashi, H., Maeda, H., Nozaki, C., Miyata, T., Fujimura, Y., and Nakagaki, T, 2003. ADAMTS-13 cysteine-rich/spacer domains are functionally essential for von Willebrand factor cleavage. *Blood* **102**:3232-3237.

Somerville, R., Longpre, J., Jungers, K., Engle, J., Ross, M., Evanko, S., Wight, T., Leduc, R., and Apte, SS, 2003. Characterization of ADAMTS-9 and ADAMTS-20 as a distinct ADAMTS subfamily related to Caenorhabditis elegans GON-1. *J. Biol. Chem.* **278**: 9503-9513.

Stanton, H., Ung, L., and Fosang, A, 2002. The 45 kDa collagen-binding fragment of fibronectin induces matrix metalloproteinase-13 synthesis by chondrocytes and aggrecan degradation by aggrecanases. *Biochem. J.* **364**:181-190.

Su, S., Grover, J., Roughley, P.J., DiBattista, J.A., Martel-Pelletier, J., Pelletier, J.P., and Zafarullah, M, 1999. Expression of the tissue inhibitor of metalloproteinases (TIMP) gene family in normal and osteoarthritic joints. *Rheumatol. Int.* **18**:183-191.

Sugimoto, K., Takahashi, M., Yamamoto, Y., Shimada, K., and Tanzawa, K, 1999. Identification of aggrecanase activity in medium of cartilage culture. *J. Biochem.* **126**:449-455.

Sylvester, J., Liacini, A., Li, W., and Zafarullah, M, 2004. Interleukin-17 signal transduction pathways implicated in inducing matrix metalloproteinase-3, -13 and aggrecanase-1 genes in articular chondrocytes. *Cell Signal* **16**:469-476.

Sztrolovics, R., White, R.J., Roughley, P.J., and Mort, J.S, 2002. The mechanism of aggrecan release from cartilage differs with tissue origin and the agent used to stimulate catabolism. *Biochem. J.* **362**:465-472.

Taylor, N., Van de Ven, W., and Creemers, J, 2003. Curbing activation: proprotein convertases in homeostasis and pathology. *FASEB J.* **17**:1215-1227.

Tortorella, M., Arner, E., Hills, R., Easton, A., Korte-Sarfaty, J., Fok, K., Wittwer, A., Liu, R., and Malfait, A.M, 2004. Alpha2-macroglobulin is a novel substrate for ADAMTS-4 and ADAMTS-5 and represents an endogenous inhibitor of these enzymes. *J. Biol. Chem.* **279**:17554-17561.

Tortorella, M.D., Burn, T.C., Pratta, M.C., Abbaszade, I., Hollis, J.M., Liu, R., Rosenfeld, S.A., Copeland, R.A., Decicco, C.P., Wynn, R., *et al.*. 1999. Purification and cloning of aggrecanase-1: a member of the ADAMTS family of proteins. *Science* **284**:1664-1666.

Tortorella, M.D., Liu, R.-Q., Burn, T., and Arner, E, 2002. Characterization of human aggrecanase 2 (ADAM-TS5): Substrate specificity studies in comparison with aggrecanase 1 (ADAM-TS4). *Matrix Biology* **21**:499-511.

Tortorella, M.D., Malfait, A.M., Decicco, C., and Arner, E, 2001. The role of ADAM-TS4 (aggrecanase-1) and ADAM-TS5 (aggrecanase-2) in a model of cartilage degradation. *Osteoarthritis and Cartilage* **9**:539-552.

Tortorella, M.D., Pratta, M.A., Fox, J.W., and Arner, E.C, 1998. The interglobular domain of cartilage aggrecan is cleaved by hemorrhagic metalloproteinase HT-d (Atrolysin-C) at the matrix metalloproteinase and aggrecanase sites. *J. Biol. Chem* **273**:5846-5850.

Tortorella, M., Pratta, M., Liu, R.Q., Abbaszade, I., Ross, H., Burn, T., and Arner, E, 2000a. The thrombospondin motif of aggrecanase-1 (ADAMTS-4) is critical for aggrecan substrate recognition and cleavage. *J. Biol. Chem.* **275**:25791-25797.

Tortorella, M.D., Pratta, M., Liu, R.Q., Austin, J., Ross, O.H., Abbaszade, I., Burn, T., and Arner, E, 2000b. Sites of aggrecan cleavage by recombinant human aggrecanase-1 (ADAMTS-4). *J. Biol. Chem.* **275**:18566-18573.

Tsuzaki, M., Guyton, G., Garrett, W., Archambault, J., Herzog, W., Almekinders, L., Bynum, D., Yang, X., and Banes, A, 2003. IL-1 beta induces COX2, MMP-1, -3 and -13, ADAMTS-4, IL-1 beta and IL-6 in human tendon cells. *J. Orthop. Res.* **21**:256-264.

Vankemmelbeke, M., Holen, I., Wilson, A., Ilic, M., Handley, C., Kelner, G., Clark, M., Liu, C., Maki, R., Burnett, D., *et al.*. 2001. Expression and activity of ADAMTS-5 in synovium. *Eur. J. Biochem.* **268**:1259-1268.

Vankemmelbeke, M., Jones, G., Fowles, C., Ilic, M., Handley, C., Day, A., Knight, C., Mort, J., and Buttle, D, 2003. Selective inhibition of ADAMTS-1, -4 and -5 by catechin gallate esters. *Eur. J. Biochem.* **270**:2394-2403.

Wang, P., Tortorella, M., England, K., Malfait, A.M., Thomas, G., Arner, E.C., and Pei, D, 2004a. Proprotein convertase furin interacts with and cleaves pro-ADAMTS4 (Aggrecanase-1) in the trans-Golgi network. *J. Biol. Chem.* **279**:15434-15440.

Wang, Z., Luo, J., and Chen, Q, 2004b. Matrilin-2: a novel extracellular matrix substrate of aggrecanase-1 (ADAMTS4). *Trans. Orthop. Res. Soc.* **29**:339.

Westling, J., Gottschall, P., Thompson, V., Cockburn, A., Perides, G., Zimmermann, D., and Sandy, J, 2004. ADAMTS4 (aggrecanase-1) cleaves human brain versican V2 at Glu405-Gln406 to generate glial hyaluronate binding protein. *Biochem. J.* **377**:787-795.

Worley, J., Baugh, M., Hughes, D., Edwards, D., Hogan, A., Sampson, M., and Gavrilovic, J, 2003. Metalloproteinase expression in PMA-stimulated THP-1 cells. Effects of peroxisome proliferator-activated receptor-gamma (PPAR gamma) agonists and 9-cis-retinoic acid. *J. Biol. Chem.* **278**:51340-51346.

Xie, D., Meyers, R., and Homandberg, G, 1992. Fibronectin fragments in osteoarthritic synovial fluid. *J. Rheumatol.* **19**:1448-1452.

Yamanishi, Y., Boyle, D.L., Clark, M., Maki, R., Tortorella, M.D., Arner, E.C., and Firestein, G.S, 2002. Expression and regulation of aggrecanase in arthritis: the role of TGF-beta. *J. Immunol.* **168**:1405-1412.

Yao, W., Wasserman, Z.K., Chao, M., Reddy, G., Shi, E., Liu, R.Q., Covington, M.B., Arner, E.C., Pratta, M.A., Tortorella, M., *et al.*. 2001. Design and synthesis of a series of (2r)-n(4)-hydroxy-2-(3-hydroxybenzyl)-n(1)-[(1s,2r)-2-hydroxy-2,3-dihydro-1h-inden-1ryl]butanediamide derivatives as potent, selective, and orally available aggrecanase inhibitors. *J. Med. Chem.* **21**:3347-3350.

Yasuda, T., and Poole, A, 2002. A fibronectin fragment induces type II collagen degradation by collagenase through an interleukin-1-mediated pathway. *Arthritis Rheum.* **46**:138-148.

Yu, W.H., Yu, S.C., Meng, Q., Brew, K., and Woessner, J, 2000. TIMP-3 binds to sulfated glycosaminoglycans of the extracellular matrix. *J. Biol. Chem.* **275**:31226-31232.

Zeng, W., Chockalingam, P.S., Shakey, Q., and Flannery, C.R, 2004. Selective cleavage of TIMP-4 by ADAMTS-4. *Trans. Orthop. Res. Soc.* **29**:610.

# Chapter 15

# ADAMTS-13
## *Thrombotic Thrombocytopenic Purpura*

Han-Mou Tsai
*Hematology Division, Montefiore Medical Center, Albert Einstein College of Medicine, New York, USA*

Abstract: von Willebrand factor (VWF) is a plasma glycoprotein that mediates platelet adhesion and aggregation at the site of vessel injury. Studies on the homeostasis of VWF led to the discovery that it is cleaved in the circulation by a zinc metalloprotease, ADAMTS-13, in a shear stress dependent manner. ADAMTS-13 is phylogenetically most different from other members of the reprolysin-type ADAMTS zinc metalloprotease family. In the presence of ADAMTS-13, shear stress on VWF promotes its cleavage by ADAMTS-13 to smaller, less active forms. In the absence of ADAMTS-13, shear stress increases the platelet-aggregating activity of VWF, resulting in accumulation of super-active forms of VWF and subsequently intravascular platelet thrombosis, as observed in patients with thrombotic thrombocytopenic purpura (TTP). In TTP, autoimmune inhibitors of the protease or genetic mutations of the *ADAMTS-13* gene cause a severe deficiency of ADAMTS-13 in plasma. Analysis of ADAMTS-13 enables the differentiation of TTP from other types of microvascular thrombosis on a pathogenetic basis and facilitates advances in the diagnosis and therapy of the disease.

Key words: ADAMTS-13, von Willebrand factor, thrombotic thrombocytopenic purpura.

## 1. INTRODUCTION

von Willebrand factor (VWF) is an endothelial cell-derived glycoprotein that mediates platelet hemostatic plug formation under high shear stress conditions (Ruggeri 2003). Although it exists in the plasma fraction of circulation as a series of multimers, these multimers are not detectable in endothelial cells or their culture media (Tsai *et al.* 1989; Kaul *et al.* 1993).



*N. M. Hooper and U. Lendeckel (eds.), The ADAM Family of Proteases,* 323-340.

Studies on the genesis of VWF multimers led to the discovery of a protease, ADAMTS-13, whose evidence of existence was previously based on the imprint it left on von Willebrand factor (Zimmerman *et al.* 1986). Investigations of the homeostasis of von Willebrand factor were instrumental in linking ADAMTS-13 deficiency to a thrombotic disorder, thrombotic thrombocytopenic purpura (TTP).

## 2. MOLECULAR BIOLOGY OF ADAMTS-13

Human ADAMTS-13 gene contains 29 exons spanning approximately 37 kb on chromosome 9q34 (Levy *et al.* 2001; Soejima *et al.* 2001; Zheng *et al.* 2001). The regulatory elements of the gene have not been defined. Alignment of human and mouse ADAMTS-13 sequences surrounding the transcription initiation site reveals no outstanding region of homology. Neither the human nor mouse putative promoter has a TATA box or a CpG island. Motif analysis has not revealed striking patterns of transcription factor binding sites.

The transcripts of ADAMTS-13 are relatively complex. At least four splicing, and several truncated variants have been described. Because some of these variants profoundly change the structure of the predicted protein, it is speculated that ADAMTS-13 function may vary in different tissues. Multiple-tissue Northern blotting demonstrated that ADAMTS-13 encodes a 4.7-kb transcript primarily in the liver, and a 2.4-kb transcript is detectable in placenta, skeletal muscle, and certain tumor cell lines (Levy *et al.* 2001; Soejima *et al.* 2001; Zheng *et al.* 2001; Cal *et al.* 2002).

The ADAMTS-13 isoform with VWF cleaving activity is almost certainly the one with all 29 exons, since it has the right molecular weight and cDNA length and since transfection studies of this cDNA demonstrated that this protein has proteolytic activity.

The full-length transcript encodes a precursor polypeptide with 1427 amino acid residues. The amino acid sequence contains 10 consensus sites for N-linked glycosylation, several potential sites for O-linked glycosylation, and one consensus site for C-mannosylation (Zheng *et al.* 2001). Expression studies revealed that ADAMTS-13 is synthesized in the cells as a 180-kD, instead of the calculated 145-kD, protein, indicating that the protein undergoes extensive glycosylation and other modifications. Smaller forms of ADAMTS-13 (170 kD, 160 kD, and 120 kD) have been isolated from the plasma. These minor forms contain identical N-terminal amino acid sequence and may derive from the shorter transcripts or proteolytic truncation.

The sequence of ADAMTS-13 exhibits a multi-domain structure, consisting of a signal peptide, a propeptide that ends with a consensus RQRR sequence, a metalloprotease domain with zinc binding catalytic sequence motif (HExGHxxGxxHD), a disintegrin-like domain, a central thrombospondin type 1 repeat (TSR-1), a cysteine-rich domain, a cysteine-free spacer region, 7 additional TSR's, and two CUB (complement, uEGF, and bone morphogenesis) domains (Figure 1).

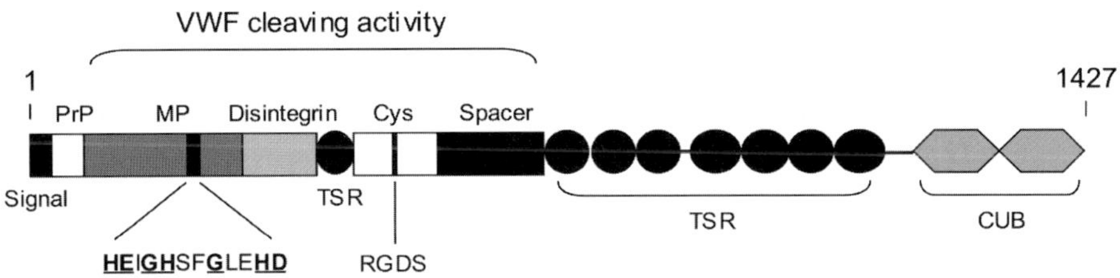


*Figure 1.* A schematic depiction of the ADAMTS-13 domain structure. PrP: propeptide; MP: metalloprotease. Cys: cysteine-rich domain; TSR; thrombospondin type 1 repeat. The $Zn^{++}$-binding catalytic motif of the metalloprotease domain and the RGDS sequence in the cysteine-rich domain are shown. A truncated protein encompassing the amino acid residues from the MP to the spacer domains, as indicated by the bracket, has been found to be proteolytically active.

An RGD sequence is present at residues 498-500 in the cysteine-free spacer domain. The CUB domains appear to be unique for ADAMTS-13, as it has not been found in other members of the ADAMTS family. Sequence alignment and homology modeling based on the structure of adamalysin II suggested that disulfide bonding connects $C^{155}$-$C^{208}$, $C^{202}$-$C^{281}$, and $C^{242}$-$C^{265}$ of the metalloprotease domain (Zheng *et al.* 2001). These structural features of ADAMTS-13 are consistent with previous observations that disulfide bond-reducing agents, tetracyclines, EDTA, EGTA, or phenanthroline abolishes VWF-cleaving activity of ADAMTS-13 and $Ca^{++}$ or $Zn^{++}$ reverses the inhibitory effect of EDTA (Tsai *et al.* 1994; Furlan *et al.* 1996; Tsai 1996; Tsai *et al.* 1997).

ADAMTS-13 contains an unusually short (41 amino acid residues) propeptide whose cleavage does not appear necessary for expression of proteolytic activity (Majerus *et al.* 2003). As a matter of fact, phylogenic analysis indicates that ADAMTS-13 diverts early from other members of the family (Apte 2004).

Enzymatic analysis of proteins expressed by mammalian cells reveals that a truncated protein comprised of the metalloprotease domain is secreted from the cells but is proteolytically inactive. When it is extended to include the spacer domain, the protein is released in the culture media in proteolytically active form (Zheng *et al.* 2003; Soejima *et al.* 2003). Other reports have proposed that the CUB domains may be involved in ligand recognition. The causes of the discrepancy are unclear and the function of the protein sequence downstream of the spacer domain remains to be defined; speculatively it may affect protein clearance in the circulation. The RGD sequence does not appear essential for expression of VWF-cleaving activity, since a substitution of the D residue by E does not affect the expression of the protease activity (Soejima *et al.* 2003).

## 3. VON WILLEBRAND FACTOR, ADAMTS-13 AND HEMOSTASIS

von Willebrand factor (VWF) is the only known substrate of ADAMTS-13. A deficiency in VWF causes mucocutaneous bleeding diathesis and decreases factor VIII procoagulant activity (Ginsburg 1999). Initially thought to represent the antigenic component of coagulation factor VIII (hence the term VIII-related antigen in old literature), VWF serves as a carrier of factor VIII in the circulation and supports platelet adhesion and aggregation at sites of vessel injury. VWF, encoded by a gene located on chromosome 12p, is a highly glycosylated, multi-functional plasma protein that exists in the circulation as a series of disulfide-bonded multimers with molecular weights ranging from $0.5 \times 10^6$ to $20 \times 10^6$ Daltons. VWF is also present in the storage granules of vascular endothelial cells (Weibel-Palade bodies), the α-granules of megakaryocytes and platelets, and the subendothelial matrix of the vasculature, where it is bound to type VI collagen (Rand *et al.* 1991). Endothelial cells account for > 95% of the plasma pool of VWF (Nichols *et al.* 1995). Platelet-derived VWF differs from plasma VWF in that it does not contain ABO blood type antigens and appears to be less effective in supporting platelet aggregation (Sarode *et al.* 2000; Williams *et al.* 1994).

In endothelial cells, VWF is synthesized as a precursor form consisting of a 22-amino acid signal peptide, a 741-amino acid propeptide (also known as VWF antigen II), and the mature 2050-amino acid polypeptide (Handin & Wagner 1989). Many copies of the mature polypeptide are linked by disulfide bonds to form a very large polymer (Tsai *et al.* 1989). The VWF polypeptide consists of repeats of homologous domains in which its various functional epitopes are located. The epitope involved in the binding with platelet surface glycoprotein receptor Ib/V/IX is located in the A1 domain, while the RGD sequence near the carboxyl terminus binds platelet glycoprotein αIIbβ3. ADAMTS-13 cleaves the $Y^{1605}$-$M^{1606}$ bond located in the central A2 domain (Figure 2) (Dent *et al.* 1990).

To maintain circulatory fluidity, glycoprotein αIIbβ3 of platelets normally exists in an inactive conformation and is transformed to an active form for ligand binding when platelets are activated. In contrast, the mechanism preventing VWF-Ib/IX/V interaction is less well understood. There is no experimental evidence that glycoprotein receptor Ib/IX/V requires activation.

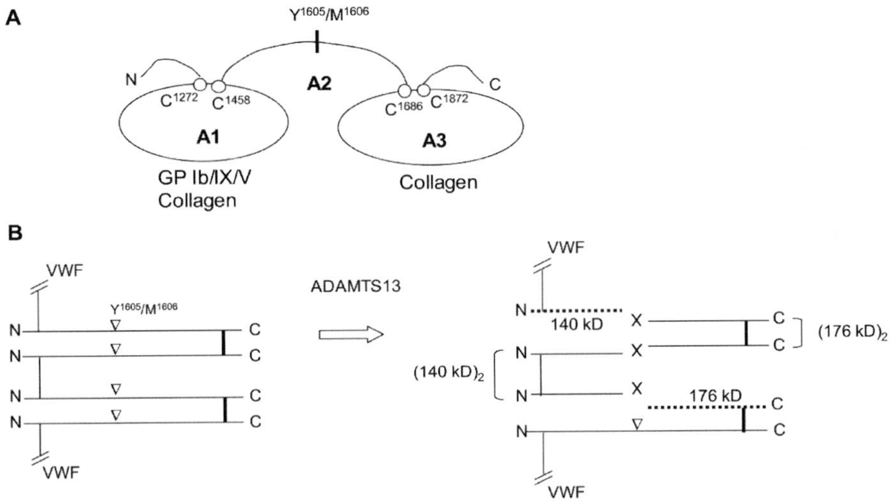


*Figure 2.* Structure of VWF and its cleavage by ADAMTS-13. **A**. A schematic depiction of the three central A domains of the VWF polypeptide. The A1 domain contains the epitope for binding platelet surface glycoprotein complex Ib/IX/V, as well as epitopes for other ligands. The A3 domain is critical for binding with collagen. The $Y^{1605}$-$M^{1606}$ cleavage site of ADAMTS-13 is located in the A2 domain. Both A1 and A3 domains, but not the A2 domain, contain an intra-chain disulfide loop. **B**. Endothelial cell-derived VWF polymer consists of many copies of the VWF polypeptide linked by alternating disulfide bonds (vertical bars) near the carboxyl or amino terminus. Four copies of the VWF polypeptide in a VWF polymer are depicted. Each cleavage of the $Y^{1605}$-$M^{1606}$ bond by ADAMTS-13 decreases the size of VWF and creates multimers with either a 140-kD and/or 176-kD fragment (dashed lines) at the end of the strand. The smallest fragments resulting from the cleavage include dimers of the 140-kD and the 176-kD fragments.

Previously, no models explained why the size of VWF is a major determinant of its hemostatic activity, i.e. large VWF multimers are hemostatically more effective than small multimers. Nor was it understood why shear stress promoted VWF-supported platelet adhesion and aggregation (Weiss *et al.* 1978). As will be further discussed later, recent studies suggested that responsiveness of VWF to shear stress is intimately linked to these peculiar features of VWF and that ADAMTS-13 is critical in preventing VWF-platelet glycoprotein Ib/IX/V interaction.

In SDS PAGE analysis, plasma VWF consists of a full-length 225 kD polypeptide of 2050 amino acid residues and at least three proteolytic fragments of 189 kD, 176 kD, and 140 kD, which together account for approximately 20% of the VWF protein (Zimmerman *et al.* 1986). The 140 kD and the 176 kD fragments result from cleavage of the full-length VWF subunit at the $Y^{1605}$-$M^{1606}$ bond (Dent *et al.* 1990). The cleavage site leading to the generation of the 189 kD fragment has not been defined. Two dimensional gel analysis, in which VWF multimers are first separated by SDS agarose gel electrophoresis, followed by electrophoresis in a perpendicular direction under reducing conditions, reveals that most likely all multimers contain proteolytic fragments (Dent *et al.* 1991; Tsai *et al.* 1991). This suggests that the cleavage of VWF at the $Y^{1605}$-$M^{1606}$ bond is involved in generating VWF multimers. The present scheme of VWF multimer generation (Figure 2B) is also consistent with results of EM studies that vWF multimer strands have at least two morphologically distinct types of ends (Fowler *et al.* 1985).

# 4. SHEAR STRESS AND VON WILLEBRAND FACTOR

## 4.1 Shear stress and proteolysis of von Willebrand factor

The laminar flow pattern in the circulation is responsible for creating shear stress that is at its highest level at the endothelial boundary, and the lowest at the center of the lumen. Since shear stress is inversely proportional at the $3^{rd}$ order to the internal radius of the lumen, the highest levels of wall shear stress are encountered in the arterioles and capillaries of the vasculature. Shear stress is critically involved in modulating the biological behavior of VWF.

ADAMTS-13 and VWF co-exist in the plasma fraction of blood, and yet incubation of blood or plasma in a test tube does not result in demonstrable cleavage of VWF by ADAMTS-13. After VWF is briefly exposed to levels of shear stress as encountered in the microcirculation, it becomes susceptible to cleavage by ADAMTS-13 (Tsai *et al.* 1994; Tsai 1996). Since shear stress also promotes cleavage of VWF by other enzymes such as leukocyte serine proteases, its effect is not specific for ADAMTS-13 (Tsai *et al.* 1994). Nevertheless, ADAMTS-13 is the only constitutively active protease with constant access to VWF. In the plasma, only the fragments produced by ADAMTS-13 are detectable. Thus ADAMTS-13 is considered the physiologic protease of VWF.

Studies using atomic force microscopy revealed that shear stress causes conformational unfolding of VWF (Marchant *et al.* 1997). The conformational change of VWF may explain why shear stress enhances the susceptibility of VWF to proteolysis; presumably unfolding of VWF makes the molecule more accessible to the protease. In support of this concept, chaotropic agents such as guanidine HCl also cause conformational unfolding of VWF and increase its proteolytic susceptibility (Tsai 1996). Nevertheless, excessive unfolding of VWF decreases its susceptibility, suggesting that an intermediate form of VWF is optimal for interaction with ADAMTS-13 (Tsai 2003a).

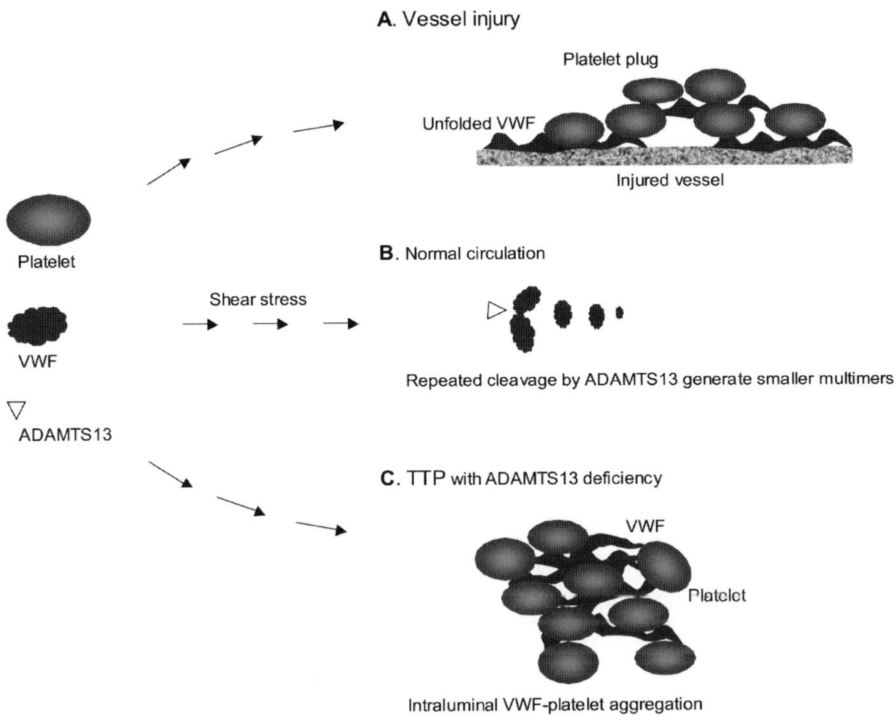


*Figure 3.* The interplay of shear stress, ADAMTS-13 and von Willebrand factor. **A**. At a site of vessel injury, VWF binds to the exposed matrix and is quickly unfolded by shear stress to form the substrate for supporting platelet adhesion and aggregation. **B**. In normal circulation, VWF is cleaved by ADAMTS-13 whenever it is partially unfolded by shear stress in the microcirculation. This process generates VWF multimers. **C**. In TTP, the absence of ADAMTS-13 causes accumulation of unfolded, super-active forms of VWF, resulting in intravascular platelet aggregation and thrombosis in the arterioles and capillaries.

The biphasic pattern of response to conformational unfolding may be essential for VWF to perform adhesive function in the presence of

ADAMTS-13. Presumably, at a site of blood vessel injury, VWF, once attached to the vascular wall matrix, is quickly unfolded by shear stress to a resistant form that supports platelet adhesion before it is cleaved by ADAMTS-13 (Figure 3A). In the circulation, a partially unfolded VWF is immediately cleaved by ADAMTS-13 (Figure 3B). This scheme allows VWF to function under high shear stress conditions and ensures that VWF does not become fully unfolded and cause platelet aggregation in the circulation (Figure 3C).

A subset of von Willebrand disease (type 2A) mutations cause VWF to be spontaneously susceptible to cleavage by ADAMTS-13 in the absence of shear stress (Tsai *et al.* 1997; O'Brien *et al.* 2004). It is conceivable that while the wild-type VWF is cleaved by ADAMTS-13 only when high levels of shear stress in the microcirculation unfold its conformation, the type 2A VWF variants are susceptible to cleavage by ADAMTS-13 throughout the circulation. Thus patients with this type of mutation have smaller VWF multimers in their plasma. When their plasma samples are left in a test tube, the VWF is further proteolyzed unless a metal cation chelator such as EDTA is present to suppress the ADAMTS-13 activity (Gralnick *et al.* 1985; Batlle *et al.* 1986).

## 4.2 Shear stress and platelet thrombus formation

In inverted vessels, cone-and-plate viscometers, and parallel-plate perfusion chambers, shear stress enhances platelet adhesion and thrombus formation in the presence of VWF (Weiss *et al.* 1978; Moake *et al.* 1986; Ikeda *et al.* 1991; Alevriadou *et al.* 1993). The process of platelet thrombus formation is initiated by binding between VWF and platelet glycoprotein complex Ib/IX/V, followed by platelet activation and VWF-$\alpha$IIb/$\beta$3 binding. Collagen and fibrinogen/fibrin are believed to help stabilize VWF-platelet thrombus (Ruggeri 2003).

The molecular mechanism of how shear stress affects platelet thrombus formation has not been adequately determined. In aggregation studies, shear stress promotes platelet aggregation without increasing VWF binding. Conceivably, by causing conformational unfolding of VWF, shear stress exposes more binding epitopes to platelet receptors without increasing the fraction of VWF that is bound to platelet. This process brings platelets to proximity, which favors platelet-platelet interaction, thereby facilitating the growth of platelet thrombus. Unfolded VWF may also provide the matrix that captures detaching platelets before they are swept away by blood flow. Thus, conformational unfolding may explain why shear stress promotes platelet adhesion and thrombus formation (Figure 3A). Furthermore, the size is a major determinant of the responsiveness of VWF to shear stress (Tsai

2002). In the presence of ristocetin (an antibiotic that induces VWF-platelet binding), shear stress increases platelet aggregation caused by large multimers but does not demonstrably affect platelet-aggregating activity of small multimers.

Those observations provide a unified explanation for the interplay of shear stress, molecular size, and VWF adhesive function: VWF is uniquely capable of supporting platelet adhesion under high shear stress conditions due to its flexible conformation and responsiveness to shear stress; high levels of shear stress promotes platelet thrombus formation due to more VWF unfolding; and large multimers are hemostatically more effective because they are more responsive to shear stress.

### 4.3 Shear stress on VWF in the presence of ADAMTS-13

In the circulation, VWF is exposed to shear stress in the presence of ADAMTS-13. When shear stress is applied on normal plasma, which contains VWF and ADAMTS-13, VWF is cleaved to smaller forms (Figure 3B), and its adhesive activity is decreased (Tsai and Lien 1998). In contrast, when the ADAMTS-13 activity is suppressed by doxycycline, which binds to the metal cation moiety of the protease, application of shear stress increases the platelet-aggregating activity of VWF. Thus, these observations suggested that the role of ADAMTS-13 is to prevent the relentless increase of VWF activity under high shear stress environments. In the absence of ADAMTS-13, VWF multimers, particularly the large forms, will be unfolded by shear stress and become super-active molecules, creating a pro-thrombotic state in which the elongated forms of VWF cause platelet aggregation, microvascular thrombosis and depletion of large VWF multimers in the circulation, as occurs in thrombotic thrombocytopenic purpura (Figure 3C) (Tsai and Lien 1998).

## 5. ADAMTS-13 AND THROMBOTIC THROMBOCYTOPENIC PURPURA

First described in 1924, TTP is a mysterious systemic disorder that typically affects previously healthy adolescents or adult individuals, manifesting with thrombocytopenia, anemia with schistocytes on blood smears, and, frequently but not invariably, focal neurological deficits or mental changes (Bukowski 1982). TTP may also involve other organs, such as kidney, heart, and pancreas. Pathologically the disease is characterized by widespread hyaline thrombi in the arterioles and capillaries of multiple organs. Although relatively uncommon, affecting approximately 1-5 cases per million annually, TTP is a serious disease that frequently resulted in a fatal outcome within a few days. Since the 1980's, plasma exchange has reduced the case fatality rate from > 90% to 10% - 30% (Rock *et al.* 1991; Bell *et al.* 1991). The remarkable accomplishment was achieved without knowledge of the pathogenesis of TTP and the mechanism of plasma therapy.

Immunohistochemical and electron microscopic studies demonstrate that the thrombi of TTP consist of platelets and VWF (Asada *et al.* 1985). Flow cytometry also reveal VWF-platelet binding occurring in patients with TTP (Chow *et al.* 1998). Those observations suggest that thrombosis as observed in TTP might be caused by ADAMTS-13 deficiency. This consideration led to the discovery that patients with TTP have ADAMTS-13 deficiency caused by autoimmune inhibitors of the protease or genetic mutations in the ADAMTS-13 gene.

## 5.1 Inhibitors of ADAMTS-13 in TTP

Patients with TTP have deficiency of ADAMTS-13 due to inhibitors of the protease (Furlan *et al.* 1998; Tsai & Lian 1998). The prevalence of ADAMTS-13 deficiency among patients with TTP varies from 13% to 100%, according to the criteria used for definition of the study cases (Furlan *et al.* 1998; Tsai and Lian 1998; Veyradier *et al.* 2001; Bianchi *et al.* 2003; Vesely *et al.* 2003;). Since a clear distinction between TTP and HUS or other types of microvascular thrombosis is not always clinically feasible, it is inevitable that some TTP case series include patients with other types of microvascular thrombosis. However, if a set of strict criteria is applied to define patients with unequivocal, idiopathic TTP, a profound deficiency in ADAMTS-13 is detected in each case (Tsai 2003b).

IgG molecules isolated from patients with TTP suppressed ADAMTS-13 activity (Furlan *et al.* 1998; Tsai & Lian 1998). In our experience, inhibitory activity mediated by IgG is detectable in every patient of TTP investigated. Nevertheless, this does not exclude the possibility that some patients may also have non-inhibitory antibodies. The IgG molecules isolated from patients with TTP also recognized ADAMTS-13 on Western blots (Fujikawa *et al.* 2001; Plaimauer *et al.* 2002).

When a TTP patient is treated with plasma exchange, the ADAMTS-13 activity levels are increased. It is believed that plasma exchange replenishes the missing ADAMTS-13; plasma exchange may also help remove the inhibitors. However, not uncommonly the protease activity level is not normalized and inhibitors of the protease remain detectable when the patients are investigated during clinical remission, suggesting that the autoimmune reaction against ADAMTS-13 may persist sub-clinically.

The target epitopes of the ADAMTS-13 inhibitors have not been determined. Western blotting studies of recombinant ADAMTS-13 or its truncated forms revealed that IgG molecules isolated from TTP patients react with recombinant ADAMTS-13 sequences that include the cysteine-rich and spacer domains (Soejima *et al.* 2003; Klaus *et al.* 2004). These observations suggest that the cysteine-rich and spacer domains are potential targets of TTP inhibitors. A systemic, prospective investigation is needed to determine the prevalence and duration of ADAMTS-13 inhibitors among patients with TTP and the nature and etiology of the autoimmune reaction against ADAMTS-13.

## 5.2 Genetic mutations of the ADAMTS-13 gene

Upshaw-Schulman syndrome, characterized by thrombocytopenia and microangiopathic hemolysis presenting soon after birth, represents the congenital form of TTP. The patients have a profound deficiency in ADAMTS-13 activity due to either homozygous or double heterozygous mutations in the *ADAMTS-13* gene (Levy *et al.* 2001). Parents of the patients are carriers of single mutant alleles, and have partial ADAMTS-13 activity level in the range of 0.4 – 0.8 U/mL, but they are otherwise phenotypically unremarkable. Studies of family members with ADAMTS-13 deficiency were instrumental in cloning of the gene and establishing its role in causing TTP (Levy *et al.* 2001).

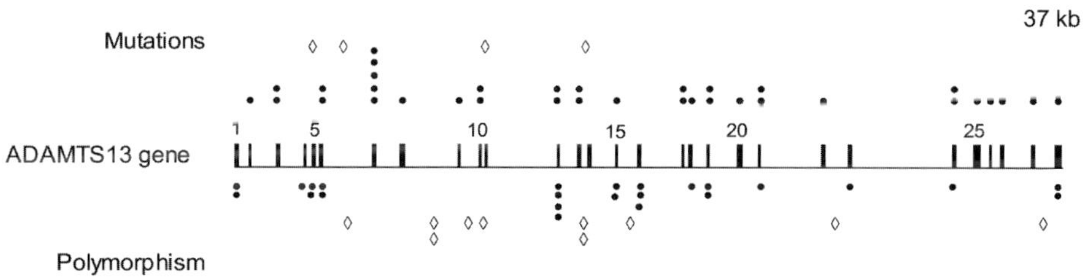


*Figure 4.* Distribution of ADAMTS-13 mutations and polymorphisms (above and below the line respectively). |: exon.•: a substitution in an exon. ◊: a substitution in an intron.

Forty-one different mutations of the ADAMTS-13 gene have been reported (Table 1) (Levy *et al.* 2001; Kokame *et al.* 2002; Schneppenheim *et al.* 2002; Antoine *et al.* 2003; Savasan *et al.* 2003; Matsumoto *et al.* 2003; Assink *et al.* 2003; Pimanda *et al.* 2003; Uchida *et al.* 2004). The mutations, which include mis-senses, non-senses, frame-shifting deletions or insertions, and intronic splicing mutations, distribute throughout the various domains of ADAMTS-13. Twenty-five mutations affect the sequence of metalloprotease-spacer domains that are necessary for expression of proteolytic activity (Figure 4). Mutations have been detected in individuals with various racial descents including African, American Indian, Asian, and Caucasian. There are 14 recurrent mutations, including 7 mutations detected in seemingly unrelated patients (Table 2). Three reports have described the 4143insA mutation in multiple individuals. Nevertheless, it remains to be determined whether any of the recurrent mutations occurred independently.

*Table 1. ADAMTS-13* gene mutations responsible for hereditary TTP

| Exon | Nucleotide | Amino acid | Domain | Exon | Nucleotide | Amino acid | Domain |
|---|---|---|---|---|---|---|---|
| Mis-sense | | | | Non-sense | | | |
| 3 | 286C>G | H96D | Propeptide | 2 | 130C>T | Q44X | Propeptide |
| 3 | 304C>T | R102C | MP | 10 | 1169G>A | W390X | TSR-1 |
| 6 | 577C>T | R193W | MP | 12 | 1345C>T | Q449X | Cys |
| 6 | 587C>T | T196I | MP | 21 | 2728C>T | R910X | TSR-5 |
| 7 | 695T>A | L232Q | MP | 24 | 3100A>T | R1034X | TSR-7 |
| 7 | 703G>C | D235H | MP | | | | |
| 7 | 749C>T | A250V | MP | | | | |
| 7 | 788C>G | S263C | MP | Deletion/insertion | | | |
| 7 | 803G>C | R268P | MP | 7 | 718-724del | T245X | MP |
| 8 | 932G>A | C311Y | Dis | 15 | 1783delTT | P612X | Spacer |
| 9 | 1058C>T | P353L | Dis | 19 | 2279delG | S776X | TSR-3 |
| 10 | 1193G>A | R398H | TSR-1 | 19 | 2376-2401del | D834X | TSR-4 |
| 12 | 1370C>T | P457L | Cys | 20 | 2549delAT | P855X | TSR-4 |
| 13 | 1523G>A | C508Y | Cys | 27 | 3769insT | S1292X | CUB-1 |
| 13 | 1582A>G | R528G | Cys | 29 | 4143insA | G1386X | CUB-2 |
| 17 | 2017 A>T | I673F | Spacer | | | | |
| 17 | 2074C>T | R692C | TSR-2 | | | | |
| 18 | 2195C>T | A732V | TSR-2 | Splicing | | | |
| 21 | 2723 G>A | C908Y | TSR-5 | 3 | 331+1 G>A | - | MP |
| 22 | 2851T>G | C951G | TSR-5 | 4 | 414+1 G>A | - | MP |
| 24 | 3070T>G | C1024G | TSR-7 | 6 | 686+1 G>A | - | MP |
| 25 | 3367 C>T | R1123C | TSR-8 | 10 | 1244+2 T>G | - | TSR-1 |
| 26 | 3638G>A | C1213Y | CUB-1 | 13 | 1584+5 G>A | - | Cys |
| 28 | 4006C>T | R1336W | CUB-2 | | | | |

*MP: metalloprotease; Dis: disintegrin-like; TSR: thrombospondin repeat; Cys: cysteine-rich domain.

Eight mutations have been investigated in expression studies: one mutation creates a proteolytically inactive form, while the other seven mutations impede secretion of the protein (Table 3). All five intronic mutations have been investigated by RT PCR and were confirmed to be associated with abnormal splicing.

In addition to mutations, multiple polymorphisms of the gene are detected in individuals from different geographic areas. Overall, each of the exons contains at least one genetic variation. The data suggest that variation in the ADAMTS-13 gene is not uncommon. Mutations that compromise the expression of ADAMTS-13 protease activity may persist because a carrier of one mutant allele is not phenotypically disadvantaged.

Table 2. Recurrent *ADAMTS-13* mutations

| Exon | Nucleotide | Amino acid | Domain |
|---|---|---|---|
| 4 | 414+1 G>A* | MP | Splice |
| 6 | 587C>T | T196I | MP |
| 9 | 1058C>T | P353L | Dis |
| 21 | 2728C>T | R910X | TSR-5 |
| 29 | 4143insA* | G1386X | CUB-2 |
| Homozygotes | | | |
| 4 | 414+1 G>A* | MP | Splice |
| 7 | 695T>A | L232Q | MP |
| 7 | 703G>C | D235H | MP |
| 8 | 932G>A | C311Y | Dis |
| 12 | 1345C>T | Q449X | Cys |
| 15 | 1783delTT | P612X | Spacer |
| 17 | 2074C>T | R692C | TSR-2 |
| 19 | 2279delG | S776X | TSR-3 |
| 29 | 4143insA* | G1386X | CUB-2 |

Table 3. Defects by ADAMTS-13 mutations

| Exon | Nucleotide | Amino acid | Domain |
|---|---|---|---|
| Secretion | | | |
| 6 | 577C>T | R193W | MP |
| 7 | 803G>C | R268P | MP |
| 13 | 1523G>A | C508Y | Cys |
| 17 | 2017A>T | I673F | Spacer |
| 21 | 2723G>A | C908Y | TSR-5 |
| 25 | 3367C>T | R1123C | TSR-8 |
| 29 | 4143insA | G1386X | CUB-2 |
| VWF-cleaving activity | | | |
| 12 | 1345C>T | Q449X | Cys |

## 6. CONCLUSIONS

Hemostasis is essential for survival after infliction of vessel injury. The down side of an effective hemostatic system is a higher risk of thrombosis. In human subjects, thrombosis, or inappropriate hemostatic reaction, contributes to the development of most cardiovascular events. In this regard, anti-thrombotic molecules such as ADAMTS-13 are essential for maintaining a delicate balance between hemostasis and thrombosis.

The role of ADAMTS-13 is perhaps better appreciated by comparing with another anti-thrombotic protease, protein C. Both ADAMTS-13 and protein C prevent thrombosis by cleaving hemostatic proteins. For protein C the substrate is factor V. Since factor V is critical for thrombin activation and fibrin generation, deficiency in protein C primarily causes venous thrombosis. In contrast, the substrate of ADAMTS-13, VWF, supports platelet adhesion and aggregation. Thus, ADAMTS-13 deficiency causes thrombosis in the arterioles and capillaries.

Both ADAMTS-13 and protein C co-exist in the circulation with their respective substrates. However, ADAMTS-13 does not exist in a zymogenic form, and no natural inhibitors are known to exist for the protease. Regulation of protease-substrate interaction relies on VWF itself. VWF is protected from excessive proteolysis because it exists in a resistant conformation to ADAMTS-13. This scheme is critical for regulation of two opposite processes. At a vessel injury site, high levels of shear stress unfold the conformation of VWF, allowing it to support platelet adhesion. In the normal circulation, ADAMTS-13 cleaves VWF whenever it is partially unfolded by shear stress. This type of regulation appears to be unique for VWF-ADAMTS-13.

Deficiency of ADAMTS-13 results from genetic mutations or, more commonly, autoimmune inhibitors of the protease. It remains uncertain whether a transient decrease in the ADAMTS-13 activity level may occur in sepsis or other pathological conditions. It is conceivable that certain mutations in the *VWF* gene may create proteolytically resistant VWF variants. Such resistant mutations have not been detected, presumably because the presence of even one completely resistant VWF allele will cause widespread microvascular thrombosis and early death.

The discovery of ADAMTS-13 helps unravel the mystery associated with TTP, a hitherto intriguing disease. It has provided new directions for improving the diagnosis and management of thrombotic thrombocytopenic purpura (Kokame *et al.* 2004; Zhou *et al.* 2004; Yomtovian *et al.* 2004). More advances are anticipated to result from ongoing investigation of ADAMTS-13.

## REFERENCES

Alevriadou,B.R., Moake,J.L., Turner,N.A., Ruggeri,Z.M., Folie,B.J., Phillips,M.D., Schreiber,A.B., Hrinda,M.E., and McIntire,L.V. 1993. Real-time analysis of shear-dependent thrombus formation and its blockade by inhibitors of von Willebrand factor binding to platelets. *Blood* **81:** 1263-76.

Antoine,G., Zimmermann,K., Plaimauer,B., Grillowitzer,M., Studt,J.D., Lammle,B., and Scheiflinger,F. 2003. ADAMTS-13 gene defects in two brothers with constitutional thrombotic thrombocytopenic purpura and normalization of von Willebrand factor-cleaving protease activity by recombinant human ADAMTS-13. *Br J Haematol* **120:** 821-824.

Apte,S.S. 2004. A disintegrin-like and metalloprotease (reprolysin type) with thrombospondin type 1 motifs: the ADAMTS family. *Int J Biochem Cell Biol* **36:** 981-985.

Asada Y, Sumiyoshi A, Hayashi T, Suzumiya J, Kaketani K. 1985. Immunohistochemistry of vascular lesion in thrombotic thrombocytopenic pupura, with special reference to factor VIII related antigen. *Thromb Res* **38:** 467-479.

Assink,K., Schiphorst,R., Allford,S., Karpman,D., Etzioni,A., Brichard,B., van de,K.N., Monnens,L., and van den,H.L. 2003. Mutation analysis and clinical implications of von Willebrand factor-cleaving protease deficiency. *Kidney Int* **63:** 1995-1999.

Batlle,J., Lopez Fernandez,M.F., Campos,M., Justica,B., Berges,C., Navarro,J.L., Diaz Cremades,J.M., Kasper,C.K., Dent,J.A., Ruggeri,Z.M., and Zimmermann,T.S. 1986. The heterogeneity of type IIA von Willebrand's disease: studies with protease inhibitors. *Blood* **68:** 1207-12.

Bell,W.R., Braine,H.G., Ness,P.M., and Kickler,T.S. 1991. Improved survival in thrombotic thrombocytopenic purpura-hemolytic uremic syndrome. Clinical experience in 108 patients. *N Engl J Med* **325:** 398-403.

Bianchi,V., Robles,R., Alberio,L., Furlan,M., & Lammle,B. 2002. Von Willebrand factor-cleaving protease (ADAMTS-13) in thrombocytopenic disorders: a severely deficient activity is specific for thrombotic thrombocytopenic purpura. *Blood* **100:** 710-713.

Bukowski,R.M. 1982. Thrombotic thrombocytopenic purpura. A review. *Prog Hemost Thromb* **6:** 287-337.

Cal,S., Obaya,A.J., Llamazares,M., Garabaya,C., Quesada,V., and Lopez-Otin,C. 2002. Cloning, expression analysis, and structural characterization of seven novel human ADAMTSs, a family of metalloproteinases with disintegrin and thrombospondin-1 domains. *Gene* **283:** 49-62.

Chow,T.W., Turner,N.A., Chintagumpala,M., McPherson,P.D., Nolasco,L.H., Rice,L., Hellums,J.D., and Moake,J.L. 1998. Increased von Willebrand factor binding to platelets in single episode and recurrent types of thrombotic thrombocytopenic purpura. *Am J Hematol* **57:** 293-302.

Dent,J.A., Berkowitz,S.D., Ware,J., Kasper,C.K., and Ruggeri,Z.M. 1990. Identification of a cleavage site directing the immunochemical detection of molecular abnormalities in type IIA von Willebrand factor. *Proc Natl Acad Sci USA* **87:** 6306-10.

Dent,J.A., Galbusera,M., and Ruggeri,Z.M. 1991. Heterogeneity of plasma von Willebrand factor multimers resulting from proteolysis of the constituent subunit. *J Clin Invest* **88:** 774-82.

Fowler,W.E., Fretto,L.J., Hamilton,K.K., Erickson,H.P., and McKee,P.A. 1985. Substructure of human von Willebrand factor. *J Clin Invest* **76:** 1491-500.

Fujikawa,K., Suzuki,H., McMullen,B., and Chung,D. 2001. Purification of human von Willebrand factor-cleaving protease and its identification as a new member of the metalloproteinase family. *Blood* **98:** 1662-1666.

Furlan,M., Robles,R., and Lamie,B. 1996. Partial purification and characterization of a protease from human plasma cleaving von Willebrand factor to fragments produced by in vivo proteolysis. *Blood* **87:** 4223-34.

Furlan,M., Robles,R., Galbusera,M., Remuzzi,G., Kyrle,P.A., Brenner,B., Krause,M., Scharrer,I., Aumann,V., Mittler,U., Solenthaler,M., and Lammle,B. 1998. von Willebrand factor-cleaving protease in thrombotic thrombocytopenic purpura and the hemolytic-uremic syndrome. *N Engl J Med* **339:** 1578-1584.

Ginsburg,D. 1999. Molecular genetics of von Willebrand disease. *Thromb Haemost* **82:** 585-591.

Gralnick,H.R., Williams,S.B., McKeown,L.P., Maisonneuve,P., Jenneau,C., Sultan,Y., and Rick,M.E. 1985. In vitro correction of the abnormal multimeric structure of von Willebrand factor in type IIa von Willebrand's disease. *Proc Natl Acad Sci USA* **82:** 5968-5972.

Handin,R.I. and Wagner,D.D. 1989. Molecular and cellular biology of von Willebrand factor. *Prog Hemost Thromb* **9:** 233-59.

Ikeda,Y., Handa,M., Kawano,K., Kamata,T., Murata,M., Araki,Y., Anbo,H., Kawai,Y., Watanabe,K., Itagaki,I., and . 1991. The role of von Willebrand factor and fibrinogen in platelet aggregation under varying shear stress. *J Clin Invest* **87:** 1234-1240.

Kaul,D.K., Nagel,R.L., Chen,D., and Tsai,H.M. 1993. Sickle erythrocyte-endothelial interactions in microcirculation: the role of von Willebrand factor and implications for vasoocclusion. *Blood* **81:** 2429-2438.

Klaus,C., Plaimauer,B., Studt,J.D., Dorner,F., Lammle,B., Mannucci,P.M., and Scheiflinger,F. 2004. Epitope mapping of ADAMTS-13 autoantibodies in acquired thrombotic thrombocytopenic purpura. *Blood* **103:** 4514-4519.

Kokame,K., Matsumoto,M., Soejima,K., Yagi,H., Ishizashi,H., Funato,M., Tamai,H., Konno,M., Kamide,K., Kawano,Y., Miyata,T., and Fujimura,Y. 2002. Mutations and common polymorphisms in ADAMTS-13 gene responsible for von Willebrand factor-cleaving protease activity. *Proc Natl Acad Sci USA* **99:** 11902-11907.

Kokame,K., Matsumoto,M., Fujimura,Y., and Miyata,T. 2004. VWF73, a region from D1596 to R1668 of von Willebrand factor, provides a minimal substrate for ADAMTS-13. *Blood* **103:** 607-612.

Levy,G.G., Nichols,W.C., Lian,E.C., Foroud,T., McClintick,J.N., McGee,B.M., Yang,A.Y., Siemieniak,D.R., Stark,K.R., Gruppo,R., Sarode,R., Shurin,S.B., Chandrasekaran,V., Stabler,S.P., Sabio,H., Bouhassira,E.E., Upshaw,J.D., Jr., Ginsburg,D., and Tsai,H.M. 2001. Mutations in a member of the ADAMTS gene family cause thrombotic thrombocytopenic purpura. *Nature* **413:** 488-494.

Majerus,E.M., Zheng,X., Tuley,E.A., and Sadler,J.E. 2003. Cleavage of the ADAMTS-13 propeptide is not required for protease activity. *J Biol Chem* **278:** 46643-46648.

Marchant,R.E., Barb,M.D., Shainoff,J.R., Eppell,S.J., Wilson,D.L., and Siedlecki,C.A. 1997. Three dimensional structure of human fibrinogen under aqueous conditions visualized by atomic force microscopy. *Thromb Haemost* **77:** 1048-51.

Matsumoto,M., Kokame,K., Soejima,K., Miura,M., Hayashi,S., Fujii,Y., Iwai,A., Ito,E., Tsuji,Y., Takeda-Shitaka,M., Iwadate,M., Umeyama,H., Yagi,H., Ishizashi,H., Banno,F., Nakagaki,T., Miyata,T., and Fujimura,Y. 2003. Molecular characterization of ADAMTS-13 gene mutations in Japanese patients with Upshaw-Schulman syndrome. *Blood* **103:** 1305-1310.

Moake,J.L., Turner,N.A., Stathopoulos,N.A., Nolasco,L.H., and Hellums,J.D. 1986. Involvement of large plasma von Willebrand factor (vWF) multimers and unusually large vWF forms derived from endothelial cells in shear stress-induced platelet aggregation. *J Clin Invest* **78:** 1456-61.

Nichols,T.C., Samama,C.M., Bellinger,D.A., Roussi,J., Reddick,R.L., Bonneau,M., Read,M.S., Bailliart,O., Koch,G.G., Vaiman,M., and . 1995. Function of von Willebrand factor after crossed bone marrow transplantation between normal and von Willebrand disease pigs: effect on arterial thrombosis in chimeras. *Proc Natl Acad Sci USA* **92:** 2455-2459.

O'Brien,L.A., Sutherland,J.J., Hegadorn,C., Benford,K., Racz,H., Rapson,D., Hough,C., and Lillicrap,D. 2004. A novel type 2A (Group II) von Willebrand disease mutation (L1503Q) associated with loss of the highest molecular weight von Willebrand factor multimers. *J Thromb Haemost* **2:** 1135-1142.

Pimanda,J.E., Maekawa,A., Wind,T., Paxton,J., Chesterman,C.N., and Hogg,P.J. 2003. Congenital thrombotic thrombocytopenic purpura in association with a mutation in the second CUB domain of ADAMTS-13. *Blood* **103:** 627-629.

Plaimauer,B., Zimmermann,K., Voelkel,D., Antoine,G., Kerschbaumer,R., Jenab,P., Furlan,M., Gerritsen,H., Laemmle,B., Schwarz,H.P., and Scheiflinger,F. 2002. Cloning, Expression and Functional Characterization of the von Willebrand Factor-Cleaving Protease (ADAMTS-13). *Blood* **100:** 3626-3632.

Rand,J.H., Patel,N.D., Schwartz,E., Zhou,S.L., and Potter,B.J. 1991. 150-kD von Willebrand factor binding protein extracted from human vascular subendothelium is type VI collagen. *J Clin Invest* **88:** 253-259.

Rock,G.A., Shumak,K.H., Buskard,N.A., Blanchette,V.S., Kelton,J.G., Nair,R.C., and Spasoff,R.A. 1991. Comparison of plasma exchange with plasma infusion in the treatment of thrombotic thrombocytopenic purpura. Canadian Apheresis Study Group. *N Engl J Med* **325:** 393-7.

Ruggeri,Z.M. 2003. Von Willebrand factor, platelets and endothelial cell interactions. *J Thromb Haemost* **1:** 1335-1342.

Sarode,R., Goldstein,J., Sussman,I.I., Nagel,R.L., and Tsai,H.M. 2000. Role of A and B blood group antigens in the expression of adhesive activity of von Willebrand factor. *Br J Haematol* **109:** 857-864.

Savasan,S., Lee,S.K., Ginsburg,D., and Tsai,H.M. 2003. ADAMTS-13 gene mutation In congenital thrombotic thrombocytopenic purpura with previously reported normal VWF cleaving protease activity. *Blood* **101:**4449-4451.

Schneppenheim,R., Budde,U., Oyen,F., Angerhaus,D., Aumann,V., Drewke,E., Hassenpflug,W.A., Haberle,J., Kentouche,K., Kohne,E., Kurnik,K., Mueller-Wiefel,D.E., Obser,T., Santer,R., and Sykora,K.W. 2002. Von Willebrand factor cleaving protease and ADAMTS-13 mutations in childhood TTP. *Blood* **101:** 1845-1850.

Soejima,K., Mimura,N., Hirashima,M., Maeda,H., Hamamoto,T., Nakagaki,T., & Nozaki,C. 2001. A novel human metalloprotease synthesized in the liver and secreted into the blood: possibly, the von Willebrand factor-cleaving protease? *J Biochem (Tokyo)* **130:** 475-480.

Soejima,K., Matsumoto,M., Kokame,K., Yagi,H., Ishizashi,H., Maeda,H., Nozaki,C., Miyata,T., Fujimura,Y., and Nakagaki,T. 2003. ADAMTS-13 cysteine-rich/spacer domains are functionally essential for von Willebrand factor cleavage. *Blood* **102:** 3232-3237.

Tsai,H.M., Nagel,R.L., Hatcher,V.B., and Sussman,I.I. 1989. Multimeric composition of endothelial cell-derived von Willebrand factor. *Blood* **73:** 2074-2076.

Tsai,H.M., Nagel,R.L., and Sussman,I.I. 1991. Subunit composition of plasma von Willebrand factor multimers: evidence for a non-proteolytic mechanism resulting in apparent increase in proteolytic fragments. *Thromb Res* **63:** 179-188.

Tsai,H.M., Sussman,I.I., and Nagel,R.L. 1994. Shear stress enhances the proteolysis of von Willebrand factor in normal plasma. *Blood* **83:** 2171-2179.

Tsai,H.M. 1996. Physiologic cleavage of von Willebrand factor by a plasma protease is dependent on its conformation and requires calcium ion. *Blood* **87:** 4235-4244.

Tsai,H.M., Sussman,I.I., Ginsburg,D., Lankhof,H., Sixma,J.J., and Nagel,R.L. 1997. Proteolytic cleavage of recombinant type 2A von Willebrand factor mutants R834W and R834Q: inhibition by doxycycline and by monoclonal antibody VP-1. *Blood* **89:**1954-1962.

Tsai,H.M. and Lian,E.C. 1998. Antibodies to von Willebrand factor-cleaving protease in acute thrombotic thrombocytopenic purpura. *N Engl J Med.* **339:** 1585-1594.

Tsai,H.M. 2002. Von Willebrand factor, ADAMTS-13, and thrombotic thrombocytopenic purpura. *J Mol Med* **80:** 639-647.

Tsai,H.M. 2003a. Shear stress and von Willebrand factor in health and disease. *Semin Thromb Hemost* **29:** 479-488.

Tsai,H.M. 2003b. Advances in the pathogenesis, diagnosis, and treatment of thrombotic thrombocytopenic purpura. *J Am Soc Nephrol* **14:** 1072-1081.

Uchida,T., Wada,H., Mizutani,M., Iwashita,M., Ishihara,H., Shibano,T., Suzuki,M., Matsubara,Y., Soejima,K., Matsumoto,M., Fujimura,Y., Ikeda,Y., and Murata,M. 2004. Identification of novel mutations in ADAMTS-13 in an adult patient with congenital thrombotic thrombocytopenic purpura. *Blood* (in press).

Vesely,S.K., George,J.N., Lammle,B., Studt,J.D., Alberio,L., El Harake,M.A., and Raskob,G.E. 2003. ADAMTS-13 activity in thrombotic thrombocytopenic purpura-hemolytic uremic syndrome: relation to presenting features and clinical outcomes in a prospective cohort of 142 patients. *Blood* **102:** 60-68

Veyradier,A., Obert,B., Houllier,A., Meyer,D., and Girma,J.P. 2001. Specific von Willebrand factor-cleaving protease in thrombotic microangiopathies: a study of 111 cases. *Blood* **98:** 1765-1772.

Weiss,H.J., Turitto,V.T., and Baumgartner,H.R. 1978. Effect of shear rate on platelet interaction with subendothelium in citrated and native blood. I. Shear rate--dependent decrease of adhesion in von Willebrand's disease and the Bernard-Soulier syndrome. *J Lab Clin Med* **92:** 750-764.

Williams,S.B., McKeown,L.P., Krutzsch,H., Hansmann,K., and Gralnick,H.R. 1994. Purification and characterization of human platelet von Willebrand factor. *Br J Haematol* **88:** 582-591.

Yomtovian,R., Niklinski,W., Silver,B., Sarode,R., and Tsai,H.M. 2004. Rituximab for chronic recurring thrombotic thrombocytopenic purpura: a case report and review of the literature. *Br.J Haematol* **124:** 787-795.

Zheng,X., Chung,D., Takayama,T.K., Majerus,E.M., Sadler,J.E., and Fujikawa,K. 2001. Structure of von Willebrand factor-cleaving protease (ADAMTS-13), a metalloprotease involved in thrombotic thrombocytopenic purpura. *J Biol Chem* **276:** 41059-41063.

Zheng,X., Nishio,K., Majerus,E.M., and Sadler,J.E. 2003. Cleavage of von Willebrand factor requires the spacer domain of the metalloprotease ADAMTS-13. *J Biol Chem* ***278:***46643-46648.

Zhou,W. and Tsai,H.M. 2004. An enzyme immunoassay of ADAMTS-13 distinguishes patients with thrombotic thrombocytopenic purpura from normal individuals and carriers of ADAMTS-13 mutations. *Thromb Haemost* **91:** 806-811.

Zimmerman,T.S., Dent,J.A., Ruggeri,Z.M., and Nannini,L.H. 1986. Subunit composition of plasma von Willebrand factor. Cleavage is present in normal individuals, increased in IIA and IIB von Willebrand disease, but minimal in variants with aberrant structure of individual oligomers (types IIC, IID, and IIE). *J Clin Invest* **77:** 947-51.

# INDEX